INTERNATIONAL CENTRE FOR MECHANICAL SCIENCES

COURSES AND LECTURES - No. 262

RECENT DEVELOPMENTS IN THERMOMECHANICS OF SOLIDS

EDITED BY

G. LEBON
UNIVERSITY OF LIEGE

P. PERZYNA
POLISH ACADEMY OF SCIENCES

SPRINGER - VERLAG WIEN GMBH

ISBN 978-3-211-81597-7 ISBN 978-3-7091-3351-4 (eBook)
DOI 10.1007/978-3-7091-3351-4

P R E F A C E

The main objective of the contributions contained in this volume is to present the thermodynamic foundations of the response of elastic and dissipative materials.

In particular, the governing equations of nonlinear thermoelasticity and thermoinelasticity as well as the basic properties of these equations as resulting from the primary assumptions of continuum thermodynamics are derived.

The global formulation of thermodynamics of continua is discussed. A special attention is paid to the properties of the balance equations on a singular surface. The possible forms of the second law of thermodynamics are discussed within the framework of axiomatic thermodynamics.

Furthermore, the thermodynamic requirements for different kinds of materials are examined.

The secondary purpose of the Course was to discuss some connections between rational and classical formulations of the principles of thermodynamics.

The present volume contains the texts of three (of the four delivered) Course lectures. I hope it will constitute a useful source of information on the problems presented and discussed in Udine.

Special thanks are due to the International Centre for Mechanical Sciences whose direction encouraged us to prepare and to deliver the lectures.

P. Perzyna

Udine, August 1980.

List of Contributors

G. Lebon, Liège University, Department of Mechanics, B-4000 Sart Tilman, Liège (Belgium)

P. Perzyna, Institute of Fundamental Technological Research, Polish Academy of Sciences, ul. Swietokrzyska 21, Warsaw (Poland)

K. Wilmanski, Institute of Fundamental Technological Research, Polish Academy of Sciences, ul. Swietokrzyska 21, Warsaw (Poland).

CONTENTS

Page

Thermodynamic Foundations of Thermoelasticity
by K. Wilmanski

Preface 1
1. The Notion of a Thermodynamic Process in Continua 3
2. Balance Laws 15
3. Thermoelastic Materials 41
4. Heat Conduction in Thermoelastic Materials 61
5. Waves in Thermoelastic Materials 73
References 89
Contents 93

Thermodynamics of Dissipative Materials
by P. Perzyna

Preface 95
Contents 99
1. Global Formulation of Thermodynamics of Continua 101
2. General Material Structure 123
3. Internal State Variable Material Structure 145
4. Rate Type Material Structure 153
5. Isomorphic Material Structures (Isothermal Processes) 156
6. Thermo-Viscoplasticity for Finite Strains 160
7. Thermoplasticity for Finite Deformations 194
References 205
Figures 217

Variational Principles in Thermomechanics
by G. Lebon

Introduction 221
I. Basic Concepts of the Calculus of Variation 224
II. Variational Principles in Classical Mechanics and in Elasticity 264
III. Variational Theory of Heat Conduction 330
IV. Coupled Thermoelasticity 366
General Appendix: Convergence and Error Estimates of the Variational and Galerkin Methods 397
Contents 412

THERMODYNAMIC FOUNDATIONS OF THERMOELASTICITY

KRZYSZTOF WILMANSKI
Institute of Fundamental Technological Research, Warsaw, Poland

PREFACE

Many attempts have been made to formulate the phenomenological thermodynamics in terms of precise mathematical axioms and statements. In spite of many successes during the last two decades, most of all within the framework of the thermodynamic theory of materials, the main goal has not been achieved. Starting from the simplest thermomechanics of continuum, we show in this paper both the results and the difficulties of such a formulation of thermodynamics. In the first part of the paper we introduce such notions as thermodynamic processes, balance equations for subsystems, their localization in space and the resulting field equations, restricting the class of admissible constitutive relations. We discuss also some recent formulations of the second law of thermodynamics.

The second part of the paper contains the revue of the application of thermodynamics in the theory of thermoelastic materials. We pay a special attention to the theory of heat conduction in such materials and to the theory of wave propagation.

The references are far from being complete. We point usually at the leading research in the field. Some details concerning the contributions to the presented branch of thermodynamics can be found in the cited papers.

Udine, July 1978

CHAPTER 1
THE NOTION OF A THERMODYNAMIC PROCESS IN CONTINUA

We consider a continuous medium B to be the bounded closed subset of the Euclidean space E^3. In the theory of materials, the medium B is considered to be a differentiable manifold and is called a material space. However, for our purposes it is not necessery to use this notion. Our set B corresponds to the reference configuration of the medium. We assume that the behavior of the medium B is described by the following fields

$$\underset{\sim}{\chi}: B \rightarrow E^3, \qquad \theta: B \rightarrow R. \tag{1.1}$$

The first mapping describes the *configuration* of B and the second one - an *empirical temperature* distribution in B. To simplify the notation, we denote each pair of these fields as follows

$$s := (\underset{\sim}{\chi}(\cdot), \theta(\cdot)). \tag{1.2}$$

Such a pair is called a s t a t e of the medium B. It is obvious that in each physical systems may appear only certain ordered sequencies of states. If two states s_1 and s_2 can appear in the medium B in the same sequence and the state s_1 precedes the state s_2 , we say that the state s_2 is a c c e s s i b l e from the state s_1 and we write

$$s_1 \rightleftarrows s_2 . \tag{1.3}$$

This relation must satisfy certain conditions, which can be put in the form of the axiom

(1.4) A x i o m: Let S be the set of all states appearing in the medium B. The accessibility relation $\rightleftarrows$, given on the set S, satisfies the following conditions

i/ $\forall_{s \in S} \; s \rightleftarrows s ,$ (reflexivity)

ii/ $\forall_{s_1, s_2, s_3 \in S} \; s_1 \rightleftarrows s_2 \;\&\; s_2 \rightleftarrows s_3 \Rightarrow s_1 \rightleftarrows s_3 ,$ (transitivity)

iii/ $\exists_{S_0 \subset S} \; \forall_{s_1, s_2 \in S_0} \; s_1 \leftrightarrow s_2 \quad \text{or} \quad s_1 \not\leftrightarrow s_2$

and

$$\forall_{s \in S} \; \exists_{s_0 \in S_0} \; s \rightleftarrows s_0 \;\&\; \forall_{s'_0 \in S_0} \; s'_0 \not\leftrightarrow s_0 \Rightarrow s \not\leftrightarrow s'_0$$

(existence of equilibrium states)

where

$s_1 \leftrightarrow s_2 \Leftrightarrow s_1 \rightleftarrows s_2 \;\&\; s_2 \rightleftarrows s_1$ (reversible accessibility)

$s_1 \not\leftrightarrow s_2 \Leftrightarrow \text{no} \; (s_1 \rightleftarrows s_2 \text{ or } s_2 \rightleftarrows s_1)$ (inaccessibility)

Usually this axiom is supplemented by certain additional conditions, such as the metrizability of some subsets of S. We will not discuss those problems in this paper. The details, concerning different formulations of such an axiom can be found in the papers of R.Giles[1964], J.L.B.Cooper[1967],

K.Wilmanski[1972,1976], B.D.Coleman and D.R.Owen[1974], F,J.Zeleznik[1976]. Further we strengthen this axiom to justify the operations to be performed.

It has to be pointed out that the above axiom can be fulfilled only by the material bodies of the very limited type of interactions with the external fields. Namely, certain interactions, such as the heat supply from the external world, can produce the repetitions of the same state in an irreversible process and such repetitions violate the transitivity of the accessibility relation. For instance, the order of appearance of states $\{s_2, s_3, s_1, s_2\}$ without repetitions of states s_1 and s_3 is possible in the case of the body interacting with the external world but it means that the relations $s_1 \rightrightarrows s_2$ and $s_2 \rightrightarrows s_3$ do not imply $s_1 \rightrightarrows s_3$. For this reason, we say that the above axiom describes the *isolated* bodies.

The main aim of all thermodynamic theories is to determine the state space and the accessibility relation for a given body B. We can procede in two ways:

i/ assuming certain state space and accessibility relation we can seek the class of materials, which admit the assumed structure,

ii/ assuming the properties of a certain class of materials we can seek the admissible state space and the accessibility relation.

The first procedure is commonly used in the modern thermodynamic theory of materials. In this paper, we also base our considerations on the assumptions fitting the first approach. The second approach is used in some particular cases such as the theory of ideal gases, the linear theory of elastic materials etc.

The axiom (1.4) yields the natural definition of the thermodynamic process. Namely, for any given state s, we define the left and right cross-sections as follows

$$L(s) := \{s' \mid s' \rightleftarrows s\} \quad , \quad R(s) := \{s' \mid s \rightleftarrows s'\}. \tag{1.5}$$

The set $L(s)$ contains all states from which the given state s is accessible and the set $R(s)$ contains all states accessible from the state s . For any two states s_1, s_2 such that $s_1 \rightleftarrows s_2$, let us define the set

$$M := R(s_1) \cap L(s_2). \tag{1.6}$$

It is easy to see that for each state s belonging to the set M, we have

$$s_1 \rightleftarrows s \rightleftarrows s_2 .$$

Taking the chains

$$s_1 \rightleftarrows s' \rightleftarrows s_2 , \quad s_1 \rightleftarrows s' \rightleftarrows s'' \rightleftarrows s_2 , \quad s_1 \rightleftarrows s''' \rightleftarrows s' \rightleftarrows s'' \rightleftarrows s_2 , \dots$$

where $s', s'', s''', \dots \in M$, we can construct maximum chains starting from the state s_1 and ending at the state s_2 . The existence of such maximum chains is assured by the Hausdorff theorem. Every maximum chain constructed as above is called a direct thermodynamic process between the states s_1 and s_2 . We denote this chain as follows

$$p_d := (\mathfrak{T}, \rightleftarrows) \, , \qquad \mathfrak{T} \subset M \subset \mathcal{S} \, , \tag{1.7}$$

where $\mathfrak{T}$ is the collection of all states of this chain. The state

$$l p_d := s_1 \tag{1.8}$$

is called an initial state of the process p_d, and the state

$$r p_d := s_2 \tag{1.9}$$

is called a f i n a l s t a t e of the process p_d.

The theory of thermodynamic processes at this level of generality is far from being simple. We do not need to go into the discussion of this problem and, therefore, we introduce the assumption

(1.10) A x i o m: For every two states s_1, s_2 such that $s_1 \rightleftarrows s_2$, there is a direct thermodynamic process p_d such that

$$l p_d = s_1 , \quad r p_d = s_2$$

and there is a closed interval $T = \langle t_l, t_r \rangle$ of the real line such that

$$\underset{t \in T}{\forall} \; \underset{s \in \mathfrak{S}}{\exists} \; t \mapsto s , \qquad t_l \mapsto l p_d , \; t_r \mapsto r p_d$$

and

$$\underset{t', t'' \in T}{\forall} \; t' \neq t'' \Rightarrow s(t') \neq s(t'')$$

The above mapping is called a p a r a m e t r i z a t i o n of the process p_d. Making use of the previous definition of the state (1.2) we see that the direct process in the body B under consideration is of the form

$$\{ \underset{\sim}{\chi}(\cdot, t), \theta(\cdot, t) \}_{t \in T} , \qquad T = \langle t_l, t_r \rangle \subset R . \tag{1.11}$$

Obviously, the parametrization is interpreted as the timing of the process. It can be different in different bodies according to the change of initial conditions and the material properties. For instance, two bodies kept at the same configuration $\underset{\sim}{\chi}(\cdot ; t_0)$ may go through the same sequence of temperature distributions with the different rate $\dot{\theta}(\cdot ; t)$. Then the collection of states in both cases will be the same but due to the change of the accessibility relation the corresponding process will be different.

The definition (1.7) of the process leads to the basic classification of the thermodynamic processes. For a given direct process $p_d = (\pi, \rightleftarrows)$ it may happen that there exists a direct process $\tilde{p}_d = (\pi, \leftrightarrows)$ containing the same states but in the inverse order, given by the same accessibility relation $\rightleftarrows$. In such a case we say that the process p_d is reversible. Otherwise we call the process irreversible. It is easy to notice that for any reversible process p_d

$$l\, p_d \leftrightarrow r\, p_d, \tag{1.12}$$

and, for any irreversible process p_d'

$$l p_d' \rightarrow r p_d', \qquad \text{i.e.} \quad l p_d' \rightleftarrows r p_d' \quad \& \quad no\,(r p_d' \rightleftarrows l p_d'). \tag{1.13}$$

As we have mentioned the theory, sketched above, models only very rare systems isolated from the external world. Simultaneously, the procedure cannot be repeated for non-isolated systems and the reason has been explained after the axiom (1.4). Therefore we procede in a different manner. We assume that the body B and its state space S is given. The question is if, basing on this assumption, we can determine the "state" of an arbitrary subset P of B. In general, the answer to this question is only trivial: for a chosen set $P \subset B$ its state is defined as a pair (s,P). It is obvious that we are interested in more restricted notion of the state of P. However it cannot be done by the simplest restriction of functions, defining the state of B, i.e. the pair

$$\{\underset{\sim}{\chi}|_P(\cdot)\,,\theta|_P(\cdot)\}\,, \qquad P \subset B \tag{1.14}$$

usually is not sufficient to describe the state of P. It can be seen on an example of the energy function. If $E(P; s)$ is the energy of the subset P in the state s of the body B, then we cannot expect it to be given by the relation

$$E(P; s) = \underset{\underset{\sim}{X} \in P}{\mathcal{E}}\,(\underset{\sim}{\chi}(X)\,, \theta(\underset{\sim}{X})), \tag{1.15}$$

where $\mathcal{E}$ is the functional

$$\mathcal{E}: \{\underset{\sim}{\chi}|_P(\cdot),\ \theta|_P(\cdot)\} \longmapsto E(P;s) \in R. \tag{1.16}$$

The value of energy $E(P;s)$ depends also on the configuration and the empirical temperature distribution in the remaining part of the body $B \setminus P$ due to the long-range interactions.

Although, we can achieve a considerable simplification neglecting the long-range interactions but even in this case the state s_P of $P \subset B$ is not defined by the restriction of the state s to the set P. Such a restriction is justified only in undeformed systems in thermodynamic equilibrium, i.e. for

$$\forall_{\underset{\sim}{X} \in B} \quad \theta(\underset{\sim}{X}) = \theta^\circ = const. \tag{1.17}$$

Such situations are considered in classical thermostatics.

It is evident from the above considerations that one of the problems of the definition of the state of subsets of the body B is connected with the structure of those subsets. The problem has been treated extensively in the literature of the subject. We present here only these aspects which are going to be needed in this paper.

First of all, we limit the family of subsets of B to the collection of a sufficient regularity.

(1.18) D e f i n i t i o n: The collection $\mathbb{B}$ of subsets of the body B contains the sets P with the following properties

i/ $\forall_{P \in \mathbb{B}}$ P is the regular region in E^3, i.e. it is a closure of the interior of the bounded subset of E^3: $P = \overline{\text{Int } P}$,

ii/ $\forall_{P_1, P_2 \in \mathbb{B}} \quad P_1 \vee P_2 := P_1 \cup P_2 \in \mathbb{B}$, (the sum)

iii/ $\forall_{P_1, P_2 \in \mathbb{B}} \quad P_1 \wedge P_2 := \overline{\text{Int } P_1 \cap P_2} \in \mathbb{B}$, (the intersection)

iv/ $\exists_{\emptyset \in \mathbb{B}} \forall_{P \in \mathbb{B}} \quad \emptyset \vee P = P$,

v/ $\forall_{P \in \mathbb{B}} \exists_{P^e \in \mathbb{B}} \quad P \vee P^e = B \ \ \& \ \ P \wedge P^e = \emptyset$, (the exterior of B) vi/ if $\emptyset$ is the surface, being the closure of the two-dimensional differentiable C^1-submanifold of E^3, and

$\partial P_1 \cap P_2 = \emptyset$, where $P_1, P_2 \in \mathbb{B}$, then

$$\exists_{\{A_i\}_{i=1}^{\infty} \subset \mathbb{B}} \bigcap_{i=1}^{\infty} A_i = \emptyset .$$

Such surfaces are called material. We assume that all material surfaces are oriented

In the above definition Int(.) denotes the interior in the topology of B relative to E^3, and $\overline{(.)}$ is the closure in this topology. The surface of P is denoted by ∂P. The orientation of material surfaces, assumed in the above definition is chosen to be exterior to the corresponding subset P. For instance, if P_1 and P_2 are non-overlapping, i.e.

$$P_1 \wedge P_2 = \emptyset \tag{1.19}$$

and

$$\partial P_1 \cap P_2 \neq \emptyset$$

then, for $\partial P_1 \cap P_2$ being material, it has an orientation exterior to P_1 and

$$\text{orientation } (\partial P_1 \cap P_2) = -\text{orientation } (\partial P_2 \cap P_1). \tag{1.20}$$

This structure was proposed by W.Noll 1966 and discussed later by M.E.Gurtin and W.O.Williams e.g. 1967.

It is convenient to introduce the partial ordering in $\mathbb{B}$. Namely

$$\forall_{P_1, P_2 \in \mathbb{B}} \quad P_1 < P_2 \iff P_1 \wedge P_2 = P_1 . \tag{1.21}$$

The definition (1.21) is obviously equivalent to the following relation

$$\forall_{P_1, P_2 \in \mathbb{B}} \quad P_1 < P_2 \iff P_1 \vee P_2 = P_2 . \tag{1.22}$$

The following relations follow at once from the definition (1.21)

$$\forall_{P \in \mathbb{B}} \ \emptyset < P \ , \quad \forall_{P \in \mathbb{B}} \ P < B \, . \tag{1.23}$$

Many useful formulae, following from the definition (1.18) can be found in the paper: K.Wilmanski[1972].

If the body B is endowed with the properties described by the axiom (1.4), then the members of $\mathbb{B}$ will be called subsystems or s u b b o d i e s of B.

Finally, it is convenient to introduce the collection of pairs of non-overlapping subbodies which will be denoted by

$$\operatorname{sep} \mathbb{B} \times \mathbb{B} := \{ (P_1, P_2) \mid P_1, P_2 \in \mathbb{B} \ \& \ P_1 \wedge P_2 = \emptyset \} \tag{1.24}$$

In addition to the above limitations on the class of subbodies, we assume that all sets belonging to $\mathbb{B}$ are volume measurable. The Lebesgue volume measure in E^3 is denoted by $\mathcal{v}$. The assumption means that the value of $\mathcal{v}(P)$ is defined for each member of $\mathbb{B}$. Moreover, we assume

$$\mathcal{v}(B) < +\infty \, . \tag{1.25}$$

On the other hand, we assume that all material surfaces are Lebesgue surface measurable. We denote this measure by $\mathcal{s}$

The above structure together with the state space S of the body B yields the construction of a balance equation for each subbody P, and that, in turn, is leading to the notion of thermodynamic processes in subsystems. We present this construction in the next Chapter.

Throughout the paper we use the Cartesian reference frames. In the reference configuration we use the notation X_K for the Cartesian coordinates of the particle $\underset{\sim}{X} \in B$. We call them m a t e r i a l c o o r d i n a t e s of the point $\underset{\sim}{X} \in B$. In the present configuration the Cartesian coordinates of the point $\underset{\sim}{x}$ of the configuration space are denoted by x_k. Hence the motion of the generic point $\underset{\sim}{X} \in B$ is described by the relation

$$x_k = \chi_k(X_K, t). \tag{1.26}$$

The numbers x_k for a given point $\underset{\sim}{X} \in B$ and a given instant of time are called s p a t i a l c o o r d i n a t e s of this particle.

We assume that the spatial reference frame is inertial. The infinitesimal transformation of this frame

$$x_k \to x_k + \varepsilon_k + \varepsilon_{kl} x_l + \alpha_k t, \qquad \varepsilon_{kl} = -\varepsilon_{lk} \tag{1.27}$$

where $\varepsilon_k, \varepsilon_{kl}, \alpha_k$ are constants, is called an Euclidean transformation. Such transformations form a group. The constant ε_k describes the shifting of the origin, ε_{kl} - the infinitesimal rotation, α_k - the velocity of one inertial frame with respect to another. All fields, appearing in the description of the material are required to be invariant with respect to the transformation

$$x_k \to x_k + \varepsilon_k + \varepsilon_{kl} x_l. \tag{1.28}$$

This assumption is tantamount to the homogeneity and isotropy of the configuration space.

On the other hand, the equations of motion are assumed to be invariant with respect to the Galilean transformation

$$x_k \to x_k + \alpha_k t. \tag{1.29}$$

We discuss the details of these assumptions in the sequel.

Both material and spatial coordinates enable to give the operational interpretation of the motion $\underset{\sim}{\chi}$. This function can be verified experimentally through the observation of the sequence of configurations of the body. The interpretation of the empirical temperature Θ - the second element of the state (1.2) - is not so simple. The rules of interpretation of this notion in equilibrium states are well established through the zero law of thermodynamics. The problem is far more complicated for systems passing the non--equilibrium states. In such a case, we usually introduce the notion of an empirical temperature through some quantities measurable directly in non-equilibrium states and connected with the thermal properties. It may be, for instance, the intensity of infra-red radiation. This quantity in a certain range should change in such a way that its gradient is opposite to the direction of the non-mechanical energy flow in the body. Such a property is not universal for empirical temperatures, i.e. it can vary from one material to another. In Fig.1 we show the estimation of empirical temperature

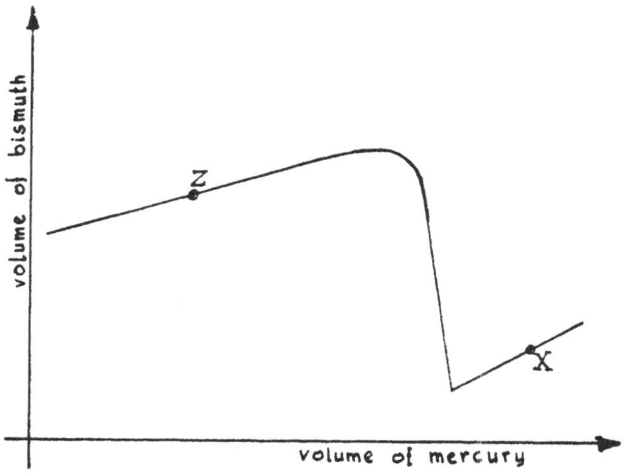

Fig.1

through the volume changes of mercury and bismuth (see:Marvan[1966]). According to the diagram the empirical temperature corresponding to the point X is higher than the empirical temperature corresponding to the point Z when the estimation is carried out through the measurements of the volume changes of mercury. However, the estimation made through the volume changes of bismuth points at the converse relation of the values of empirical temperature. It means that one of these substances cannot be used as a measure of empirical temperature in the considered range of volume changes.

Further in the paper we do not consider the means of temperature measurements assuming only that in any case an empirical temperature is an operational notion.

CHAPTER 2
Balance Laws

2.1. Continuity assumptions

All physical theories are based on a certain set of balance equations. Their derivation varies from the variational methods to the a priori statements. We adopt in this paper an approach suggested by M.E.Gurtin, W.Noll and W.O.Williams (see, for instance, M.E.Gurtin, W.O.Williams[1967]) and based on the axioms of continuity, As an example, we consider the scalar balance equation. Let p_d be a chosen direct thermodynamic process with the given parametrization

$$t \mapsto s \in \pi \,, \quad p_d = (\pi, \rightleftarrows) \qquad t \in T \tag{2.1}$$

For any closed subinterval $T' \subset T$, the collection of states

$$\pi' := \left\{ s \in \pi \,\middle|\, \underset{t \in T'}{\exists} \, t \mapsto s \right\} \tag{2.2}$$

defines a direct process

$$p'_d = (\pi', \rightleftarrows) \tag{2.3}$$

for which the parametrization is compatible with that given for the process p_d . Such a process is called a subprocess of p_d . Let

$$E : \mathbb{B} \times T \to R \tag{2.4}$$

be a given state function, defined for each subbody $P \in \mathbb{B}$ and each state s, $s \in \pi$, parametrized by $t \in T$. We say that the above function satisfies the b a l a n c e e q u a t i o n if there exists a function Γ

$$\forall_{(P_1, P_2) \in \mathrm{sep}\, \mathbb{B} \times \mathbb{B}} \quad ((P_1, P_2), T) \mapsto \Gamma(P_1, P_2; T) \tag{2.5}$$

called a f l u x from the subbody P_2 to the subbody P_1 and defined for each subprocess p'_d of the process p_d , such that

$$\forall_{P \in \mathbb{B}} \; E(P, t_r) - E(P, t_l) = \Gamma(P, P^e; T')$$
$$t_r \mapsto r p'_d \; , \quad t_l \mapsto l p'_d . \tag{2.6}$$

We dispense with the detailed discussion of this global form of the balance equation. Under certain smoothness assumptions, we can shrink down the family of subprocesses to the state s , obtaining the local form of the balance equation

$$\forall_{P \in \mathbb{B}} \; \forall_{t \in T} \; \dot{E}(P, t) = Q(P, P^e; t) ; \quad \dot{E}(P, t) := \frac{dE}{dt}(P, t) \tag{2.7}$$

where $Q(P, P^e; t)$ is the limit of $\frac{1}{t_r - t_l} \Gamma(P, P^e; T')$ for $t_l \nearrow t \swarrow t_r$

It is obvious that neither $\dot{E}(P, t)$ nor $Q(P, P^e; t)$ are the state functions.

Dealing with the continuous body, we expect the set functions in balance equations to satisfy certain continuity assumptions. We present here the simplest form of such conditions, excluding the surface concentrations. Some generalizations can be found in the papers: G.M.C.Fisher, M.J.Leitman[1968], W.O.Williams[1971,1972].

(2.8) A x i o m: There exist positive constants α, β, τ such that

$$\forall_{(P_1,P_2)\in \operatorname{sep} \mathcal{B}\times\mathcal{B}} \quad |Q(P_1,P_2;t)| \leq \alpha\, \vartheta(P_1)\, \vartheta(P_2) + \beta\, \mathfrak{s}(\partial P_1 \cap P_2),$$

$$\forall_{P\in\mathcal{B}} \quad |\dot{E}(P,t)| \leq \tau\, \vartheta(P);$$

we assume $\alpha, \hat{\beta}, \tau$ to be the least Lipschitz constants

The first inequality in this axiom yields the decomposition of the flux Q into two parts: the first one- Lipschitz continuous with respect to the surface measure and the second one - lipschitz continuous with respect to the volume measure. However, this decomposition, though expectable, is far from being unique. We achieve the uniqueness assuming in addition the biadditivity (i.e. neglecting many-body interactions):

(2.9) A x i o m: For any three mutually non-overlapping subbodies P_1, P_2, P_3

$$Q(P_1 \vee P_2, P_3; t) = Q(P_1,P_3;t) + Q(P_2,P_3;t),$$

$$Q(P_1, P_2 \vee P_3; t) = Q(P_1,P_2;t) + Q(P_1,P_3;t)$$

Let us consider two non-overlapping subbodies P_1, P_2 such that $\partial P_1 \cap P_2 \neq \emptyset$. Simultaneously, let us introduce the family of subbodies $\{A_i\}_{i=1}^{\infty}$ possesing the properties (Fig.2)

$$\text{i/} \quad \forall_{1\leq i<\infty} \quad \partial P_1 \cap A_i = \partial P_1 \cap P_2,$$

$$\text{ii/} \quad \forall_{1\leq i<j<\infty} \quad A_j < A_i < P_2, \tag{2.10}$$

$$\text{iii/} \quad \bigcap_{i=1}^{\infty} A_i = \partial P_1 \cap P_2 .$$

Such a family exists due to the point vi/ of the definition (1.18). For any two members A_i, A_j, $i > j$, of the above family we have

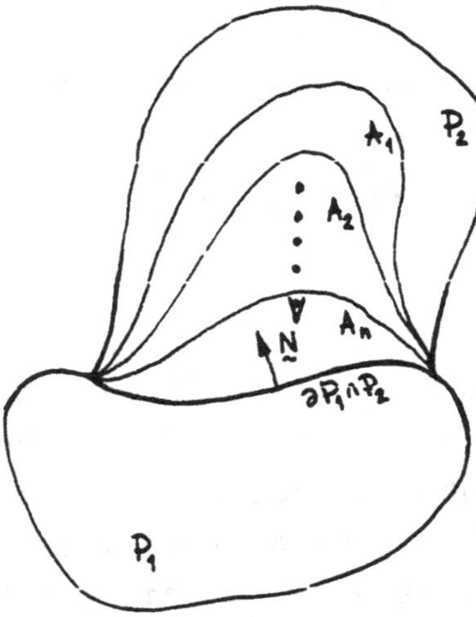

Fig. 2

$$Q(P_1, A_j; t) - Q(P_1, A_i; t) = Q(P_1, A_j - A_i; t), \tag{2.11}$$

where

$$A_j - A_i := \overline{\operatorname{Int} A_j \setminus A_i}. \tag{2.12}$$

Obviously

$$\partial P_1 \cap (A_j - A_i) = \emptyset, \quad \text{i.e.} \quad s(\partial P_1 \cap (A_j - A_i)) = 0. \tag{2.13}$$

Bearing in mind the axiom (2.8) and (2.11), we have

$$|Q(P_1, A_j; t) - Q(P_1, A_i; t)| \leq \alpha\, v(P_1)\, |v(A_j) - v(A_i)|. \tag{2.14}$$

On the other hand, the Lebesgue volume measure in the Euclidean space E^3 is complete, i.e.

$$\lim_{i \to \infty} v(A_i) = v\left(\bigcap_{i=1}^{\infty} A_i\right) = 0 \tag{2.15}$$

for the previously defined family of subbodies. According to the Bolzano-Cauchy's theorem we have then

$$\forall_{\varepsilon>0} \exists_{N} \; i,j>N \Rightarrow |\vartheta(A_i) - \vartheta(A_j)| < \frac{\varepsilon}{\alpha \vartheta(P_1)} . \tag{2.16}$$

Substituting (2.16) in (2.14), we obtain

$$\forall_{\varepsilon>0} \exists_{N} \; i,j>N \Rightarrow |Q(P_1,A_j;t) - Q(P_1,A_i;t)| < \varepsilon . \tag{2.17}$$

According to the same theorm, it means that the limit

$$Q_s(P_1,P_2;t) := \lim_{i\to\infty} Q(P_1,A_i;t) \tag{2.18}$$

exists.

Due to the completeness of the volume measure and the relation

$$\forall_{A_i} \; |Q(P_1,A_i;t)| \leq \alpha \vartheta(P_1) \vartheta(A_i) + \beta s(\partial P_1 \cap P_2)$$

we get

$$|Q_s(P_1,P_2;t)| \leq \beta s(\partial P_1 \cap P_2). \tag{2.19}$$

Hence, the surface part of the flux Q, defined by the formula (2.18) possesses one of the properties expected from the surface flux - it vanishes for two non-overlapping sub-bodies P_1, P_2 having no common part of the boundaries.

It is also easy to check that the definition (2.18) does not depend on the choice of the family $\{A_i\}_{i=1}^{\infty}$, i.e. the surface flux Q_s is well-defined. Simultaneously, it can be proved (see: K.Wilmanski[1977,2]) that for three sub-bodies P_1, P_2, P_3 endowed with the properties (Fig.3)

$$P_1 \wedge P_2 = \emptyset \;, \quad P_1 \wedge P_3 = \emptyset \;, \quad \partial P_1 \cap P_2 = \partial P_1 \cap P_3 \,, \tag{2.20}$$

the following relation holds

$$Q_s(P_1,P_2;t) = Q_s(P_1,P_3;t). \tag{2.21}$$

It means that indeed this part of the flux describes the influx to P_1 through its surface.

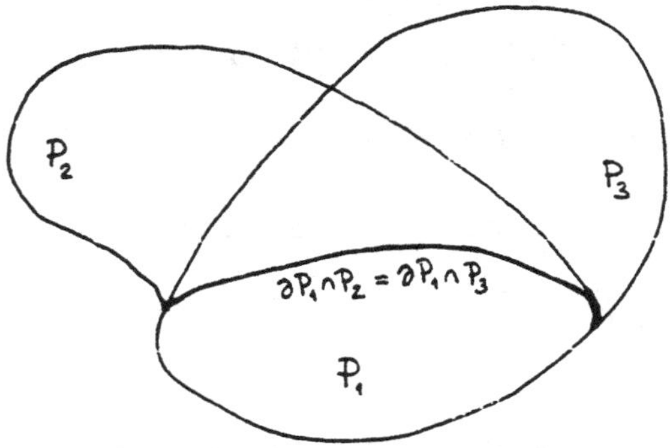

Fig.3

The remaining part of the flux is called the volume flux and it describes the long-range spatial interactions in the body B. Such a term may appear in the case of the internal radiation, mutual body forces etc. It is denoted

$$Q_v(\mathcal{P}_1, \mathcal{P}_2; t) := Q(\mathcal{P}_1, \mathcal{P}_2; t) - Q_s(\mathcal{P}_1, \mathcal{P}_2; t). \tag{2.22}$$

Taking into account the definition of the surface flux, we get

$$\begin{aligned} Q_v(\mathcal{P}_1, \mathcal{P}_2; t) &= Q(\mathcal{P}_1, \mathcal{P}_2; t) - \lim_{i \to \infty} Q(\mathcal{P}_1, A_i; t) = \\ &= \lim_{i \to \infty} \left[Q(\mathcal{P}_1, \mathcal{P}_2; t) - Q(\mathcal{P}_1, A_i; t) \right] = \\ &= \lim_{i \to \infty} Q(\mathcal{P}_1, \mathcal{P}_2 - A_i; t), \end{aligned} \tag{2.23}$$

and, according to the continuity axiom (2.8)

$$|Q_v(\mathcal{P}_1, \mathcal{P}_2; t)| \leq \alpha\, v(\mathcal{P}_1)\, v(\mathcal{P}_2), \tag{2.24}$$

due to the relation

$$\partial\mathcal{P}_1 \cap (\mathcal{P}_2 - A_i) = \emptyset \Rightarrow s(\partial\mathcal{P}_1 \cap (\mathcal{P}_2 - A_i)) = 0. \tag{2.25}$$

Summing up the above considerations we arrive at the following formula

$$\begin{aligned} &\underset{(P_1,P_2)\in \operatorname{sep}\mathcal{B}\times\mathcal{B}}{\forall} \quad Q(P_1,P_2;t) = Q_s(P_1,P_2;t) + Q_v(P_1,P_2;t), \\ &|Q_s(P_1,P_2;t)| \leq \beta\, \mathcal{J}(\partial P_1 \cap P_2), \\ &|Q_v(P_1,P_2;t)| \leq \alpha\, v(P_1)\, v(P_2). \end{aligned} \tag{2.26}$$

Further in the paper we neglect the long-range interactions assuming

$$\underset{(P_1,P_2)\in \operatorname{sep}\mathcal{B}\times\mathcal{B}}{\forall} \quad Q_v(P_1,P_2;t) \equiv 0. \tag{2.27}$$

This assumption leads at once to the further simplifications of the form of the surface flux. Taking the family $\{C_i\}_{i=1}^{\infty}$ of subbodies, satisfying the conditions similar to (2.10), but for the subbody P_1, i.e.

$$\begin{aligned} &i/ \quad \underset{1\leq i<\infty}{\forall} \partial C_i \cap P_2 = \partial P_1 \cap P_2, \\ &ii/ \quad \underset{1\leq i<j<\infty}{\forall} \quad C_j < C_i < P_1, \\ &iii/ \quad \bigcap_{i=1}^{\infty} C_i = \partial P_1 \cap P_2, \end{aligned} \tag{2.28}$$

and bearing in mind the biadditivity of the flux, we obtain

$$\underset{i}{\forall} \; Q(P_1,P_2;t) = Q(C_i,P_2;t) + Q_v(P_1 - C_i, P_2;t).$$

On the other hand, we have

$$\lim_{i\to\infty} Q_v(P_1 - C_i, P_2;t) = Q_v(P_1,P_2;t).$$

Hence

$$Q_s(P_1,P_2;t) = \lim_{i\to\infty} Q(C_i,P_2;t). \tag{2.29}$$

Taking into account both the definition (2,18) and the above formula, we see that the continuity and biadditivity assumptions lead to the notion of the surface flux depending on the contact surface $\partial P_1 \cap P_2$ and not on the subbodies P_1 and P_2. It means that we can introduce the function H(.;t), defined on the collection of all material surfaces in the body B:

$$\forall_{\mathcal{S} \subset B} \quad \mathcal{S} \longmapsto \mathcal{H}(\mathcal{S};t). \tag{2.30}$$

For any two subbodies P_1, P_2 satisfying the relation

$$\partial P_1 \cap P_2 = \mathcal{S} \tag{2.31}$$

we define

$$\mathcal{H}(\partial P_1 \cap P_2;t) := Q_s(P_1, P_2;t). \tag{2.32}$$

Let us check the behavior of the function H(.;t) under the change of the orientation of the surface. The balance equation (2.7) takes now the following form

$$\dot{E}(P;t) = \mathcal{H}(\partial P;t) + Q_v(P, P^e;t). \tag{2.33}$$

If we write down this equation for two non-overlapping subbodies P_1, P_2 and their sum $P_1 \vee P_2$, we have

$$\begin{aligned}\dot{E}(P_1 \vee P_2;t) - \dot{E}(P_1;t) - \dot{E}(P_2;t) &= Q(P_1 \vee P_2, P_1^e \wedge P_2^e;t) - \\ &- Q(P_1, P_1^e;t) - Q(P_2, P_2^e;t) = -Q(P_1, P_2;t) - Q(P_2, P_1;t) = \\ &= -\mathcal{H}(\partial P_1 \cap P_2;t) - \mathcal{H}(\partial P_2 \cap P_1;t) - \\ &- Q_v(P_1, P_2;t) - Q(P_2, P_1;t)\end{aligned} \tag{2.34}$$

due to the biadditivity of the flux. Making use of this relation for any pair (C_i, A_k) of subbodies belonging to the families (2.10) and (2,28), respectively, and shrinking down these sets we arrive at

$$\lim_{\substack{i\to\infty\\ k\to\infty}}[\dot{E}(C_i\vee A_k;t)-\dot{E}(C_i;t)-\dot{E}(A_k;t)]=-H(\partial P_1\cap P_2;t)-$$
$$-H(\partial P_2\cap P_1;t)-\lim_{\substack{i\to\infty\\ k\to\infty}}[Q_v(C_i,A_k;t)+Q_v(A_k,C_i;t)].$$

The continuity assumption (2.8), the property (2.24) and the completeness of the volume measure finally lead to the relation

$$H(\partial P_1\cap P_2;t)=-H(\partial P_2\cap P_1;t). \tag{2.35}$$

It means that the change of the orientation of the surface yields the change of sign of the surface flux. In such a case we say that the surface flux is b a l a n c e d.

For the surface flux endowed with the above property, we can prove a very important representation theorem. It has been proved by M.E.Gurtin and W.O.Williams[1967] (Theorem 7; see also W.O.Williams[1970], Proposition 6) that for each unit vector field over B there exists a scalar function

$$B\ni \underset{\sim}{X}\mapsto q^{\circ}(\underset{\sim}{X},\underset{\sim}{N}(\underset{\sim}{X});t)\in R \tag{2.36}$$

such that

$$H(\mathscr{S};t)=\int_{\mathscr{S}} q^{\circ}(\underset{\sim}{X},\underset{\sim}{N}(\underset{\sim}{X});t)\,d\mathscr{S}_X\,,\quad q^{\circ}(\underset{\sim}{X},\underset{\sim}{N}(\underset{\sim}{X});t)=-q^{\circ}(\underset{\sim}{X},-\underset{\sim}{N}(\underset{\sim}{X});t), \tag{2.37}$$

for each material surface $\mathscr{S}\subset B$. The vector field $\underset{\sim}{N}(\underset{\sim}{X})$ is ortogonal to the surface $\mathscr{S}$ and establishes its orientation.

The standard procedure (e.g. see M.E.Gurtin, V.J.Mizel, W.O.Williams[1968], C.Truesdell[1972]) yields now the Cauchy's theorem on the linear dependence of q on the vector $\underset{\sim}{N}$, i.e.

$$q^{\circ}(\underset{\sim}{X},\underset{\sim}{N}(\underset{\sim}{X});t)=\underset{\sim}{q}^{\circ}(\underset{\sim}{X};t)\cdot\underset{\sim}{N}(\underset{\sim}{X}), \tag{2.38}$$

where $\underset{\sim}{q}^{o}(.;t)$ is the vector field over B and the dot stands for the scalar product. Hence, the balance equation for any subbody of B can be written in the form

$$\dot{E}(P;t) = \oint_{\partial P} \underset{\sim}{q}^{\circ}\cdot \underset{\sim}{N}\, ds + Q_v(P,P^e;t). \tag{2.39}$$

As we have already mentioned, we neglect in this paper the long-range interactions, i.e. we assume (2.27) to hold. Then for any two non-overlapping subbodies P_1, P_2 we obtain from the equation (2.39)

$$\dot{E}(P_1 \vee P_2;t) - \dot{E}(P_1;t) - \dot{E}(P_2;t) = \oint_{\partial(P_1\vee P_2)} \underset{\sim}{q}^{\circ}\cdot \underset{\sim}{N}\, ds - \oint_{\partial P_1} \underset{\sim}{q}^{\circ}\cdot \underset{\sim}{N}\, ds - \oint_{\partial P_2} \underset{\sim}{q}^{\circ}\cdot \underset{\sim}{N}\, ds.$$

Assuming that there are no singular surfaces in the body B (we discuss this problem in the sequel), we can use the Stokes theorem

$$\dot{E}(P_1 \vee P_2;t) - \dot{E}(P_1;t) - \dot{E}(P_2;t) = \int_{P_1\vee P_2} \mathrm{Div}\, \underset{\sim}{q}^{\circ}\, dv - \int_{P_1} \mathrm{Div}\, \underset{\sim}{q}^{\circ}\, dv - \int_{P_2} \mathrm{Div}\, \underset{\sim}{q}^{\circ}\, dv = 0,$$

where Div is the divergence operator in the reference configuration B. Hence

$$\dot{E}(P_1 \vee P_2;t) = \dot{E}(P_1;t) + \dot{E}(P_2;t). \tag{2.40}$$

The above proved additivity of the rate E(.;t) and its volume continuity, assumed in (2.8), yield the following theorem, proved by M.E.Gurtin and W.O.Williams[1967]: There exists a scalar function

$$e^{\circ}(.;t): B \rightarrow R \tag{2.41}$$

such that for each subbody P

$$\dot{E}(P;t) = \int_P \dot{e}^{\circ}(\underset{\sim}{X};t)\, dv_x. \tag{2.42}$$

Let us mention that both relations (2.37) and (2.42) are the consequences of the Radon-Nikodym theorem on the existence of derivatives with respect to a measure.

Finally, in the absence of long-range interactions and singular surfaces, the balance equation under considerations takes the form

$$\underset{P \in \mathbb{B}}{\forall} \frac{d}{dt} \int_{P} e^{\circ}(\underset{\sim}{X};t)\, dv_X = \oint_{\partial P} \underset{\sim}{q}^{\circ}(\underset{\sim}{X};t) \cdot \underset{\sim}{N}(\underset{\sim}{X})\, ds_X . \tag{2.43}$$

For the purpose of this paper it is convenient to write the equation (2.43) for the current configuration of the subbody P. We use the following notation

$$\chi(P;t) := \{\underset{\sim}{x} \in E^3 \mid \underset{\sim}{x} = \chi(\underset{\sim}{X};t),\ \underset{\sim}{X} \in P\},$$
$$\partial\chi(P;t) := \{\underset{\sim}{x} \in E^3 \mid \underset{\sim}{x} = \chi(\underset{\sim}{X};t),\ \underset{\sim}{X} \in \partial P\}. \tag{2.44}$$

Then the balance equation (2.43) takes the form

$$\frac{d}{dt} \int_{\chi(P;t)} e(\underset{\sim}{x};t)\, dv = \oint_{\partial\chi(P;t)} \underset{\sim}{q}(\underset{\sim}{x};t) \cdot \underset{\sim}{n}(\underset{\sim}{x};t)\, ds \tag{2.45}$$

where

$$e(\underset{\sim}{x};t) := J\, e^{\circ}[\underset{\sim}{\chi}^{-1}(\underset{\sim}{x};t);t] \tag{2.46}$$

and similarly for the other quantities, while the volume and surface measure are refered to the current configuration of the body B.

We discuss the particular balance equations for thermo-mechanical processes in the body B in the next Section of the paper.

2.2. Local Balance Laws

Now we consider the balance laws occuring in the case of the body undergoing the thermomechanical processes. In this section, we neglect the phenomena being described in continuous models of singular surfaces. We present the simplest theory of balance laws for such surfaces in the next Section. Under these assumptions, we deal with the following laws

i/ balance of mass

ii/ balance of momentum

iii/ balance of moment of momentum

iv/ balance of energy

v/ balance of entropy.

The last law differs slightly from that considered in the previous Section due to the presence of the sources. However it is easy to extend the procedure to cover also this case. The equation (2.45) takes for these five balance laws the following form

$$\frac{d}{dt}\int_{\chi(P;t)} \varrho\, dv = 0 \qquad \text{(balance of mass)} \qquad (2.47)$$

$$\frac{d}{dt}\int_{\chi(P;t)} \varrho v_k\, dv = \oint_{\partial\chi(P;t)} t_{kl} n_l\, ds \qquad \text{(balance of momentum)} \qquad (2.48)$$

$$\frac{d}{dt}\int_{\chi(P;t)} \varrho x_{[m} v_{k]}\, dv = \oint_{\partial\chi(P;t)} x_{[m} t_{k]l} n_l\, ds \qquad \text{(balance of moment of momentum)} \qquad (2.49)$$

$$\frac{d}{dt}\int_{\chi(P;t)} \varrho(\tfrac{1}{2} v_k v_k + \varepsilon)\, dv = \oint_{\partial\chi(P;t)} (v_k t_{kl} n_l - q_k n_k)\, ds \qquad \text{(balance of energy)} \qquad (2.50)$$

$$\frac{d}{dt}\int_{\chi(P;t)} \varrho\eta\, dv = -\oint_{\partial\chi(P;t)} h_k n_k\, ds + \int_{\chi(P;t)} \varrho\eta^*\, dv \qquad \text{(balance of entropy)} \qquad (2.51)$$

where $\varrho \equiv \varrho(\underset{\sim}{x};t)$ -mass density in the current configuration;

$v_k \equiv v_k(\underset{\sim}{x};t)$ -the components of the particle velocity in the current configuration;

$$v_k(\underset{\sim}{x};t) := \frac{\partial \chi_k}{\partial t}[\underset{\sim}{\chi}^{-1}(\underset{\sim}{x};t);t] \qquad (2.52)$$

t_{kl} -the components of the Cauchy's stress tensor;

ε -internal specific energy per unit mass in the current configuration;

q_k -the components of the heat flux vector (per unit material surface in the current configuration);

η -specific entropy per unit mass in the current configuration;

h_k -the components of the entropy flux vector (per unit material surface in the current configuration);

η^* -the intensity of entropy source per unit time and unit mass in the current configuration.

For the detailed discussion of the above notions we refer the reader to any standard textbook on continuum mechanics (e.g. see: C.Truesdell[1966,1972]). Certain aspects of the entropy balance law are presented in the Section 2.4.

In the absence of singular surfaces passing through the subbody P, we can easily replace the above balance laws by their local counterparts holding at each point $\underset{\sim}{X}$ of the subbody P. Namely, due to the assumption that these laws hold for every subbody $P' \subset P$, we have, for instance

$$\lim_{P' \searrow \{\underset{\sim}{X}\} \subset P} \frac{1}{v(P')} \frac{d}{dt} \int_{\chi(P';t)} \varrho v_k \, dv = \lim_{P' \searrow \{\underset{\sim}{X}\} \subset P} \frac{1}{v(P')} \oint_{\partial\chi(P';t)} t_{kl} n_l \, ds . \tag{2.53}$$

With respect to the relation

$$\frac{d}{dt} \int_{\chi(P';t)} \varrho v_k \, dv = \int_{\chi(P';t)} \frac{\partial}{\partial t}(\varrho v_k) \, dv + \oint_{\partial\chi(P';t)} \varrho v_k v_l n_l \, ds \tag{2.54}$$

and the Stokes theorem, we obtain

$$\forall_{\underset{\sim}{x} \in \chi(P;t)} \quad \frac{\partial}{\partial t}(\varrho v_k) + (\varrho v_k v_l)_{,l} = t_{kl,l} . \tag{2.55}$$

The similar manipulations for the remaining balance laws lead to the following set of equations

$\dot{\varrho} + \varrho v_{l,l} = 0$ (local mass balance) (2.56)

$\varrho \dot{v}_k = t_{kl,l}$ (local momentum balance) (2.57)

$t_{kl} = t_{lk}$ (local moment of momentum balance) (2.58)

$$\rho\dot{\varepsilon} = v_{k,l}t_{kl} - q_{k,k} \qquad \text{(local energy balance)} \tag{2.59}$$

$$\rho\dot{\eta} = -h_{k,k} + \rho\eta^{*} \qquad \text{(local entropy balance)} \tag{2.60}$$

where the dot denotes the material time derivative. For example

$$\dot{v}_k := \frac{\partial v_k}{\partial t} + v_{k,l}v_l . \tag{2.61}$$

The direction of the heat flux vector has been chosen in the above formula, as is customary, to be positive in the case of the thermal emission of the energy from the subbody P to the exterior, i.e. if $q_k n_k(\underset{\sim}{x},t) > 0$, then the flow of the energy is at the point $\underset{\sim}{x}$ directed from P to P^{e}. The same remark concerns the entropy flux h_k.

Finally, let us mention that the mass balance equation (2.56) can be solved independently of the material and of the remaining equations. If ρ_o is the mass density in the reference configuration and J is the jacobian of the deformation gradient from the reference configuration to the current one, i.e.

$$J(\underset{\sim}{X};t) := |\det \chi_{k,K}(\underset{\sim}{X};t)| \tag{2.62}$$

then

$$\rho(\underset{\sim}{x};t)\, J(\underset{\sim}{\chi}^{-1}(\underset{\sim}{x};t);t) = \rho_o(\underset{\sim}{\chi}^{-1}(\underset{\sim}{x};t)). \tag{2.63}$$

2.3. Singular Surfaces

The brief presentation of the theory of balance equation in continuum just delivered neglects one of the most interesting features of the continuous model - the influence of the singular surface. Such surfaces model the boundaries of the body, interfaces, acoustic and shock waves and many other physical phenomena, whose structural description, if possible is usually very complicated. On the other hand, these phenomena furnish tools for the experimental measurements of dynamic properties of materials. For this reason, we discuss the properties of singular surfaces in many places in this paper.

Let us start with the simplest case of the singular surface passing through the body B and described by the balance equation of the form (2.15).

We say that the field $e(.;t)$ is s m o o t h in $\chi(B;t)$ if

i/ it is continuously differentiable in $\chi(B;t)\setminus\sigma(t)$, where $\sigma(t)$ is an oriented C^2 manifold of the dimension 2;

ii/ it approaches finite limits $e^+(x;t)$ and $e^-(x;t)$ for $\mathcal{J}$ -almost every point of $\sigma(t)$;

iii/ the flux normal to $\sigma(t)$: $\underset{\sim}{q}\cdot\underset{\sim}{n}(.;t)$ approaches finite limits $\underset{\sim}{q}^+(\underset{\sim}{x};t)\cdot\underset{\sim}{n}(\underset{\sim}{x};t)$ and $\underset{\sim}{q}^-(\underset{\sim}{x};t)\cdot\underset{\sim}{n}(\underset{\sim}{x};t)$ for $\mathcal{J}$ -almost every point of $\sigma(t)$;

iv/ the velocity field $\underset{\sim}{v}(.;t)$ approaches finite limits $\underset{\sim}{v}^+(x;t)$ and $\underset{\sim}{v}^-(x;t)$ for $\mathcal{J}$ -almost every point of $\sigma(t)$;

v/ the motion $\underset{\sim}{\chi}(.;t)$ is continuous through the surface $\sigma(t)$.

The surface $\sigma(t)$ is said to be s i n g u l a r with respect to the balance equation (2.45) if at least one of the following quantities

$$[\![e]\!]:=e^+-e^-,$$

$$[\![\underset{\sim}{q}\cdot\underset{\sim}{n}]\!]:=\underset{\sim}{q}^+\cdot\underset{\sim}{n}-\underset{\sim}{q}^-\cdot\underset{\sim}{n}, \tag{2.64}$$

$$[\![\underset{\sim}{v}]\!]:=\underset{\sim}{v}^+-\underset{\sim}{v}^-,$$

do not vanish identically on the surface $\sigma(t)$.

Deriving the balance equation for a singular surface we use the notation shown in Fig.4.

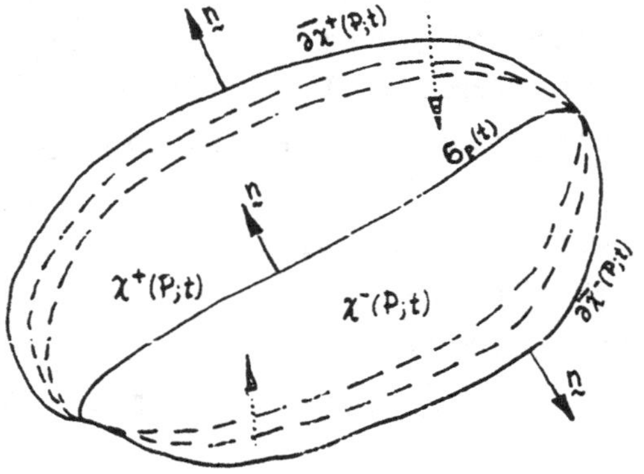

Fig.4

The derivation is based on the following

(2.65) A x i o m: The balance equation (2.45) holds for every subbody $P \in \mathbb{B}$ independently of the location of points of the singular surface $\sigma(t)$ in $\chi(B;t)$

In the case of the subbody P, whose current configuration is containing the part of the singular surface $\sigma(t)$, we cannot repeat the manipulations performed in the preceding Section. We proceed as follows. We divide the region $\chi(P;t)$ into two parts as pointed out in Fig.4. Then

$$\frac{d}{dt}\int_{\chi(P;t)} e\, dv = \frac{d}{dt}\int_{\chi^+(P;t)} e\, dv + \frac{d}{dt}\int_{\chi^-(P;t)} e\, dv =$$

$$= \int_{\chi^+(P;t)} \frac{\partial e}{\partial t} dv + \int_{\partial\chi^+(P;t)} e\, \underset{\sim}{v}\cdot\underset{\sim}{n}\, ds - \int_{\sigma_P(t)} e\, c\, ds + \tag{2.66}$$

$$+ \int_{\chi^-(P;t)} \frac{\partial e}{\partial t} dv + \int_{\partial\chi^-(P;t)} e\, \underset{\sim}{v}\cdot\underset{\sim}{n}\, ds + \int_{\sigma_P(t)} e\, c\, ds, \quad \sigma_P(t) := \sigma(t) \cap \chi(P;t).$$

In the above formulae c denotes the normal speed of displacement of the surface $\sigma(t)$ with respect to the same frame as that used for the body motion. The difference in sign of integrals over the surface $\sigma_p(t)$ results from the different orientation of this surface with respect to the regions $\chi^+(P;t)$ and $\chi^-(P;t)$.

On the other hand

$$\oint\limits_{\partial\chi(P;t)} \underset{\sim}{q}\cdot\underset{\sim}{n}\,ds = \int\limits_{\overline{\partial\chi^+}(P;t)} \underset{\sim}{q}\cdot\underset{\sim}{n}\,ds + \int\limits_{\overline{\partial\chi^-}(P;t)} \underset{\sim}{q}\cdot\underset{\sim}{n}\,ds. \tag{2.67}$$

Shrinking down the regions $\chi^+(P;t)$ and $\chi^-(P;t)$ to the surface $\sigma_p(t)$ as shown by the dotted lines in Fig.4, making use of the definition of the singular surface and the axiom (2.65) we arrive at

$$\int\limits_{\sigma_p(t)} \{e^+\underset{\sim}{v}^+\cdot\underset{\sim}{n} - e^+c - e^-\underset{\sim}{v}^-\cdot\underset{\sim}{n} + e^-c\}\,ds = \int\limits_{\sigma_p(t)} \{\underset{\sim}{q}^+\cdot\underset{\sim}{n} - \underset{\sim}{q}^-\cdot\underset{\sim}{n}\}\,ds.$$

According to the notation (2.64), we finally have

$$\int\limits_{\sigma_p(t)} [\![e(c-\underset{\sim}{v}\cdot\underset{\sim}{n}) + \underset{\sim}{q}\cdot\underset{\sim}{n}]\!]\,ds = 0. \tag{2.68}$$

With respect to the frequent use, we introduce the notation

$$U^{\pm} := c - \underset{\sim}{v}^{\pm}\cdot\underset{\sim}{n}. \tag{2.69}$$

The quantities U^+ and U^- are called instanteneous speeds of propagation of the surface $\sigma(t)$ and they measure the speeds of this surface relative to the particles instanteneously approaching and leaving the surface $\sigma(t)$. The arguments similar to these used in the localization of the balance equation (Sec.2.2.) yield the following local counterpart of the relation (2.68)

$$[\![eU]\!] + [\![\underset{\sim}{q}\cdot\underset{\sim}{n}]\!] = 0 \qquad s\text{-almost everywhere on } \sigma(t). \tag{2.70}$$

This is the sought balance equation for the singular surface and it is called a K o t c h i n e ' s condition.

It is worthy to mention that in the case of the material singular surface we have

$$U^+ = U^- \equiv 0 \tag{2.71}$$

and the Kotchine's condition takes the form

$$[\![\mathbf{q} \cdot \mathbf{n}]\!] = 0. \tag{2.72}$$

Making use of the relation (2.70), we obtain for the fields appearing in thermomechanical processes the following set of conditions

$$
\begin{aligned}
&[\![\varrho U]\!] = 0 && \text{(surface balance of mass)}\\
&[\![\varrho U v_k]\!] + [\![t_{kl} n_l]\!] = 0 && \text{(surface balance of momentum)}\\
&[\![\varrho U (\tfrac{1}{2} v_k v_k + \varepsilon)]\!] + [\![v_k t_{kl} n_l - q_k n_k]\!] = 0 && \text{(surface balance of energy)}
\end{aligned} \tag{2.73}
$$

We return to the surface balance of entropy further in this Section. In the considered case, the surface balance of moment of momentum is identically satisfied.

If we introduce the notation

$$\langle a \rangle := \tfrac{1}{2}(a^+ + a^-) \tag{2.74}$$

then it is easy to prove the following identity

$$[\![a b]\!] = [\![a]\!] \langle b \rangle + \langle a \rangle [\![b]\!]. \tag{2.75}$$

Bearing in mind the identity (2.75), we can write the above set of conditions in the form

$$
\begin{aligned}
&[\![\varrho U]\!] = 0,\\
&m [\![v_k]\!] - [\![t_{kl} n_l]\!] = 0, \qquad m := -\langle \varrho U \rangle\\
&m [\![\varepsilon]\!] - \langle t_{kl} n_l \rangle [\![v_k]\!] + [\![q_k n_k]\!] = 0.
\end{aligned} \tag{2.76}
$$

These relations are often called the Rankine--Hugoniot's conditions.

If the surface is not material and the jump of velocity $[\![v_k]\!]$ is not identically equal to zero we call the surface $\sigma(t)$ the shock wave. On the other hand, if the surface is not material but $[\![v_k]\!] \equiv 0$ then the surface $\sigma(t)$ is called the acoustic wave. Sometimes we refer to those two cases as to the surfaces of strong and weak discontinuities.

In the case of weak discontinuity the conditions (2.76) simplify considerably. Namely, we have

$$[\![\varrho]\!] = 0,$$

$$[\![t_{kl}]\!] n_l = 0, \tag{2.77}$$

$$m [\![\varepsilon]\!] + [\![q_k]\!] n_k = 0.$$

The relation $(2.77)_2$ is called the Poisson's condition and is basic for the theory of propagation of acoustic waves.

If, in addition, the surface $\sigma(t)$ does not carry the jump of specific energy $[\![\varepsilon]\!] \equiv 0$, we have, instead of the relation $(2.77)_3$, the following one

$$[\![q_k]\!] n_k = 0. \tag{2.78}$$

It is called the Fourier's condition. Both Poisson and Fourier's conditions hold also on the material surface of weak discontinuity due to the relation $m \equiv 0$.

The Kotchine's condition, derived on the basis of the balance equation (2.45), is usually violated in most interesting cases of phase transition and is obviously wrong in thecase of the entropy balance due to the influence of entropy sources. However, it is not difficult to extend the procedure of the Section 2.1. to cover also the surface concentrations and sources. This extension requires minor changes in the form of the axiom (2.8). Such changes, as we have already mentioned, were proposed by G.M.C.Fisher and M.J.Leitman[1968] and by W.O.Williams[1972]. We do not present their derivation in this paper. The final result of these considerations is the following balance equation

$$\frac{d}{dt}\Big(\int_{\chi(P;t)} e(x;t)\, dv + \int_{\sigma_P(t)} e_s(x;t)\, ds\Big) = \oint_{\partial\chi(P;t)} q(x;t)\cdot n(x;t)\, ds + \int_{\chi(P;t)} e^*(x;t)\, dv + \int_{\sigma_P(t)} e_s^*(x;t)\, ds. \qquad (2.79)$$

The new fields, appearing in the relation (2.79), are called as follows

e_s - surface concentration of the quantity e,

e_s^* - surface source of the quantity e.

We return to the physical interpretations of these fields in the sequel.

The balance equation of the form (2.79) with some additional terms has been investigated by G.P.Moeckel[1974] and, in a slightly different form, by K.Wilmanski[1974,1975,1977,1].

The modification of the balance law certainly yields the change of the definition of the singular surface and of the form of the Kotchine's condition.

The former requires the supplementation of the list of quantities (2.64) by e_s, $\frac{\partial e_s}{\partial t}$ and e_s^*. At the same time we assume that e^* is continuous everywhere in $\chi(B;t)$.

The latter follows on the same way as the relation (2.70). We have to perform in addition the differentiation of the second term on the left-hand side of the balance law (2.79). The integral counterpart of the relation (2.68) is in the case under consideration of the following form

$$\frac{d}{dt}\int_{\sigma_P(t)} e_s \, ds - \int_{\sigma_P(t)} \{ [\![eU + q\cdot n]\!] + e_s^* \} ds = 0. \tag{2.80}$$

The detailed manipulations to be performed on the first term of this relation are presented, for example, in C.Truesdell, R.A.Toupin[1960],Sec.80 and G.P.Moeckel[1975]. We present only the sketch of this procedure. Taking into account the motion of the surface $\sigma(t)$, we have

$$\frac{d}{dt}\int_{\sigma_P(t)} e_s \, ds = \int_{\sigma_P(t)} \frac{\partial e_s}{\partial t} ds + \int_{\sigma_P(t)} e_s \dot{\overline{ds}} . \tag{2.81}$$

It can be proved that for the appropriate choice of the surface coordinates (moving with the points of the surface) we have

$$\dot{\overline{ds}} = -2c K_M ds \tag{2.82}$$

where K_M is the mean curvature of the surface $\sigma(t)$. Substituting (2.82) in (2.81) and, subsequently, in (2.80), we obtain

$$\int_{\sigma_P(t)} \left\{ \frac{\partial e_s}{\partial t} - 2cK_M e_s - [\![eU + q\cdot n]\!] - e_s^* \right\} ds = 0. \tag{2.83}$$

Consequently, the g e n e r a l i z e d Kotchine's condition takes the following form

$$\frac{\partial e_s}{\partial t} - 2cK_M e_s = [\![eU + q\cdot n]\!] + e_s^* . \tag{2.84}$$

For the particular fields, appearing in the processes considered in this paper, we have

$$\frac{\partial \rho_s}{\partial t} - 2cK_M \rho_s = [\![\rho U]\!] + \rho^* ,$$ (generalized surface balance of mass)

$$\frac{\partial p_k}{\partial t} - 2cK_M p_k = [\![\rho U v_k + t_{kl} n_l]\!] + p_k^* ,$$ (generalized surface balance of momentum)

$$n_{[k} p_{l]} := \frac{1}{2}(n_k p_l - n_l p_k) = 0, \quad \text{(generalized surface balance of moment of momentum)}$$

$$\frac{\partial \varepsilon_s}{\partial t} - 2\mathcal{C} K_M \varepsilon_s = [\![\varrho U(\tfrac{1}{2}v_k v_k + \varepsilon) + v_k t_{kl} n_l - q_k n_k]\!] + \varepsilon_s^*, \quad \text{(generalized surface balance of energy)} \tag{2.85}$$

$$\frac{\partial \eta_s}{\partial t} - 2\mathcal{C} K_M \eta_s = [\![\varrho U \eta - h_k n_k]\!] + \eta_s^*, \quad \text{(generalized surface balance of entropy).}$$

The physical interpretetion of the new quantities related by the conditions (2.85) is as follows:

ϱ_s describes the surface concetration of mass; such a quantity appears, for instance, in the description of strong shock waves in gases (e.g. see: K.Wilmanski[1977,1]);

ϱ_s^* describes the surface sources of mass, which may appear, for example, in the theory of phase transition;

p_k are the components of the surface momentum concentration; they appear again in the theory of strong shock waves;

p_k^* are the components of the surface sources of momentum; they are related to the Maxwell's tension of the theory of capillarity;

ε_s is the surface concentration of the specific energy; such a quantity appears also in the theory of capillarity;

ε_s^* is the surface source of energy; it is again the quantity important in the theory of phase transition;

η_s is the surface concentration of entropy; we investigate the properties of this surface field in the connection with the Galilean invariance;

η_s^* is the surface source of entropy; we discuss it in the next Section.

Let us mention that the generalized surface balance of moment of momentum follows under the additional assumption of the absence of surface couples.

We return to the properties of the set (2.85) in the Section 3.1.2. However, it is worthy to mention a certain bothering feature of the set (2.85), apparent without any further relations. The classical Rankine - Hugoniot's conditions (2.76) are used to find the motion of the surface of strong discontinuity. Hence, they should be invariant with respect to the Galilean transformation of the reference frame. However, due to the presence of the speed of displacement c in the conditions (2.85), in general these conditions are not invariant. The invariance condition for these relations yields a special structure of the concentrations and sources, which we discuss in the Section 3.1.2.

2.4. Second Law of Thermodynamics$

As we have mentioned in the first Section, the construction of the thermodynamic theory can be performed in two ways. The first one, commonly used in the contemporary theory of materials, requires the formulation of analytical conditions, limitting the admissible class of materials. In the case of thermomechanical processes, such conditions form the set of balance equations. However, it is possible to provide many examples, satisfying this set of equations and still impossible in nature. The condition for the admissibility of the process is called the s e c o n d l a w o f t h e r m o d y n a m i c s. Together with balance laws it forms the sufficient condition of admissibility. Since almost the beginning of thermodynamics, this law was formulated in terms of an entropy function. Its physical meaning in thermostatics has been checked in many ways and the measurability of entropy in equilibrium has been proved in numerous experiments. The formal extension of this notion beyond thermostatics has been performed by Carathéodory in 1909 and the experimental verification has justified such a procedure in the case of so-called systems in local equilibrium. However, the general case of non-equilibrium systems is not solved yet. Consequently, the analytical form of the second law of thermodynamics in which the entropy function and its flux are used, should be considered as an ad hoc assumption still open for the criticism. Some progress in this field can be expected after completing the investigation of the geometrical properties of the state space. Such considerations formulate the second law of thermodynamics as an assumption on the stability of thermodynamic processes and the procedure is similar to that of the theory of dynamical systems.

For our purposes, it is sufficient to assume the second law of thermodynamics in the form proposed by I.Müller[1971,1,2].

(2.86) A x i o m: For any given direct thermodynamic process $p_d = (\pi, \chi)$ and given subbody $P \in \mathcal{B}$, the entropy source is non-negative, i.e. for any instant of time $t \in T$ the following inequality holds

$$\int_{\chi(P,t)} \eta^* \, dv + \int_{\sigma_P(t)} \eta_s^* \, ds \geq 0 .$$

Making use of the balance equation (2.79), we obtain the following Clausius - Duhem - Müller inequality

$$\frac{d}{dt}\left(\int_{\chi(P,t)} \varrho \eta \, dv + \int_{\sigma_P(t)} \eta_s \, ds \right) \geq - \oint_{\partial\chi(P,t)} h_k n_k \, ds, \tag{2.87}$$

where η_s is the entropy surface concentration and the remaining symbols were explained after the formula (2.52).

After easy calculations, we obtain the following local counterparts of the inequality (2.87)

$$\forall_{\underset{\sim}{x} \in \chi(\mathcal{B},t)\setminus\sigma(t)} \quad \varrho \dot{\eta} + h_{k,k} \geq 0, \tag{2.88}$$

$$\forall_{x \in \sigma(t)} \quad \frac{\partial \eta_s}{\partial t} - 2cK_M \eta_s - [\![\varrho \eta U - h_k n_k]\!] \geq 0 \tag{2.89}$$

(see: the formulae (2.60) and $(2.85)_5$).

It should be pointed out that so far we have not assumed the relation

$$h_k = \frac{q_k}{\Theta}$$

appearing in earlier formulation of the Clausius - Duhem inequality. The generalisation of this relation has been proposed by I.Müller and it is of the very important consequences, presented further in the paper.

The simplified form of the inequality (2.89) has appeared already in the papers on the propagation of shock waves in gases. Assuming the absence of the surface concentration of entropy η_s, and the continuity of the entropy flux (adiabatic shock), we have

$$[\![\varrho \eta U]\!] \leqslant 0. \tag{2.90}$$

The balance of mass on the singular surface $(2.76)_1$ yields

$$m\,[\![\eta]\!] \geqslant 0. \tag{2.91}$$

It means that the entropy increases ($m > 0 \Rightarrow \eta^+ \geqslant \eta^-$) after the transition of the considered weak adiabatic shock. The inequality (2.91) is called the s t a b i l i t y c o n d i t i o n of the shock wave.

2.5. Summary

The derived balance equations for the thermomechanical processes in continuum without long-range interactions constitute the following set for each point $\underset{\sim}{x} \in \chi(B;t) \setminus \sigma(t)$ and each instant of time from the given interval $T \subset R$:

$$\begin{aligned} &\varrho J = \varrho_0 , && J := \det \chi_{k,K}(\underset{\sim}{\chi}^{-1}(\underset{\sim}{x},t);t), \\ &\varrho \dot{v}_k = t_{kl,l} , && v_k(\underset{\sim}{x},t) := \frac{\partial \chi_k}{\partial t}(\underset{\sim}{\chi}^{-1}(\underset{\sim}{x},t);t), \end{aligned} \tag{2.92}$$

$$
\begin{aligned}
& t_{kl} = t_{lk}, \\
& \varrho\dot{\varepsilon} = v_{k,l} t_{kl} - q_{k,k}, \\
& \varrho\dot{\eta} + h_{k,k} \geqslant 0.
\end{aligned}
\tag{2.92}
$$

At the same time, we have the following set of relations for each point of the singular surface $\sigma(t)$:

$$
\begin{aligned}
& \frac{\partial \varrho_s}{\partial t} - 2c K_M \varrho_s = [\![\varrho U]\!] + \varrho_s^*, \qquad U^{\pm} := c - v_k^{\pm} n_k, \\
& \frac{\partial p_k}{\partial t} - 2c K_M p_k = [\![\varrho U v_k + t_{kl} n_l]\!] + p_k^*, \\
& n_k p_l = n_l p_k, \\
& \frac{\partial \varepsilon_s}{\partial t} - 2c K_M \varepsilon_s = [\![\varrho U (\tfrac{1}{2} v_k v_k + \varepsilon) + v_k t_{kl} n_l - q_k n_k]\!] + \varepsilon_s^*, \\
& \frac{\partial \eta_s}{\partial t} - 2c K_M \eta_s - [\![\varrho \eta U - h_k n_k]\!] \geqslant 0.
\end{aligned}
\tag{2.93}
$$

These sets of relations cannot be solved without the supplementary conditions, called the c o n s t i t u t i v e r e l a t i o n s and specifying the material. Such relations are presented for the class of thermoelastic materials in the next Chapter.

CHAPTER 3
THERMOELASTIC MATERIALS

3.1. Constitutive relations

3.1.1. The structure of the constitutive functionals

The definition of the class of thermoelstic materials has already appeared in many monographs and textbooks, where also its basic properties have been extensively investigated. Let us just mention a few references commonly available. Some courses have been held on this subject in Udine: C.C. Wang[1971], R.Stojanović[1972]. The detailed and careful presentation can be found in the book of C.Truesdell[1972],Sec.XV (see also: C.Truesdell, W.Noll[1965],Sec.96).

We present only a rough sketch sufficient for our purposes. Let us start with the localization of the state and of the process. We show the procedure on the example of the rigid heat conductor. For such a material the state at the chosen instant of time t is defined by the distribution of the empirical temperature

$$\theta(\cdot,t): B \to R. \tag{3.1}$$

Neglecting the memory effects, we can also introduce the notion of the rate of thermodynamic process

$$\dot{\theta}(\cdot,t) := B \times \left\{ \frac{\partial \theta}{\partial t}(\underset{\sim}{X},t) \middle| \underset{\sim}{X} \in B \right\}. \tag{3.2}$$

For a given state $\theta(\cdot,t)$ the rate $\dot{\theta}(\cdot,t)$ defines the direction of the thermodynamic process passing through this state (fig.5).

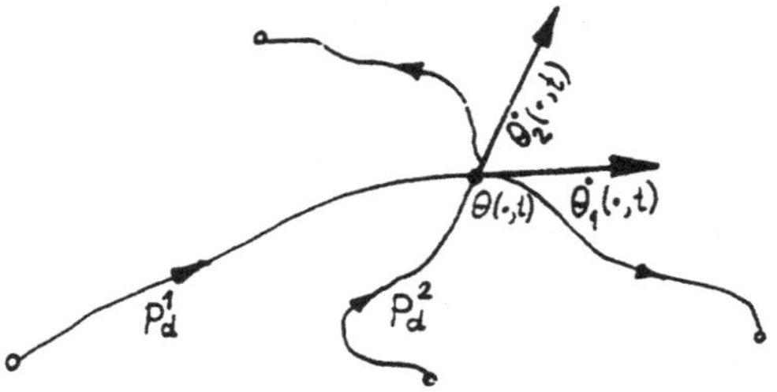

Fig.5

It means that the quantities appearing in the energy balance law for the rigid heat conductor

$$\rho\dot{\varepsilon} = -q_{k,k} \tag{3.3}$$

are given by the following constitutive functionals

$$\varepsilon(\underset{\sim}{X},t) = \mathcal{E}(\underset{\sim}{X},t;\theta(\cdot,t)) \tag{3.4}$$

$$q_k(\underset{\sim}{X},t) = \mathfrak{S}_k(\underset{\sim}{X},t;\theta(\cdot,t),\dot{\theta}(\cdot,t)) \tag{3.5}$$

ε being the function of state and q_k being the function of the process.

Due to the spatial localization carried out in the previous Chapter we would like to have also these functional replaced by the local counterparts. It requires assumptions on the range of interactions and we will not go into this problem. However on a very simple example we show that such a localization violates the first order theory for the energy given by the following functional

$$\varepsilon(\underset{\sim}{X},t)=\int_B K(\underset{\sim}{X},\underset{\sim}{Z},t,\dot{\theta}(\underset{\sim}{X},t))\,\theta(\underset{\sim}{Z},t)\,dv_Z. \tag{3.6}$$

The energy ε is in such a case the state function if

$$\int_B \frac{\partial}{\partial\dot{\theta}(\underset{\sim}{X},t)} K(\underset{\sim}{X},\underset{\sim}{Z},t,\dot{\theta}(\underset{\sim}{X},t))\,\theta(\underset{\sim}{Z},t)\,dv_Z=0. \tag{3.7}$$

In particular, for a homogeneous distribution $\forall_{\underset{\sim}{X}\in B}\ \theta(\underset{\sim}{X},t)=\theta_0(t)$

$$\int_B \frac{\partial}{\partial\dot{\theta}(\underset{\sim}{X},t)} K(\underset{\sim}{X},\underset{\sim}{Z},t,\dot{\theta}(\underset{\sim}{X},t))\,dv_Z=0. \tag{3.8}$$

Obviously, it does not mean that $\frac{\partial K}{\partial\dot{\theta}}=0$ for every $\underset{\sim}{X}\in B$. Hence, the transition from the state space of infinite dimension $\{\theta(\cdot,\cdot)\}$ to the state space of the finite dimension $\{\theta(\underset{\sim}{X},\cdot)\}$ for each point $X\in B$ may yield the change of the state function into the function of the local process.

Let us construct the first gradient theory, i.e. we substitute in the formula (3.6) the following approximate expressions

$$\begin{aligned}\theta(\underset{\sim}{Z},t)&\cong\theta(\underset{\sim}{X},t)+\nabla_X\theta(\underset{\sim}{X},t)\cdot(\underset{\sim}{Z}-\underset{\sim}{X}),\\ \dot{\theta}(\underset{\sim}{X},t)&\cong\dot{\theta}(\underset{\sim}{Z},t),\end{aligned} \tag{3.9}$$

Then

$$\varepsilon(\underset{\sim}{X},t)\cong\theta(\underset{\sim}{X},t)\int_B K(\underset{\sim}{X},\underset{\sim}{Z},t,\dot{\theta}(\underset{\sim}{X},t))\,dv_Z+\nabla_X\theta(\underset{\sim}{X},t)\cdot\int_B(\underset{\sim}{Z}-\underset{\sim}{X})K(\underset{\sim}{X},\underset{\sim}{Z},t,\dot{\theta}(\underset{\sim}{X},t))\,dv_Z \tag{3.10}$$

Such a result is typical for the above described transition from the global to local states and processes. Functions appearing in the thermodynamic theory of materials and being the state functions in thermodynamic equilibrium remain also the state functions in non-equilibrium processes but the state space for these functions is usually of the higher dimension than that used in practice.

Those remarks partially justify Truesdell's equipresence principle in spite of all its drawbacks.

We proceed now to the considerations concerning the constitutive functionals for thermoelastic materials. We base this part of the paper on I.Müller's results[1971,2]. The inspection of the relations (2.92) shows that we must construct the constitutive relations for the following fields: stress tensor t_{kl}, specific energy ε, heat flux q_k, specific entropy η and entropy flux h_k. In the thermomechanical theory, the local process is described by the following set of quantities

$$\{\chi_k(\underset{\sim}{X},t), \chi_{k,K}(\underset{\sim}{X},t), \dot{\chi}_k(\underset{\sim}{X},t), \theta(\underset{\sim}{X},t), \theta_{,K}(\underset{\sim}{X},t), \dot{\theta}(\underset{\sim}{X},t)\} =: \mu_\alpha(\underset{\sim}{X},t), \qquad \alpha=1,\dots,20 \tag{3.11}$$

for each $\underset{\sim}{X}$ and t.

Hence, in the case of thermoelastic materials, we have to supplement the relations (2.92) by the following set of functions

$$\begin{aligned}
t_{kl}(\underset{\sim}{X},t) &= \hat{t}_{kl}(\mu_\alpha(\underset{\sim}{X},t)),\\
\varepsilon(\underset{\sim}{X},t) &= \hat{\varepsilon}(\mu_\alpha(\underset{\sim}{X},t)),\\
q_k(\underset{\sim}{X},t) &= \hat{q}_k(\mu_\alpha(\underset{\sim}{X},t)),\\
\eta(\underset{\sim}{X},t) &= \hat{\eta}(\mu_\alpha(\underset{\sim}{X},t)),\\
h_k(\underset{\sim}{X},t) &= \hat{h}_k(\mu_\alpha(\underset{\sim}{X},t)).
\end{aligned} \tag{3.12}$$

Among other limitations, these functions have to be invariant with respect to the Euclidean transformation (Chap.1). We show the consequences of this requirement on the example of the specific energy. The infinitesimal Euclidean transformation yields the following variations of the quantities, listed in the formula (3.11)

$$\begin{aligned} &\delta\chi_k = \varepsilon_k + \varepsilon_{kl}\chi_l + \alpha_k t, \qquad \varepsilon_{kl} = -\varepsilon_{lk}, \\ &\delta\chi_{k,K} = \varepsilon_{kl}\chi_{l,K}, \\ &\delta\dot{\chi}_k = \varepsilon_{kl}\dot{\chi}_l + \alpha_k, \\ &\delta\theta = 0, \\ &\delta\theta_{,K} = 0, \\ &\delta\dot{\theta} = 0. \end{aligned} \tag{3.13}$$

The variation of energy, resulting from the above variation is as follows

$$\delta\varepsilon = \left(\frac{\partial\hat{\varepsilon}}{\partial\chi_k}\right)\varepsilon_k + \left(\frac{\partial\hat{\varepsilon}}{\partial\chi_k}\chi_l + \frac{\partial\hat{\varepsilon}}{\partial\chi_{k,K}}\chi_{l,K} + \frac{\partial\hat{\varepsilon}}{\partial\dot{\chi}_k}\dot{\chi}_k\right)\varepsilon_{kl} + \left(\frac{\partial\hat{\varepsilon}}{\partial\chi_k}t + \frac{\partial\hat{\varepsilon}}{\partial\dot{\chi}_k}\right)\alpha_k. \tag{3.14}$$

Due to the assumed invariance we have

$$\delta\varepsilon = 0 \tag{3.15}$$

Hence

$$\frac{\partial\hat{\varepsilon}}{\partial\chi_k} = 0, \quad \frac{\partial\hat{\varepsilon}}{\partial\dot{\chi}_k} = 0, \quad \frac{\partial\hat{\varepsilon}}{\partial\chi_{k,K}}\chi_{l,K} = \frac{\partial\hat{\varepsilon}}{\partial\chi_{l,K}}\chi_{k,K}. \tag{3.16}$$

The similar relations follow for all remaining fields. It means that the functions (3.12) cannot depend explicitely on the motion χ_k and on the velocity v_k (Noll's theorem; see,e.g. C.Truesdell[1972],Chap.IV). Their dependence on the deformation gradient $\chi_{k,K}$ is limitted by $(3.16)_3$. Then, it can be proved that the quantities (3.12) depend on the deformation gradient through the right Cauchy-Green tensor

$$C_{KL} := x_{k,K} x_{k,L}. \tag{3.17}$$

Bearing in mind the above results, we can replace the relations (3.12) by the following set

$$\begin{aligned} t_{kl}(\underset{\sim}{X},t) &= \tilde{t}_{kl}(C_{KL};\theta;\theta_{,K};\dot{\theta}), \\ \varepsilon(\underset{\sim}{X},t) &= \tilde{\varepsilon}(C_{KL};\theta;\theta_{,K};\dot{\theta}), \\ q_k(\underset{\sim}{X},t) &= \tilde{q}_k(C_{KL};\theta;\theta_{,K};\dot{\theta}), \\ \eta(\underset{\sim}{X},t) &= \tilde{\eta}(C_{KL};\theta,\theta_{,K};\dot{\theta}), \\ h_k(\underset{\sim}{X},t) &= \tilde{h}_k(C_{KL};\theta,\theta_{,K};\dot{\theta}). \end{aligned} \tag{3.18}$$

The equations (2.92) and the relations (3.18) form the governing set of equations for thermoelastic materials. They should be satisfied everywhere except of the points of singular surfaces. We present some properties of this set further in this Section.

3.1.2. Constitutive functionals for singular surface

It is easy to notice in the relations (2.93) that all quantities appearing in square brackets are defined by the relations (3.18) and kinematics of the surface. However the remaining surface fields, namely

$$\varrho_s, \varrho_s^*, p_K, p_K^*, \varepsilon_s, \varepsilon_s^*, \eta_s \tag{3.19}$$

should be described by the additional surface constitutive relations. The local process at the point of the singular surface is determined by the following limits

$$C_{KL}^+, C_{KL}^-, v_k^+, v_k^-, \theta^+, \theta^-, \dot{\theta}^+, \dot{\theta}^-, \theta_{,K}^+, \theta_{,K}^- \tag{3.20}$$

and by the speed of displacement c. It is convenient to use the definition (2.69) and replace the set (3.20) by the following one

$$[\![v_k]\!]\,,\ \langle v_k\rangle\,,\ \langle U\rangle,\ [\![C_{KL}]\!]\,,\langle C_{KL}\rangle,[\![\theta]\!]\,,\langle\theta\rangle,\ [\![\dot\theta]\!]\,,\langle\dot\theta\rangle,\ [\![\theta_{,K}]\!],\langle\theta_{,K}\rangle\,,$$

equivalent to (3.20) due to the relations (3.21)

$$\begin{aligned} v_k^+ &= \langle v_k\rangle + \tfrac{1}{2}[\![v_k]\!],\\ v_k^- &= \langle v_k\rangle - \tfrac{1}{2}[\![v_k]\!],\\ c &= \langle U\rangle + \langle v_k\rangle n_k. \end{aligned} \tag{3.22}$$

We will not investigate the possible forms of surface constitutive relations for the quantities (3.19) but we check the consequences of Galilean invariance condition. As we have already pointed out, the classical Rankine-Hugoniot's conditions (2.76) are invariant with respect to the change of the inertial reference frame

$$x_k \longrightarrow x_k + \alpha_k t\,. \tag{3.23}$$

For arbitrary surface fields the set (2.93) does not have to be invariant due to the following rules of transformation for variables appearing in these relations

$$\delta[\![v_k]\!]=0,\quad \delta\langle v_k\rangle=\alpha_k,\quad \delta\langle U\rangle=0,\quad \delta c=\alpha_k n_k,$$

$$\delta[\![C_{KL}]\!]=0,\ \delta\langle C_{KL}\rangle=0,\ \delta[\![\theta]\!]=\delta\langle\theta\rangle=\delta[\![\dot\theta]\!]=\delta\langle\dot\theta\rangle=\delta[\![\theta_{,K}]\!]=0. \tag{3.24}$$

Let us check the change of the relation (2.84) due to the transformation (3.23). To this aim, we write this relation in the form

$$\frac{\partial \varepsilon_s}{\partial t} - 2(\langle U\rangle + \langle v_k\rangle n_k)K_M\varepsilon_s = [\![eU]\!] + [\![q]\!]\cdot\underset{\sim}{n} + \varepsilon_s^*. \tag{3.25}$$

Then the only variable undergoing the variation is the mean value of the velocity $\langle v_K\rangle$. The corresponding change of the relation (3.25) is of the form

$$\delta_{<v>}\left(\frac{\partial \varrho_s}{\partial t}\right) - 2n_k K_M \varrho_s - 2(<U> + <v_k>n_k) K_M \delta_{<v>}\varrho_s = \delta_{<v>} \varrho_s^* \tag{3.26}$$

where we have made use of the constitutive relations (3.18).

We solve the relation (3.26) in the particular case, when two additional conditions are satisfied

i/ $\delta_{<v>} \varrho_s \equiv 0$,

ii/ ϱ_s^* is an analytical function of $<v_k>$, i.e.

$$\varrho_s^* = \varrho_0^* + \underset{\sim}{\varrho}_1^* \cdot <\underset{\sim}{v}> + \frac{1}{2} <\underset{\sim}{v}> \cdot \underset{\sim}{\varrho}_2^* <\underset{\sim}{v}> + \ldots, \qquad \underset{\sim}{\varrho}_1^* \neq 0.$$

Let us start with the surface balance of mass $(2.93)_1$. The relation (3.26) takes in this case the following form

$$-2\underset{\sim}{n} K_M \varrho_s = \underset{\sim}{\varrho}_1^* + \underset{\sim}{\varrho}_2^* <\underset{\sim}{v}> + \ldots \tag{3.27}$$

for each value of $<v>$. Hence

$$\begin{aligned} \underset{\sim}{\varrho}_1^* &= -2\underset{\sim}{n} K_M \varrho_s, \\ \underset{\sim}{\varrho}_2^* &= \ldots = 0. \end{aligned} \tag{3.28}$$

It means that the surface mass sources should have the following structure

$$\varrho_s^* = \varrho_0^* - 2<v_k>n_k K_M \varrho_s \tag{3.29}$$

where ϱ_0^* is independent of $<v_k>$. The corresponding balance equation is of the form

$$\frac{\partial \varrho_s}{\partial t} - 2<U> K_M \varrho_s = [\![\varrho U]\!] + \varrho_0^*, \tag{3.30}$$

invariant with respect to the Galilean transformation.

The surface balance of momentum, written in the form convenient for our considerations is as follows

$$\frac{\partial p_k}{\partial t} - 2(\langle U\rangle + \langle v_l\rangle n_l) K_M p_k = [\![\varrho U]\!]\langle v_k\rangle + \langle \varrho U\rangle [\![v_k]\!] + [\![t_{kl}]\!] n_l + p_k^*;$$

or, bearing in mind the balance law (3.30),

$$\frac{\partial p_k}{\partial t} - 2(\langle U\rangle + \langle v_l\rangle n_l) K_M p_k = \left\{ \frac{\partial \varrho_s}{\partial t} - 2\langle U\rangle K_M \varrho_s - \varrho_o^* \right\} \langle v_k\rangle +$$

$$+ \langle \varrho U\rangle [\![v_k]\!] + [\![t_{kl}]\!] n_l + p_k^*. \tag{3.31}$$

According to our simplifying assumptions, the invariance condition for this relation follows in the form

$$-2 n_l K_M p_k = \left\{ \frac{\partial \varrho_s}{\partial t} - 2\langle U\rangle K_M \varrho_s - \varrho_o^* \right\} \delta_{kl} + {}_1p_{kl}^* + {}_2p_{klm}^* \langle v_m\rangle + \ldots, \tag{3.32}$$

due to the following expansion of the surface momentum source

$$p_k^* = {}_0p_k^* + {}_1p_{kl}^* \langle v_l\rangle + \frac{1}{2}\, {}_2p_{klm}^* \langle v_l\rangle \langle v_m\rangle + \ldots; \qquad {}_1p_{kl}^* \not\equiv 0. \tag{3.33}$$

Hence, we have

$${}_1p_{kl}^* = -K_M n_l p_k - \left\{ \frac{\partial \varrho_s}{\partial t} - 2\langle U\rangle K_M \varrho_s - \varrho_o^* \right\} \delta_{kl} = -K_M n_l p_k - [\![\varrho U]\!] \delta_{kl},$$

$${}_2p_{klm}^* = \ldots = 0. \tag{3.34}$$

The invariant form of the momentum balance is finally as follows

$$\frac{\partial p_k}{\partial t} - 2\langle U\rangle K_M p_k = \langle \varrho U\rangle [\![v_k]\!] + [\![t_{kl}]\!] n_l + {}_0p_k^*. \tag{3.35}$$

Due to our assumptions, the surface moment of momentum balance $(2.93)_3$ is invariant with respect to the transformation (3.23).

The considerations similar to that performed for the momentum balance lead to the following form of the energy balance

$$\frac{\partial \varepsilon_s}{\partial t} - 2\langle U\rangle K_M \varepsilon_s = [\![\varrho U \varepsilon]\!] + \frac{1}{8} [\![\varrho U]\!] [\![v_k]\!] [\![v_k]\!] + [\![v_k]\!] \langle t_{kl}\rangle n_l - [\![q_k]\!] n_k + \varepsilon_c^* \tag{3.36}$$

where ε_o^* is independent of $\langle v_k \rangle$, and

$$\varepsilon_s^* = \varepsilon_o^* - \left\{\frac{\partial p_k}{\partial t} - 2\langle U\rangle K_M p_k - {}_o p_k^* + 2n_k K_M \varepsilon_s\right\}\langle v_k\rangle - \frac{1}{2}[\![\rho U]\!]\langle v_k\rangle\langle v_k\rangle . \tag{3.37}$$

Finally, the inspection of the inequality $(2.93)_5$ shows that, under the above assumptions, the surface entropy inequality cannot be Galilean invariant. If we want to execute this requirement we can proceed in two ways:

i/ replace the inequality (2.86) by the following one

$$\int_{\chi(P,t)} \eta^* \, dv + \int_{\sigma_P(t)} \left\{\eta_s^* + 2\langle v_k\rangle n_k K_M \eta_s\right\} ds \geq 0 ; \tag{3.38}$$

ii/ assume that η_s depends on $\langle v_k \rangle$ in such a way that $(2.93)_5$ becomes invariant.

The first possibility fits the pattern of the above presented considerations. In this case, the structure of the surface entropy sources is as follows

$$\eta_s^* = \eta_o^* - 2\langle v_k\rangle n_k K_M \eta_s$$

and only η_o^*, being independent of $\langle v_k \rangle$, contributes to the production of entropy in the body B.

Summing up the above considerations, we arrive at the following Galilean invariant set of surface relations

$$\begin{aligned}
&\frac{\partial \rho_s}{\partial t} - 2\langle U\rangle K_M \rho_s = [\![\rho U]\!] + \rho_o^* ,\\
&\frac{\partial p_k}{\partial t} - 2\langle U\rangle K_M p_k = \langle \rho U\rangle [\![v_k]\!] + [\![t_{kl}]\!] n_l + {}_o p_k^* ,\\
&n_k p_l = n_l p_k , \qquad\qquad (3.39)\\
&\frac{\partial \varepsilon_s}{\partial t} - 2\langle U\rangle K_M \varepsilon_s = [\![\rho U \varepsilon_s]\!] + \frac{1}{8}[\![\rho U]\!][\![v_k]\!][\![v_k]\!] + [\![v_k]\!]\langle t_{kl}\rangle n_l - [\![q_k]\!] n_k + \varepsilon_o^* ,\\
&\frac{\partial \eta_s}{\partial t} - 2\langle U\rangle K_M \eta_s - [\![\rho U \eta - h_k n_k]\!] \geq 0,
\end{aligned}$$

where the fields ϱ_s , ϱ_0^* , p_k, ${}_0p_k^*$, ε_s, ε_0^* , η_s may still depend on $[v_k]$, $\langle U\rangle$, but are independent of $\langle v_k\rangle$. In the case of absence of these fields, we apparently arrive at the Rankine-Hugoniot's conditions (2.76).

3.2. Thermodynamically admissible constitutive relations

According to our previous remarks, the thermodynamic method, utilized in this paper, relies upon the choice of constitutive relations (3.18) satisfying the balance equations $(2.92)_{1-4}$ and the entropy inequality $(2.92)_5$. To make this choice we can use the method of Lagrange multipliers proposed for thermodynamic applications by I-Shih Liu[1972]. To this aim we write the set (2.92) in the form

$$-\Lambda_k[\dot{v}_k - \frac{J}{\varrho_0}(\chi_{l,K})^{-1} t_{kl,K}] - \Lambda[\dot{\varepsilon} - \frac{J}{2\varrho_0}\dot{C}_{MN}(\chi_{k,M})^{-1}\cdot(\chi_{l,N})^{-1} t_{kl} +$$

$$+ \frac{J}{\varrho_0}(\chi_{k,K})^{-1} q_{k,K}] + [\dot{\eta} + \frac{J}{\varrho_0}(\chi_{k,K})^{-1} h_{k,K}] \geq 0 \tag{3.40}$$

where Λ and Λ_k are Lagrange multipliers. It is convenient to use the material densities rather than the spatial ones. Therefore we define the following fields

$$t_{KL} := J(\chi_{k,K})^{-1}(\chi_{l,L})^{-1} t_{kl},$$

$$q_K := J(\chi_{k,K})^{-1} q_k, \tag{3.41}$$

$$h_K := J(\chi_{k,K})^{-1} h_k.$$

The substitution of these fields in (3.40) and the utilization of the constitutive relations (3.18) together with the chain rule for the differentiation yield the following inequality

$$(\frac{\partial\tilde{\eta}}{\partial\theta} - \Lambda\frac{\partial\tilde{\varepsilon}}{\partial\theta})\dot{\theta} + \frac{1}{\varrho_0}(\frac{\partial\tilde{h}_K}{\partial\theta} - \Lambda\frac{\partial\tilde{q}_K}{\partial\theta} + \Lambda_k\chi_{k,L}\frac{\partial\tilde{t}_{KL}}{\partial\theta})\theta_{,K} + (\frac{\partial\tilde{\eta}}{\partial\dot{\theta}} - \Lambda\frac{\partial\tilde{\varepsilon}}{\partial\dot{\theta}})\ddot{\theta} +$$

$$+ (\frac{\partial\tilde{\eta}}{\partial\theta_{,K}} + \frac{1}{\varrho_0}\frac{\partial\tilde{h}_K}{\partial\dot{\theta}} - \Lambda\frac{\partial\tilde{\varepsilon}}{\partial\theta_{,K}} - \Lambda\frac{1}{\varrho_0}\frac{\partial\tilde{q}_K}{\partial\dot{\theta}} + \Lambda_k\frac{1}{\varrho_0}\chi_{k,L}\frac{\partial\tilde{t}_{KL}}{\partial\dot{\theta}})\dot{\theta}_{,K} + \frac{1}{\varrho_0}(\frac{\partial\tilde{h}_K}{\partial\theta_{,M}} -$$

$$- \Lambda\frac{\partial\tilde{q}_K}{\partial\theta_{,M}} + \Lambda_K\chi_{k,L}\frac{\partial\tilde{t}_{KL}}{\partial\theta_{,M}})\theta_{,KM} - \Lambda_k\ddot{\chi}_k + \frac{1}{\varrho_0}(\frac{\partial\tilde{h}_K}{\partial C_{MN}} - \Lambda\frac{\partial\tilde{q}_K}{\partial C_{MN}} + \Lambda_k\chi_{k,L}\frac{\partial\tilde{t}_{KL}}{\partial C_{MN}})C_{MN,K} +$$

$$+ (\frac{\partial\tilde{\eta}}{\partial C_{MN}} - \Lambda\frac{\partial\tilde{\varepsilon}}{\partial C_{MN}} + \frac{\Lambda}{2\varrho_0}\tilde{t}_{MN})\dot{C}_{MN} \geq 0. \tag{3.42}$$

According to our assumptions, the local process at the generic point $\underset{\sim}{X} \in B$ is described by the following quantities

$$\{C_{KL}; \theta; \dot{\theta}; \theta_{,K}\}. \tag{3.43}$$

Assuming the analyticity of these solutions we conclude that for any choice of the quantities

$$\{\ddot{\theta}; \dot{\theta}_{,K}; \theta_{,KM}; \ddot{\chi}_k; C_{MN,K}; \dot{C}_{MN}\} \tag{3.44}$$

at the point $\underset{\sim}{X} \in B$ we can find such initial conditions

$$C_{KL}(\underset{\sim}{X}); \theta(X); \dot{\theta}(X); \theta_{,K}(X) \tag{3.45}$$

that the chosen quantities will be given by the solution corresponding to these conditions. Consequently, the inequality (3.42) is satisfied for any process if the coefficients of the quantities (3.44) vanish identically, providing $\frac{\partial\tilde{\varepsilon}}{\partial\dot{\theta}} \neq 0$. Hence

$$\Lambda_k = 0,$$

$$\Lambda = (\frac{\partial\tilde{\eta}}{\partial\dot{\theta}})(\frac{\partial\tilde{\varepsilon}}{\partial\dot{\theta}})^{-1}, \tag{3.46}$$

$$\frac{\partial\tilde{\eta}}{\partial\theta_{,K}} + \frac{1}{\varrho_0}\frac{\partial\tilde{h}_K}{\partial\dot{\theta}} - \Lambda(\frac{\partial\tilde{\varepsilon}}{\partial\theta_{,K}} + \frac{1}{\varrho_0}\frac{\partial\tilde{q}_K}{\partial\dot{\theta}}) = 0,$$

$$\frac{\partial \tilde{h}_{K)}}{\partial \theta_{,(M}} - \Lambda \frac{\partial \tilde{q}_{K)}}{\partial \theta_{,(M}} = 0,$$

$$\frac{\partial \tilde{h}_{K)}}{\partial C_{M(N}} - \Lambda \frac{\partial \tilde{q}_{K)}}{\partial C_{M(N}} = 0,$$

$$\frac{\partial \tilde{\eta}}{\partial C_{MN}} - \Lambda \frac{\partial \tilde{\varepsilon}}{\partial C_{MN}} + \frac{\Lambda}{2\varrho_0} \tilde{t}_{MN} = 0, \tag{3.46}$$

$$\left(\frac{\partial \tilde{\eta}}{\partial \theta} - \Lambda \frac{\partial \tilde{\varepsilon}}{\partial \theta}\right)\dot{\theta} + \frac{1}{\varrho_0}\left(\frac{\partial \tilde{h}_K}{\partial \theta} - \Lambda \frac{\partial \tilde{q}_K}{\partial \theta}\right) \theta_{,K} \geq 0.$$

We investigate the consequences of these identities under an additional assumption of the colinearity of the heat and entropy fluxes

$$\tilde{h}_K = \Lambda \tilde{q}_K . \tag{3.47}$$

I.Müller and some other authors have proved that this relation holds for a wide class of materials. It holds, for instance, for isotropic solids (see: I.Müller[1971,2]). Then the relations $(3.46)_{3-7}$ have the following form

$$\frac{\partial \tilde{\eta}}{\partial \theta_{,K}} - \Lambda \frac{\partial \tilde{\varepsilon}}{\partial \theta_{,K}} + \frac{1}{\varrho_0} \frac{\partial \Lambda}{\partial \dot{\theta}} \tilde{q}_K = 0, \tag{3.48}$$

$$\frac{\partial \Lambda}{\partial \theta_{,(M}} \tilde{q}_{K)} = 0, \tag{3.49}$$

$$\frac{\partial \Lambda}{\partial C_{M(N}} \tilde{q}_{K)} = 0, \tag{3.50}$$

$$\tilde{t}_{MN} = 2\varrho_0 \left(\frac{\partial \tilde{\varepsilon}}{\partial C_{MN}} - \frac{1}{\Lambda} \frac{\partial \tilde{\eta}}{\partial C_{MN}}\right), \tag{3.51}$$

$$\left(\frac{\partial \tilde{\eta}}{\partial \theta} - \Lambda \frac{\partial \tilde{\varepsilon}}{\partial \theta}\right)\dot{\theta} + \frac{1}{\varrho_0} \frac{\partial \Lambda}{\partial \theta} \tilde{q}_K \theta_{,K} \geq 0. \tag{3.52}$$

These relations have to hold for all heat fluxes q_K. According to the conditions (3.49) and (3.50), it means that Λ is independent of $\theta_{,M}$ and C_{MN}. Hence

$$\Lambda = \Lambda(\theta, \dot{\theta}). \tag{3.53}$$

The multiplier Λ is called a c o l d n e s s. We show in the sequel some properties of this quantity. If we introduce the notion of the Helmholtz free energy

$$\tilde{F} := \tilde{\varepsilon} - \frac{1}{\Lambda}\tilde{\eta}, \tag{3.54}$$

we can write the above formulae in the form

$$\begin{aligned} &\tilde{q}_K = \varrho_0 \Lambda \left(\frac{\partial \Lambda}{\partial \dot{\theta}}\right)^{-1} \frac{\partial \tilde{F}}{\partial \theta_{,K}}, \\ &\tilde{t}_{MN} = 2\varrho_0 \frac{\partial \tilde{F}}{\partial C_{MN}}, \\ &\left(\frac{\partial \ln \Lambda}{\partial \theta}\tilde{\eta} - \Lambda \frac{\partial \tilde{F}}{\partial \theta}\right)\dot{\theta} + \frac{1}{\varrho_0}\frac{\partial \Lambda}{\partial \theta}\tilde{q}_K \theta_{,K} \geq 0, \\ &\Lambda = \Lambda(\theta, \dot{\theta}), \quad \tilde{F} = \hat{F}(C_{KL}; \theta; \dot{\theta}; \theta_{,K}). \end{aligned} \tag{3.55}$$

It means that the second law of thermodynamics reduces the number of required constitutive relations to three scalar formulae for the coldness Λ, the Helmholtz free energy F and the entropy η. Some further limitations for the constitutive relations follow from the integrability conditions of (3.55). We will not use them in this paper.

Let us now return to the consequences of the inequality (3.52). The function

$$\tilde{\sigma} := \left(\frac{\partial \tilde{\eta}}{\partial \theta} - \Lambda \frac{\partial \tilde{\varepsilon}}{\partial \theta}\right)\dot{\theta} + \frac{1}{\varrho_0}\frac{\partial \Lambda}{\partial \theta}\tilde{q}_K \theta_{,K} \tag{3.56}$$

describes the e n t r o p y p r o d u c t i o n at the point $\underset{\sim}{X}$ and in the instant of time t. It is obvious that σ is positive semi-definite and it reaches its minimum value for

$$\dot{\theta}(\underset{\sim}{X}, t) = 0, \quad \theta_{,K}(\underset{\sim}{X}, t) = 0. \tag{3.57}$$

The state described at the point $\underset{\sim}{X}$ by the conditions (3.57) is called a l o c a l t h e r m o d y n a m i c e q u i l i b r i u m. The necessery conditions for this state are

$$\frac{\partial\tilde{\varepsilon}}{\partial\dot{\theta}}\Big|_E = 0, \qquad \frac{\partial\tilde{\varepsilon}}{\partial\theta_{,K}}\Big|_E = 0, \tag{3.58}$$

where the index E indicates the state of equilibrium. Their explicit form is as follows

$$\frac{\partial}{\partial\dot{\theta}}\left(\frac{\partial\tilde{\eta}}{\partial\theta} - \Lambda\frac{\partial\tilde{\varepsilon}}{\partial\theta}\right)\dot{\theta}\Big|_E + \left(\frac{\partial\tilde{\eta}}{\partial\theta} - \Lambda\frac{\partial\tilde{\varepsilon}}{\partial\theta}\right)\Big|_E + \frac{\partial}{\partial\dot{\theta}}\left(\frac{1}{\varrho_0}\frac{\partial\Lambda}{\partial\theta}\tilde{q}_K\right)\theta_{,K}\Big|_E = 0,$$

$$\frac{\partial}{\partial\theta_{,L}}\left(\frac{\partial\tilde{\eta}}{\partial\theta} - \Lambda\frac{\partial\tilde{\varepsilon}}{\partial\theta}\right)\dot{\theta}\Big|_E + \frac{\partial}{\partial\theta_{,L}}\left(\frac{1}{\varrho_0}\frac{\partial\Lambda}{\partial\theta}\tilde{q}_K\right)\theta_{,K}\Big|_E + \frac{1}{\varrho_0}\frac{\partial\Lambda}{\partial\theta}\tilde{q}_L\Big|_E = 0.$$

Making use of (3.57), we obtain

$$\frac{\partial\tilde{\eta}_{|E}}{\partial\theta} - \Lambda_{|E}\frac{\partial\tilde{\varepsilon}_{|E}}{\partial\theta} = 0, \tag{3.59}$$

$$\tilde{q}_{K|E} = 0. \tag{3.60}$$

It means that both entropy and heat flux vanish in the equilibrium. This result has been obtained by Pipkin and Rivlin in 1958 and it shows the N o n - e x i s t e n c e o f a p i e z o - c a l o r i c e f f e c t.

On the other hand,

$$d\tilde{\eta}_{|E} = \frac{\partial\tilde{\eta}_{|E}}{\partial\theta}d\theta + \frac{\partial\tilde{\eta}_{|E}}{\partial C_{KL}}dC_{KL}. \tag{3.61}$$

The substitution of (3.51) and (3.59) in this relation yields

$$d\tilde{\eta}_{|E} = \Lambda_{|E}\left[\frac{\partial\tilde{\varepsilon}_{|E}}{\partial\theta}d\theta + \left(\frac{\partial\tilde{\varepsilon}_{|E}}{\partial C_{KL}} - \frac{1}{2\varrho_0}\tilde{t}_{KL|E}\right)dC_{KL}\right],$$

or, bearing in mind the formula

$$d\tilde{\varepsilon}_E = \frac{\partial\tilde{\varepsilon}_{|E}}{\partial\theta}d\theta + \frac{\partial\tilde{\varepsilon}_{|E}}{\partial C_{KL}}dC_{KL},$$

we have

$$d\tilde{\eta}_{|E} = \Lambda_{|E}\left(d\tilde{\varepsilon}_{|E} - \frac{1}{2\varrho_0}\tilde{t}_{KL|E}\,dC_{KL}\right). \tag{3.62}$$

The comparison of the relation (3.62) with the classical Gibbs identity for thermoelastic materials leads to the condition

$$\Lambda|_E = \frac{1}{T} \tag{3.63}$$

where T is the absolute temperature.

It proves that the coldness is a natural exrension of the reciprocal of equilibrium absolute temperature on a certain class of non-equilibrium states. It can be proved that Λ is a universal function for all materials, satisfying the condition (3.47). The details are presented in I.Müller's papers (e.g. 1971,2). Simultaneously it proves that $\Lambda|_E$ is monotonically decreasing function of θ, i.e.

$$\frac{\partial \Lambda|_E}{\partial \theta} < 0. \tag{3.64}$$

The conditions (3.58) are sufficient for the minimum of entropy production in equilibrium when supplemented by the following inequality

$$\left\| \begin{matrix} \dfrac{\partial^2 \tilde{\sigma}}{\partial \dot{\theta}^2} & \dfrac{\partial^2 \tilde{\sigma}}{\partial \dot{\theta}\, \partial \theta_{,L}} \\ \dfrac{\partial^2 \tilde{\sigma}}{\partial \dot{\theta}\, \partial \theta_{,L}} & \dfrac{\partial^2 \tilde{\sigma}}{\partial \theta_{,L}\, \partial \theta_{,M}} \end{matrix} \right\| \quad \text{is positive definite} \tag{3.65}$$

It means that at least the following inequality

$$2 \frac{\partial}{\partial \dot{\theta}} \left(\frac{\partial \tilde{\eta}}{\partial \theta} - \Lambda \frac{\partial \tilde{\varepsilon}}{\partial \theta} \right) \Big|_E \lambda^2 + 2 \left[\frac{\partial}{\partial \theta_{,L}} \left(\frac{\partial \tilde{\eta}}{\partial \theta} - \Lambda \frac{\partial \tilde{\varepsilon}}{\partial \theta} \right) + \right.$$
$$\left. + \frac{1}{\varrho_0} \frac{\partial \Lambda}{\partial \theta} \frac{\partial \tilde{q}_L}{\partial \dot{\theta}} \right]_E \lambda \lambda_L + \frac{1}{\varrho_0} \frac{\partial \Lambda|_E}{\partial \theta} \left(\frac{\partial \tilde{q}_M}{\partial \theta_{,L}} + \frac{\partial \tilde{q}_L}{\partial \theta_{,M}} \right) \Big|_E \lambda_L \lambda_M \geqslant 0. \tag{3.66}$$

should hold for any λ and λ_L. In particular, for $\lambda_L \equiv 0$, we have

$$\frac{\partial}{\partial \dot{\theta}} \left(\frac{\partial \tilde{\eta}}{\partial \theta} - \Lambda \frac{\partial \tilde{\varepsilon}}{\partial \theta} \right) \Big|_E \lambda^2 = \left(\frac{\partial \Lambda|_E}{\partial \theta} \frac{\partial \varepsilon}{\partial \dot{\theta}} \Big|_E - \frac{\partial \Lambda}{\partial \dot{\theta}} \Big|_E \frac{\partial \tilde{\varepsilon}|_E}{\partial \theta} \right) \lambda^2 \geqslant 0.$$

Hence

$$\frac{\partial \Lambda}{\partial \theta}\Big|_E \frac{\partial \tilde{\varepsilon}}{\partial \dot{\theta}}\Big|_E \geqslant \frac{\partial \Lambda}{\partial \dot{\theta}}\Big|_E \frac{\partial \tilde{\varepsilon}}{\partial \theta}\Big|_E . \tag{3.67}$$

Due to the inequality (3.64) and the relation (3.63), we have

$$-\frac{T^2}{\frac{dT}{d\theta}}\left(\frac{\partial \Lambda}{\partial \dot{\theta}}\right)\Big|_E \frac{\partial \tilde{\varepsilon}}{\partial \theta}\Big|_E \geqslant \frac{\partial \tilde{\varepsilon}}{\partial \dot{\theta}}\Big|_E . \tag{3.68}$$

In the case $\lambda \equiv 0$, the inequality (3.66) leads to the following condition

$$\frac{\partial \tilde{q}_L}{\partial \theta_{,M}}\Big|_E \lambda_L \lambda_M \leq 0 \tag{3.69}$$

for any choice of λ_L. We return to these inequalities in the next Section of the paper.

3.3. Elastic Heat Conductors

The above presented thermodynamic theory of thermoelastic materials relies on Müller's assumption on the appearance of the rate $\dot{\varepsilon}$ as a constitutive variable. Many contemporary papers on the heat conduction in solids (e.g. C.Truesdell, W.Noll 1965, Sec.96) are based on simpler constitutive relations without this argument. For completeness, we briefly present also the theory of such relations. It means that we replace the relations (3.18) by the following ones

$$\begin{aligned}
t_{kl}(\underset{\sim}{X},t) &= \tilde{t}_{kl}(C_{KL};\theta;\theta_{,K}),\\
\varepsilon(\underset{\sim}{X},t) &= \tilde{\varepsilon}(C_{KL};\theta;\theta_{,K}),\\
q_k(\underset{\sim}{X},t) &= \tilde{q}_k(C_{KL};\theta;\theta_{,K}),\\
\eta(\underset{\sim}{X},t) &= \tilde{\eta}(C_{KL};\theta;\theta_{,K}),\\
h_k(\underset{\sim}{X},t) &= \tilde{h}_K(C_{KL};\theta;\theta_{,K}).
\end{aligned} \tag{3.70}$$

The material, satisfying these constitutive relations, is called an e l a s t i c h e a t c o n d u c t o r.

We utilize again the method of Lagrange multipliers. It yields the following inequality

$$\left(\frac{\partial\tilde\eta}{\partial\theta} - \Lambda\frac{\partial\tilde\varepsilon}{\partial\theta}\right)\dot\theta + \frac{1}{\varrho_0}\left(\frac{\partial\tilde h_K}{\partial\theta} - \Lambda\frac{\partial\tilde q_K}{\partial\theta} + \Lambda_k\chi_{k,L}\frac{\partial\tilde t_{KL}}{\partial\theta}\right)\theta_{,K} + \left(\frac{\partial\tilde\eta}{\partial\theta_{,K}} - \Lambda\frac{\partial\tilde\varepsilon}{\partial\theta_{,K}}\right)\dot\theta_{,K} +$$

$$+ \frac{1}{\varrho_0}\left(\frac{\partial\tilde h_K}{\partial\theta_{,M}} - \Lambda\frac{\partial\tilde q_K}{\partial\theta_{,M}} + \Lambda_k\chi_{k,L}\frac{\partial\tilde t_{KL}}{\partial\theta_{,M}}\right)\theta_{,KM} - \Lambda_k\ddot\chi_k + \tag{3.71}$$

$$+ \frac{1}{\varrho_0}\left(\frac{\partial\tilde h_K}{\partial C_{MN}} - \Lambda\frac{\partial\tilde q_K}{\partial C_{MN}} + \Lambda_k\chi_{k,L}\frac{\partial\tilde t_{KL}}{\partial C_{MN}}\right)C_{MN,K} + \left(\frac{\partial\tilde\eta}{\partial C_{MN}} - \Lambda\frac{\partial\tilde\varepsilon}{\partial C_{MN}} + \frac{\Lambda}{2\varrho_0}\tilde t_{MN}\right)\dot C_{MN} \geq 0.$$

Bearing in mind the assumption (3.47) on the colinearity of the heat and entropy fluxes, the arguments presented after the inequality (3.42) yield the following relations

$$\Lambda_k = 0 \tag{3.72}$$

$$\frac{\partial\tilde\eta}{\partial\theta_{,K}} - \Lambda\frac{\partial\tilde\varepsilon}{\partial\theta_{,K}} = 0, \qquad \frac{\partial\tilde\eta}{\partial\theta} - \Lambda\frac{\partial\tilde\varepsilon}{\partial\theta} = 0, \tag{3.73}$$

$$\frac{\partial\Lambda}{\partial\theta_{,(M}}\tilde q_{K)} = 0, \tag{3.74}$$

$$\frac{\partial\Lambda}{\partial C_{M(N}}\tilde q_{K)} = 0, \tag{3.75}$$

$$\tilde t_{MN} = 2\varrho_0\left(\frac{\partial\tilde\varepsilon}{\partial C_{MN}} - \frac{1}{\Lambda}\frac{\partial\tilde\eta}{\partial C_{MN}}\right), \tag{3.76}$$

$$\frac{1}{\varrho_0}\frac{\partial\Lambda}{\partial\theta}\tilde q_K\,\theta_{,K} \geq 0. \tag{3.77}$$

The relations (3.74) and(3.75) lead to the following formula

$$\Lambda = \Lambda(\theta). \tag{3.78}$$

Due to the definition (3.57) of the equilibrium state we see from the formula (3.78) that the dependence of coldness on the temperature distribution θ is the same in an arbitrary state as in the equilibrium

$$\Lambda = \Lambda|_E = \frac{1}{T}. \tag{3.79}$$

With respect to (3.78) it is convenient to introduce again the notion of the Helmholtz free energy

$$F := \varepsilon - T\eta. \tag{3.80}$$

Then the above relations take the simple form

$$\frac{\partial \tilde{F}}{\partial \theta_{,K}} = 0, \quad \tilde{\eta} = -\left(\frac{\partial T}{\partial \theta}\right)^{-1} \frac{\partial \tilde{F}}{\partial \theta}, \tag{3.81}$$

$$\tilde{t}_{MN} = 2\varrho_0 \frac{\partial \tilde{F}}{\partial C_{MN}}, \tag{3.82}$$

$$\tilde{q}_K \theta_{,K} \leq 0. \tag{3.83}$$

According to the relations (3.81) the free energy function, the entropy and the internal energy functions do not depend on the temperature gradient. This result has been reported, for instance, by B.D.Coleman and V.J.Mizel[1964]. It means that the elastic heat conductor belongs to the class of a few materials for which the reduction of the dimension of the state space does not change the state function F into the function of the local process (see: Sec.3.11.). It means also, due to the relation (3.82), that the stress tensor does not depend on the temperature gradient. The reduced constitutive relations should be of the following form

$$F(\underset{\sim}{X}, t) = \tilde{F}(C_{MN}; \theta),$$

$$q_K(\underset{\sim}{X}, t) = \check{q}_K(C_{MN}, \theta; \theta_{,M}), \tag{3.84}$$

and, in addition, they have to satisfy the relations (3.82) and (3.83). The form of the inequality (3.83) shows that the dissipation in elastic heat conductors is solely a result of the heat conduction. Finally, the appropriate relations for equilibrium take in this case the following form

$$\tilde{q}_K\big|_E = 0\,; \qquad \frac{\partial \tilde{q}_K}{\partial \theta_{,L}}\bigg|_E \lambda_K \lambda_L \leq 0. \tag{3.85}$$

CHAPTER 4

HEAT CONDUCTION IN THERMOELASTIC MATERIALS

4.1. Classification of Heat Conductors

One of the main properties of thermoelastic bodies is their ability to transport the energy in the form of the heat flux. As we have already seen during the analysis of the second law of thermodynamics this property is connected with the appearance of the temperature gradient. The type of admissible heat conductivity strongly influences also the propagation condition of acoustic waves in the material. In this Chapter, we present the basic classification of thermoelastic materials as heat conductors. In the next Chapter, we discuss the influence of the heat conductivity properties on the propagation of waves in such materials.

Let us define the following function

$$\bar{Q}(\xi, \theta_{,M}; \theta, \dot{\theta}, C_{MN}) := n_K \check{q}_K(C_{MN}, \theta, \theta_{,M} - \xi n_M, \dot{\theta}), \tag{4.1}$$

where n_K are components of any unit vector $\underset{\sim}{n}$. The thermoelastic material is said to be a n o r m a l c o n d u c t o r (P.Chadwick, P.K.Currie[1972]) if for any $\theta(\underset{\sim}{X},t)$, $\theta_{,M}(\underset{\sim}{X},t)=0$, $\dot{\theta}(\underset{\sim}{X},t)$, $C_{MN}(\underset{\sim}{X},t)$ at the generic point $\underset{\sim}{X} \in B$

$$\underset{\xi \neq 0}{\forall} \; \Phi(\xi, 0; \theta, \dot{\theta}, C_{MN}) \neq 0 \tag{4.2}$$

for any unit vector $\underset{\sim}{n}$, i.e.

$$\underset{\xi \neq 0}{\forall} \; \underset{\underset{\sim}{n}}{\forall} \; \xi n_K \tilde{q}_K(C_{MN}, \theta, \xi n_M, \dot{\theta}) \neq 0. \tag{4.3}$$

In such a case, a non-zero temperature gradient is always accompanied by a non-zero heat flux in the same direction.

On the other hand, if

$$\underset{\xi \neq 0}{\forall} \; \Phi(\xi, 0; \theta, \dot{\theta}, C_{MN}) = 0 \tag{4.4}$$

for any unit vector $\underset{\sim}{n}$, the thermoelastic material is called a n o n - c o n d u c t o r. If we write the formula (4.4) in the explicit form

$$\underset{\xi \neq 0}{\forall} \; \underset{\underset{\sim}{n}}{\forall} \; n_K \tilde{q}_K(C_{MN}, \theta, \xi n_M, \dot{\theta}) = 0 \tag{4.5}$$

it is readily seen through the definitions

$$\xi := (\theta_{,M}\theta_{,M})^{\frac{1}{2}}, \quad n_M := \frac{\theta_{,M}}{\sqrt{\theta_{,N}\theta_{,N}}} \tag{4.6}$$

that there is no heat flux across planes perpendicular to the temperature gradient $\theta_{,M}$. Due to the arbitrariness of $\theta_{,M}$ it means that in a non-conductor temperature gradients are not accompanied with the heat flux.

There are also possible some intermediate cases. A material which is neither a non-conductor nor a normal conductor is called an a n o m a l o u s c o n d u c t o r. For such a conductor there are non-zero solutions of the equation

$$\theta_{,K}\,\tilde{q}_K(C_{MN},\theta,\theta_{,M},\dot{\theta}) = 0. \tag{4.7}$$

Such solutions are, in turn, called a n o m a l i e s (P.Chadwick, P.K.Currie[1972]). We omit the discussion of properties of the anomalous conductor refering the reader to the above mentioned paper.

It is worthy to mention that the same thermoelastic material may behave as either of the above defined conductors due to the dependence of the heat flux vector q_K on the arguments θ , $\dot{\theta}$, C_{KL}.

The strengthening of the definition (4.2) to the form

$$\underset{\xi\neq 0}{\forall}\ \underset{\theta_{,M}}{\forall}\ \Phi(\xi,\theta_{,M};\theta,\dot{\theta},C_{MN}) - \Phi(0,\theta_{,M};\theta,\dot{\theta},C_{MN}) \neq 0 \tag{4.8}$$

for any unit vector $\underline{n}$ yields the notion of a s t r o n g c o n d u c t o r (C.Truesdell[1969], P.Chadwick[1977]). With respect to the definition (4.1) the formula (4.8) can be written as follows

$$\underset{\xi\neq 0}{\forall}\ \underset{\theta_{,M}}{\forall}\ n_K\tilde{q}_K(C_{MN},\theta,\theta_{,M}+\xi n_M,\dot{\theta}) \neq n_K\tilde{q}_K(C_{MN},\theta,\theta_{,M},\dot{\theta}). \tag{4.9}$$

Physically, this condition means that the change in the temperature gradient invariably gives rise to a co-directional variation of the heat flux. It is easily seen that every strong conductor is also a normal conductor.

For n_K being the components of the unit normal vector of the singular surface and

$$\theta_{,M} \equiv \theta^-_{,M} \quad , \quad \xi n_M \equiv [\![\theta_{,M}]\!] \tag{4.10}$$

we conclude that the Fourier's condition (2.78)

$$[\![q_K]\!]N_K = 0$$

cannot be satisfied in strong conductors for $[\![\theta_{,M}]\!]N_M \neq 0$. It means that across the weak discontinuity (e.g. acoustic wave) in the strong conductor the normal component of the temperature gradient must be continuous. If the gradient suffers the jump, the surface has either to carry concentrations or it is the strong discontinuity (e.g. shock wave).

Applying the mean value theorem of the differentiation to the formula (4.8) we see that it can be written in the form

$$\xi n_K \frac{\partial q_K}{\partial \theta_{,L}}(C_{MN}, \theta, \theta_{,M} + \alpha \xi n_M, \dot{\theta}) n_L \neq 0, \qquad 0 < \alpha < 1. \tag{4.11}$$

The derivative

$$K_{KL}(\underset{\sim}{X}, t) := -\frac{\partial \tilde{q}_K}{\partial \theta_{,L}}(C_{MN}, \theta, \theta_{,M}, \dot{\theta}), \tag{4.12}$$

is called the c o n d u c t i v i t y of the material.

In the special case of the elastic heat conductor the conductivity takes the form

$$K_{KL} = \tilde{K}_{KL}(C_{MN}, \theta, \theta_{,M}). \tag{4.13}$$

For such conductors B.D.Coleman and M.E.Gurtin[1965,1,2] have introduced the notion of a d e f i n i t e c o n d u c t o r (see also: C.Truesdell[1969]). This type of conductors has the positive definite conductivity, i.e.

$$\underset{\theta_{,M}}{\forall} \; K_{KL} n_K n_L > 0 \tag{4.14}$$

for any unit vector $\underset{\sim}{n}$. It follows immediately from the relation (4.11) that every definite conductor is also a strong conductor. P.Chadwick[1977] shows that the converse does not hold. The following conductor

$$\tilde{q}_K(C_{MN}, \theta, \theta_{,M}) = K(C_{MN}, \theta)\left(1 - \frac{2g^2}{3g_0^2} + \frac{g^4}{5g_0^4}\right)\theta_{,K} \tag{4.15}$$

where

$$g := (\theta_{,M}\theta_{,M})^{\frac{1}{2}}. \tag{4.16}$$

and g_0-positive constant, $K(C_{MN}, \theta)$-a negative function, is a strong conductor but it is not a definite conductor. In fact, the conductivity satisfies in this case the inequality

$$\underset{\theta_{,M}}{\forall} \quad K_{KL} n_K n_L \geq 0 \tag{4.17}$$

so it is positive semi-definite only. The materials satisfying this inequality are called s e m i - d e f i n i t e c o n d u c t o r s. By virtue of the definition (4.1), we obtain

$$\begin{aligned} \frac{\partial \Phi}{\partial \xi}(\xi, \theta_{,M}; \theta, C_{MN}) &= n_K \frac{\partial \tilde{q}_K}{\partial \theta_{,L}}(C_{MN}, \theta, \theta_{,M} + \xi n_M) n_L = \\ &= -K_{KL}(C_{MN}, \theta, \theta_{,M} + \xi n_M) n_K n_L . \end{aligned} \tag{4.18}$$

The definition (4.8) of the strong conductor leads to the following implication

$$\xi_1 \neq \xi_2 \Rightarrow \Phi(\xi_1, \theta_{,M}; \theta, C_{MN}) \neq \Phi(\xi_2, \theta_{,M}; \theta, C_{MN}). \tag{4.19}$$

It means that $\Phi(\cdot, \theta_{,M}; \theta, C_{MN})$ is monotonous function and, hence $\frac{\partial \Phi}{\partial \xi}$ is either non-positive or non-negative. However the substitution

$$\overset{\circ}{\xi} = -\sqrt{\theta_{,N}\theta_{,N}} \quad , \qquad \overset{\circ}{n}_M = \frac{\theta_{,M}}{\sqrt{\theta_{,N}\theta_{,N}}}$$

gives

$$\begin{aligned} \frac{\partial \Phi}{\partial \xi}(\overset{\circ}{\xi}, \theta_{,M}, \theta, C_{MN}) &= \overset{\circ}{n}_K \frac{\partial \check{q}_K}{\partial \theta_{,L}}(C_{MN}, \theta, 0) \overset{\circ}{n}_L = \\ &= \overset{\circ}{n}_K \overset{\circ}{n}_L \left. \frac{\partial \check{q}_K}{\partial \theta_{,L}} \right|_E \end{aligned}$$

and, according to the inequality (3.85) we obtain

$$\frac{\partial \Phi}{\partial \xi}(\overset{\circ}{\xi}, \theta_{,M}; \theta, C_{MN}) \leq 0.$$

It means that the derivative $\frac{\partial \Phi}{\partial \xi}$ should be non-positive. Now, setting $\xi = 0$ in the formula (4.18) we get

$$K_{KL}(C_{MN},\theta,\theta_{,M})n_K n_L \geqslant 0 \tag{4.20}$$

for every unit vector $\underset{\sim}{n}$. Consequently, every strong conductor is also a semi-definite conductor. Again th converse does not hold. The counterexample is as follows

$$\tilde{q}_K(C_{MN},\theta,\theta_{,M}) = K(C_{MN},\theta)\bar{n}_L\theta_{,L}\bar{n}_K \tag{4.21}$$

where $K(C_{MN},\theta)$ is a negative function and $\bar{\underset{\sim}{n}}$ is fixed unit vector. Then

$$K_{KL} = -K(C_{MN},\theta)\bar{n}_L\bar{n}_K. \tag{4.22}$$

This tensor is, certainly, positive semi-definite and, consequently, the relation (4.21) defines a semi-definite conductor. However, the substitution of (4.21) in (4.9) yields

$$n_K K(C_{MN},\theta)\bar{n}_L\theta_{,L}\bar{n}_K = n_K K(C_{MN},\theta)\bar{n}_L(\theta_{,L}-\xi n_L)\bar{n}_K = 0 \tag{4.23}$$

in the case of the ortogonality of vectors $\underset{\sim}{n}$ and $\bar{\underset{\sim}{n}}$. It means that the considered material is not a strong conductor.

Summing up the above classification, we can establish the following relations among different types of conductors (P.Chadwick[1977])

$$\begin{aligned} &\mathfrak{D} \subset \mathfrak{S} \subset \mathcal{N}, \\ &\mathfrak{S} \subset s\mathfrak{D}, \end{aligned} \tag{4.24}$$

where

$\mathfrak{D}$-the class of definite conductors

$\mathfrak{S}$-the class of strong conductors

$\mathcal{N}$-the class of normal conductors

$s\mathfrak{D}$-the class of semi-definite conductors.

Finally, let us mention the procedure commonly used in the analysis of the heat conduction. It is assumed that the heat flux q_K can be expanded into the Taylor series with respect to the temperature gradient at the point $\theta_{,K}=0$. The analysis of this series has been carried out in the paper of B.D.Coleman and V.J.Mizel[1963] for a rigid heat conductor. We limit our considerations to the first two terms

$$q_K(\underset{\sim}{X},t)=\tilde{q}_K(C_{MN},\theta,\theta_{,M}=0,\dot{\theta})+\frac{\partial\tilde{q}_K}{\partial\theta_{,L}}(C_{MN},\theta,\theta_{,M}=0,\dot{\theta})+o(|\theta_{,M}|). \tag{4.25}$$

It is easy to see that the relation

$$\theta_{,K}\tilde{q}_K=0 \tag{4.7}$$

has a non-trivial solution

$$\tilde{q}_K(C_{MN},\theta,\theta_{,M}=0,\dot{\theta})\neq 0 \tag{4.26}$$

for $\theta_{,K}=0$ and $\dot{\theta}\neq 0$. It means that the thermoelastic material in the non-equilibrium state can be an anomalous conductor. Due to the relation (3.60) this solution vanishes in the equilibrium. As expected, the anomaly does not violate the non-existence of the piezo-caloric effect.

Let us consider the case

$$\tilde{q}_K(C_{KL},\theta,\theta_{,M}=0,\dot{\theta})\equiv 0. \tag{4.27}$$

Then

$$q_K(\underset{\sim}{X},t)=-\tilde{K}_{KL}(C_{MN},\theta,\dot{\theta})\theta_{,L}+o(|\theta_{,K}|). \tag{4.28}$$

This form of the expansion holds true for every elastic heat conductor due to the constitutive relation $(3.70)_3$ and the condition $(3.85)_2$. The only difference is in the form of the constitutive relation for the conductivity

$$K_{KL}(\underset{\sim}{X},t)=\tilde{K}_{KL}(C_{MN},\theta), \tag{4.29}$$

i.e. in this case the conductivity does not depend on the rate of temperature $\dot{\theta}$.

For sufficiently small temperature gradients $\theta_{,M}$, we can neglect the higher order terms. In such a case, we obtain the Fourier-Duhamel relation for the heat flux

$$q_K(\underset{\sim}{X},t) = -\tilde{K}_{KL}(C_{MN},\theta,\dot{\theta})\,\theta_{,L} \tag{4.30}$$

for thermoelastic materials and

$$q_K(\underset{\sim}{X},t) = -\tilde{K}_{KL}(C_{MN},\theta)\,\theta_{,L} \tag{4.31}$$

for elastic heat conductors.

According to the inequality (3.69), the conductivity appearing in (4.30) should satisfy the condition

$$\tilde{K}_{KL}(C_{MN},\theta,\dot{\theta}=0)\lambda_K\lambda_L \geqslant 0. \tag{4.32}$$

Otherwise this tensor is not limited by thermodynamic conditions. In the next Section we return to some properties of this object but, in general, the problem of physically acceptable form of it has not been investigated.

On the other hand, due to the independence of the conductivity in (4.31) on the temperature gradient $\theta_{,M}$, it should satisfy the inequality $(3.55)_3$, i.e.

$$\tilde{K}_{KL}(C_{MN},\theta)\lambda_K\lambda_L \geqslant 0. \tag{4.33}$$

Hence, every elastic heat conductor is a semi-definite conductor. We can distinguish four types of such conductors (P.Chadwick, P.K.Currie 1972)

i/ $K_{KL}(C_{MN},\theta)$ is positive definite; then the conductor is definite;

ii/ $K_{KL}(C_{MN},\theta)$ has one proper number equal to zero; then for any temperature gradient $\theta_{,K}=\varkappa\,\nu_K$ where ν_K is the associate proper vector, $\varkappa$ - real parameter, the heat flux is equal to zero; it means that the conductor is d i r e c t i o n a l l y anomalous;

iii/ $K_{KL}(C_{MN},\theta)$ has two proper numbers equal to zero; the material is anomalous in every direction perpendicular to the proper vector associated with the non-zero proper number;

iv/ $K_{KL}(C_{MN}, \theta)$ has all proper numbers equal to zero; then the material is a non-conductor.

4.2. Properties of the Heat Conduction Equation

The above presented analysis of heat conductors is connected with two main problems of the thermoelastic materials: the propagation of thermal disturbances and the propagation of acoustic and shock waves. The latter problem is presented in the last Chapter of this paper. Below we show some features of the former.

In general, the problem of the heat conduction is coupled with the problem of mechanical disturbances and not much can be said about the properties of this form of the energy transport without constructing the full solution. Namely, for given constitutive relations (3.18), the distribution of temperature is described by the energy balance equation. Its explicit form depends on the deformation C_{KL}, which can be found from the balance of momentum law. These equations, in turn, depend on the temperature distribution.

However, certain important features of heat conduction can be established in the analysis of the rigid heat conductor, i.e. of the material, whose deformations are negligible in the analysis of the temperature distribution. In such a case, we have to deal solely with the energy balance of the form

$$\varrho_0 \dot{\varepsilon} + q_{K,K} = 0. \tag{4.34}$$

The constitutive relations take the form

$$\begin{aligned} \varepsilon(\underset{\sim}{X},t) &= \check{\varepsilon}(\theta, \theta_{,M}, \dot{\theta}), \\ \eta(\underset{\sim}{X},t) &= \check{\eta}(\theta, \theta_{,M}, \dot{\theta}), \end{aligned} \tag{4.35}$$

and the consequences of the second law of thermodynamics assume the form

$$\frac{\partial \Lambda}{\partial \dot{\theta}} \tilde{q}_K = \varrho_0 \left(\Lambda \frac{\partial \tilde{\varepsilon}}{\partial \theta_{,K}} - \frac{\partial \tilde{\eta}}{\partial \theta_{,K}} \right), \tag{4.36}$$

$$\left(\frac{\partial \tilde{\eta}}{\partial \theta} - \Lambda \frac{\partial \tilde{\varepsilon}}{\partial \theta} \right) \dot{\theta} + \frac{1}{\varrho_0} \frac{\partial \Lambda}{\partial \theta} \tilde{q}_K \theta_{,K} \geq 0, \tag{4.37}$$

where

$$\Lambda = \Lambda(\theta, \dot{\theta}) \tag{4.38}$$

Finally, in the state of equilibrium ($\dot{\theta} = 0$, $\theta_{,M} = 0$), the following relations should be obeyed

$$\frac{\partial \tilde{\eta}|_E}{\partial \theta} - \frac{1}{T} \frac{\partial \tilde{\varepsilon}|_E}{\partial \theta} = 0, \tag{4.39}$$

$$T \equiv \frac{1}{\Lambda}\Big|_E ; \qquad \frac{\partial \Lambda}{\partial \theta}\Big|_E < 0, \tag{4.40}$$

$$\tilde{q}_K|_E = 0, \tag{4.41}$$

$$-\left(\frac{T^2}{\frac{dT}{d\theta}} \right) \left(\frac{\partial \Lambda}{\partial \dot{\theta}} \right)_E \frac{\partial \tilde{\varepsilon}|_E}{\partial \theta} \geq \frac{\partial \tilde{\varepsilon}}{\partial \dot{\theta}}\Big|_E \tag{4.42}$$

$$\frac{\partial \tilde{q}_L}{\partial \theta_{,M}}\Big|_E \lambda_L \lambda_M \leq 0. \tag{4.43}$$

The above relations form the full set of thermodynamic formulae describing the rigid heat conductor. In addition, some material symmetry conditions should be fulfilled but we do not discuss this problem in the paper.

Let us check the properties of the equation (4.34) under the above conditions. The substitution of (4.35) and (4.36) in (4.34) yields

$$\varrho_0 \frac{\partial \tilde{\varepsilon}}{\partial \dot{\theta}} \ddot{\theta} + \left(\varrho_0 \frac{\partial \tilde{\varepsilon}}{\partial \theta_{,K}} + \frac{\partial \check{q}_K}{\partial \dot{\theta}} \right) \dot{\theta}_{,K} + \frac{\partial \check{q}_K}{\partial \theta_{,L}} \theta_{,KL} + \check{G}(\theta, \theta_{,M}, \dot{\theta}) = 0, \tag{4.44}$$

where

$$\tilde{G}(\theta, \theta_{,K}, \dot{\theta}) := \varrho_0 \frac{\partial \tilde{\varepsilon}}{\partial \theta} \dot{\theta} + \frac{\partial \tilde{q}_K}{\partial \theta} \theta_{,K} . \tag{4.45}$$

Hence the relation (4.44) is a quasilinear partial differential equation describing the temperature distribution $\theta(X,t)$. If we want the thermal disturbance to have the finite speed of propagation this equation should by hyperbolic. Let us write the condition of hyperbolicity for the one-dimensional case. Then the equation (4.44) becomes

$$\varrho_0 \frac{\partial \tilde{\varepsilon}}{\partial \dot{\theta}} \ddot{\theta} + \left(\varrho_0 \frac{\partial \tilde{\varepsilon}}{\partial \dot{\theta}} + \frac{\partial \tilde{q}}{\partial \dot{\theta}}\right) \dot{\theta}' + \frac{\partial \tilde{q}}{\partial \theta'} \theta'' + \tilde{G}(\theta, \theta', \dot{\theta}) = 0, \tag{4.46}$$

where

$$\theta' := \frac{\partial \theta}{\partial X}, \quad q := q_X . \tag{4.47}$$

Its hyperbolicity condition has the form

$$\varrho_0 \frac{\partial \tilde{\varepsilon}}{\partial \dot{\theta}} \frac{\partial \tilde{q}}{\partial \theta'} - \frac{1}{4} \left(\varrho_0 \frac{\partial \tilde{\varepsilon}}{\partial \theta'} + \frac{\partial \tilde{q}}{\partial \dot{\theta}}\right)^2 < 0 . \tag{4.48}$$

As we see, this condition can be fulfilled only if the constitutive relations contain the rate of temperature $\dot{\theta}$ as an argument. Otherwise, the first two terms in the equation (4.46) vanish (see the remark after the formula (3.83)) and the equation is either parabolic or elliptic.

The speed of the propagation of temperature disturbances for the equation (4.46) is as follows

$$C_T = \frac{1}{2\varrho_0 \frac{\partial \tilde{\varepsilon}}{\partial \dot{\theta}}} \left[\left(\varrho_0 \frac{\partial \tilde{\varepsilon}}{\partial \theta'} + \frac{\partial \tilde{q}}{\partial \dot{\theta}}\right) \pm \sqrt{\left(\varrho_0 \frac{\partial \tilde{\varepsilon}}{\partial \theta'} + \frac{\partial \tilde{q}}{\partial \dot{\theta}}\right)^2 - 4\varrho_0 \frac{\partial \tilde{\varepsilon}}{\partial \dot{\theta}} \frac{\partial \tilde{q}}{\partial \theta'}} \right] \tag{4.49}$$

for $\frac{\partial \tilde{\varepsilon}}{\partial \dot{\theta}} \neq 0$, and

$$C_T = \frac{\frac{\partial \tilde{q}}{\partial \theta'}}{\varrho_0 \frac{\partial \tilde{\varepsilon}}{\partial \theta'} + \frac{\partial \tilde{q}}{\partial \dot{\theta}}} \tag{4.50}$$

for $\frac{\partial \tilde{\varepsilon}}{\partial \dot{\theta}} = 0$.

The above feature of thermoelastic materials is even more obvious in the linear theory. The linearization of the equation (4.46) leads to the following relation

$$\varrho_0 \frac{\partial \tilde{\varepsilon}}{\partial \dot{\theta}}\Big|_E \ddot{\theta} - K|_E \theta'' + \varrho_0 \frac{\partial \tilde{\varepsilon}}{\partial \theta}\Big|_E \dot{\theta} = 0, \tag{4.51}$$

due to the formulae (4.36) and (4.41). In the above relation K is the conductivity

$$K := -\frac{\partial \tilde{q}}{\partial \theta'}. \tag{4.52}$$

The equation (4.51) is hyperbolic if

$$-\varrho_0 \frac{\partial \tilde{\varepsilon}}{\partial \dot{\theta}}\Big|_E K|_E < 0. \tag{4.53}$$

According to the inequality (4.43) the conductivity is positive. Consequently

$$\frac{\partial \tilde{\varepsilon}}{\partial \dot{\theta}}\Big|_E > 0 . \tag{4.54}$$

Comparing this condition with the inequality (4.42) we see that it is essential that the coldness Λ depends on the rate $\dot{\theta}$. Otherwise the equation (4.51) would be again parabolic or elliptic.

CHAPTER 5
WAVES IN THERMOELASTIC MATERIALS

5.1. Acoustical waves; propagation condition

5.1.1. Preliminaries

The aim of this Chapter is to show on a few simple examples the influence of heat conduction on the propagation of acoustical and shock waves in thermoelastic materials. We limit our considerations to the most immediate results bearing in mind the compact form of this article. Many interesting details can be found in the paper of M.F.McCarthy[1975].

The problem of the propagation of waves in thermoelastic materials, described by the constitutive relations (3.18) has not been investigated. The number of properties of waves moving through the material with memory can be directly transfered on this class of materials but its specific properties are unknown. On the other hand, the propagation of waves in elastic heat conductors, described by the relations (3.70), has been treated quite extensively. Further we present solely the results for such materials. Let us collect the formulae, derived earlier in the paper to be used in this Chapter. We neglect the surface sources and concentrations

$$\rho_s = \rho_o^* = 0, \; p_k = {}_o p_k^* = 0, \; \varepsilon_s = \varepsilon_o^* = 0, \; \eta_s = 0, \tag{5.1}$$

and, dealing with the acoustic waves, we assume

$$[\![v_k]\!] = 0, \; [\![\theta]\!] = 0. \tag{5.2}$$

Then the set of governing equations simplifies considerably and takes the form

$$\begin{aligned}
&1. \; \rho_o \dot{v}_k = t_{kL,L}, \\
&2. \; \rho_o \dot{\varepsilon} = v_{k,L} t_{k,L} - q_{K,K}, \\
&3. \; [\![\rho]\!] = 0, \\
&4. \; [\![U]\!] = 0, \\
&5. \; [\![t_{kL}]\!] N_L = 0, \\
&6. \; [\![q_K]\!] N_K = 0, \\
&7. \; F(\underset{\sim}{X}, t) = \tilde{F}(C_{MN}, \theta), \\
&8. \; q_K(\underset{\sim}{X}, t) = \tilde{q}_K(C_{MN}, \theta, \theta_{,M}), \\
&9. \; \eta(\underset{\sim}{X}, t) = -\left(\frac{dT}{d\theta}\right)^{-1} \frac{\partial \tilde{F}}{\partial \theta}, \\
&10. \; t_{kL}(\underset{\sim}{X}, t) = 2\rho_o \frac{\partial \tilde{F}}{\partial C_{KL}} x_{k,K} = \rho_o \frac{\partial \tilde{F}}{\partial x_{k,L}}
\end{aligned} \tag{5.3}$$

where

$$t_{kL} := t_{KL} x_{k,K} , \qquad F = \varepsilon - T\eta \tag{5.4}$$

and we have made use of the identity

$$\left(J (x_{k,K})^{-1} \right)_{,K} = 0. \tag{5.5}$$

The unit vector $\underset{\sim}{N}$ is perpendicular to the image of the surface $\sigma(t)$ in the reference configuration. If the surface $\sigma(t)$ is given by the equation

$$g(\underset{\sim}{x}, t) = 0, \tag{5.6}$$

then

$$n_k = \frac{\frac{\partial g}{\partial x_k}}{\left| \frac{\partial g}{\partial x_m} \right|} \tag{5.7}$$

and, due to the relation

$$\frac{\partial g}{\partial x_k} \frac{d x_k}{dt} + \frac{\partial g}{\partial t} = 0, \tag{5.8}$$

we have ($U = \frac{dx_k}{dt} n_k$)

$$U \left| \frac{\partial g}{\partial x_m} \right| + \frac{\partial g}{\partial t} = 0. \tag{5.9}$$

On the other hand, the image $\Sigma(t)$ of the surface $\sigma(t)$ is given by the relation

$$G(\underset{\sim}{X}, t) := g(\underset{\sim}{x}(\underset{\sim}{X}, t), t) = 0 \tag{5.10}$$

and, consequently,

$$N_K = \frac{\frac{\partial G}{\partial X_K}}{\left| \frac{\partial G}{\partial X_M} \right|} , \qquad U_0 \left| \frac{\partial G}{\partial X_M} \right| + \frac{\partial G}{\partial t} = 0. \tag{5.11}$$

It is easy to establish the following relation

$$U_0 = \frac{U}{B_{(n)}} , \qquad N_K = \frac{n_k \chi_{k,K}}{B_{(n)}} \tag{5.12}$$

where

$$B^2_{(n)} := \chi_{k,K} \chi_{l,K} n_k n_l . \tag{5.13}$$

Further we use the material description of the acoustical wave, i.e. we consider such a wave to be the surface $\Sigma(t)$ carrying no jump of the velocity v_k and of the empirical temperature θ .

Except of the above relations we need also certain geometrical conditions of compatibility (cf. C.Truesdell, R.A.Toupin 1960,Sec.175), following from the smoothness assumptions on the jumps. As an example let us consider the heat flux vector q_K. We assume that both limits q^+_K, q^-_K are differentiable along an arbitrary arc on the surface $\Sigma(t)$. If l is the parameter along this arc, then (see Fig.6)

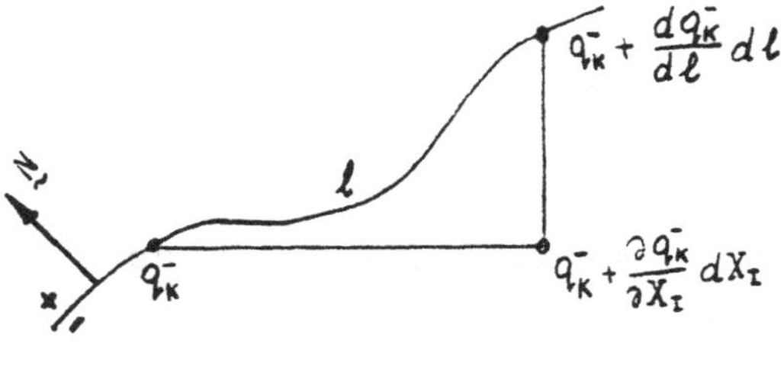

Fig.6

$$\frac{dq^+_K}{dl} = q^+_{K,L} \frac{dX_L}{dl} , \qquad \frac{dq^-_K}{dl} = q^-_{K,L} \frac{dX_L}{dl} . \tag{5.14}$$

The substitution yields the following Hadamard's lemma

$$\frac{d}{dl}[q_K] = [q_{K,L}] \tau_L \tag{5.15}$$

where τ_L are the components of the vector tangent to the arc chosen on the surface $\Sigma(t)$. In a particular case of the continuous flux, we have the Maxwell's theorem

$$[\![q_K]\!]=0 \quad \Rightarrow \quad [\![q_{K,L}]\!]=Q_K N_L \tag{5.16}$$

where Q_K is called an amplitude of discontinuity; the similar results easily follow for another fields.

In the same way we can obtain the kinematical conditions of compatibility. In this case we have to use the Hadamard's lemma (5.15) to the three-dimensional surface in the four-dimensional space-time. For the same example of the heat flux vector it follows at once

$$\frac{d}{dt}[\![q_K]\!]=U_0[\![q_{K,L}]\!]N_L+[\![\dot{q}_K]\!]. \tag{5.17}$$

Again, for the continuous field we have

$$[\![\dot{q}_K]\!]=-U_0[\![q_{K,L}]\!]N_L$$

and, according to the relation (5.16),

$$[\![\dot{q}_K]\!]=-U_0 Q_K. \tag{5.18}$$

In particular, for the acoustic wave, we have

$$[\![x_{k,K}]\!]=\lambda_k N_K$$

and

$$[\![\dot{x}_k]\!]=-U_0\lambda_k$$

but due to the continuity of the velocity we obtain

$$\lambda_k=0.$$

It means that the deformation gradient is continuous on the

acoustic wave. For this reason, we have put

$$[\![\varepsilon]\!] = 0$$

in the formula $(5.3)_6$.

5.1.2. The governing equations of acoustical waves in a strong heat conductor

We proceed now to the derivation of the equations describing the directions and speed of propagation of the acoustic waves. The theory is based on the continuity of stress tensor across the surface $\Sigma(t)$

$$[\![t_{kL}]\!] = 0, \tag{5.19}$$

following from the relation $(5.3)_{10}$ and the continuity of the deformation gradient $\chi_{k,K}$. The condition (5.19) is, certainly, sufficient for the Poisson's condition $(5.3)_5$ to hold. According to the geometrical condition of compatibility, we have then

$$[\![t_{kL,L}]\!] = T_{kL} N_L \tag{5.20}$$

where T_{kL} is the amplitude of the discontinuity. On the other hand, we obtain from the kinematical condition of compatibility the following relation

$$[\![\dot{t}_{kL}]\!] = -U_0 T_{kL}. \tag{5.21}$$

These two relations yield

$$[\![t_{kL,L}]\!] = -\frac{1}{U_0} [\![\dot{t}_{kL}]\!] N_L. \tag{5.22}$$

The equations of motion $(5.3)_1$ lead to the following jump condition on the singular surface

$$\varrho_0 [\![\dot{v}_k]\!] = [\![t_{kL,L}]\!]. \tag{5.23}$$

Again, according to the kinematical condition of compatibility, we have

$$[\![\dot{v}_k]\!] = [\![\ddot{\chi}_k]\!] = U_0^2 a_k , \tag{5.24}$$

where a_k is the amplitude of discontinuity of the acceleration. Joining the relations (5.22), (5.23), (5.24), we finally obtain the following Gurtin's lemma

$$[\![\dot{t}_{kL}]\!] N_L + \varrho_0 U_0^3 a_k = 0. \tag{5.25}$$

Now, we make use of the constitutive relations $(5.3)_7$ and $(5.3)_{10}$. Then

$$\dot{t}_{kL} = \left(2\varrho_0 \frac{\partial \tilde{F}}{\partial C_{KL}} \chi_{k,K}\right)^{\bullet} = \left(\varrho_0 \frac{\partial \tilde{F}}{\partial \chi_{k,L}}\right)^{\bullet} = \varrho_0 \frac{\partial^2 \tilde{F}}{\partial \chi_{k,L} \partial \chi_{l,M}} \dot{\chi}_{l,M} + \varrho_0 \frac{\partial^2 \tilde{F}}{\partial \chi_{k,L} \partial \theta} \dot{\theta}.$$

Let us introduce the notation

$$A_{kLlM} := \varrho_0 \frac{\partial^2 \tilde{F}}{\partial \chi_{k,L} \partial \chi_{l,M}} = 4\varrho_0 \frac{\partial^2 \tilde{F}}{\partial C_{MP} \partial C_{KL}} \chi_{k,K} \chi_{l,P} + 2 \frac{\partial \tilde{F}}{\partial C_{ML}} \delta_{kl} ,$$

$$L_{kL} := \frac{\partial^2 \tilde{F}}{\partial \theta \partial \chi_{k,L}} = 2 \frac{\partial^2 \tilde{F}}{\partial \theta \partial C_{KL}} \chi_{k,K} . \tag{5.26}$$

Then the relation (5.25) takes the form

$$[\![A_{kLlM} \dot{\chi}_{l,M} + \varrho_0 L_{kL} \dot{\theta}]\!] N_L + \varrho_0 U_0^3 a_k = 0 . \tag{5.27}$$

However both tensors A_{kLlM} and L_{kL} are continuous and

$$[\![\dot{\chi}_{l,M}]\!] = - U_0 a_l N_M$$
$$[\![\dot{\theta}]\!] = - U_0 \Theta \tag{5.28}$$

where Θ is the amplitude of discontinuity of the temperature gradient

$$[\![\theta_{,L}]\!] = \Theta N_L \Rightarrow \Theta = [\![\theta_{,L}]\!] N_L . \tag{5.29}$$

Simultaneously, we have proved in Sec.4.1. that the normal component of the temperature gradient must be continuous in strong conductors, i.e.

$$\Theta = 0 \tag{5.30}$$

in the case considered in this Section. Hence the condition (5.27) reduces to the form

$$\left(Q_{kl} - \rho_o U_o^2 \delta_{kl} \right) a_k = 0 , \qquad Q_{kl} := A_{kLlM} N_L N_M . \tag{5.31}$$

The above relation is the well-known propagation condition for hyperelastic materials. The proper numbers of the acoustical tensor Q_{kl} give the speeds of propagation of the wave, while the associate proper vectors describe the directions of propagation (the acoustical axes). These results form a classical Fresnel-Hadamard theorem. With respect to the relation (5.30) we say that the strong conductors admit solely the homothermal acoustical waves.

Now, we check the consequences of the energy balance equation $(5.3)_2$. Let us eliminate the specific energy ε from this equation. According to the definition (3.80), we have

$$\dot{\varepsilon} = \dot{F} + T\dot{\eta} + \dot{T}\eta . \tag{5.32}$$

However

$$\dot{F} = \frac{\partial \tilde{F}}{\partial \chi_{k,K}} \dot{\chi}_{k,K} + \frac{\partial \tilde{F}}{\partial \theta} \dot{\theta}$$

and, due to the relations $(5.3)_9$ and $(5.3)_{10}$, we obtain

$$\rho_o \dot{F} = t_{kK} v_{k,K} - \rho_o \dot{\theta} \frac{dT}{d\theta} \eta = t_{kK} v_{k,K} - \rho_o \dot{T} \eta . \tag{5.33}$$

The substitution of the above relation in (5.32) yields

$$\rho_o \dot{\varepsilon} = t_{kK} v_{k,K} + \rho_o T \dot{\eta} . \tag{5.34}$$

Consequently, the energy balance equation takes the form

$$\rho_0 T\dot{\eta} + q_{K,K} = 0. \tag{5.35}$$

Hence, on the singular surface $\Sigma(t)$ we have

$$\rho[\![T\dot{\eta}]\!] + [\![q_{K,K}]\!] = 0. \tag{5.36}$$

Assuming the continuity of the heat flux

$$[\![q_K]\!] = 0 \tag{5.37}$$

we can write

$$[\![q_{K,K}]\!] = -\frac{1}{U_0}[\![\dot{q}_K]\!]N_K \tag{5.38}$$

(compare the formula (5.22))and then we have ($[\![T]\!]=0$)

$$\rho_0 T U_0 [\![\dot{\eta}]\!] - [\![\dot{q}_K]\!]N_K = 0. \tag{5.39}$$

A straightforward calculation, using the constitutive relations for η and q_K as well as the continuity of the temperature gradient $\theta_{,L}$, leads to the following form of the relation (5.39)

$$\frac{\partial q_K}{\partial \theta_{,L}} N_K N_L [\![\ddot{\theta}]\!] = -\left(\rho_0 U_0^3 T \left(\frac{dT}{d\theta}\right)^{-1} L_{kK} N_K + \frac{\partial q_K}{\partial x_{k,L}} N_K N_L U_0^2\right) a_k. \tag{5.40}$$

In the derivation of this formula we have utilized the relation

$$[\![\dot{\theta}_{,L}]\!] := -\frac{1}{U_0}[\![\ddot{\theta}]\!]N_L. \tag{5.41}$$

The formula (5.40) means that the acoustic wave traversing the strong conductor creates the discontinuity of the second derivatives of temperature. The amplitude of this discontinuity can be calculated from the formula (5.40).

The propagation condition (5.31) completely describes the properties of acoustic waves in hyperelastic materials.

In such materials the amplitude a_k remains constant during the motion of the wave. It is not a case when the material is a strong conductor. The variation of the amplitude depends on the behavior of the temperature θ . The equation, describing this variation, can be derived from the equations of motion $(5.3)_1$ through their differentiation with respect to time and the evaluation of the jump of the resulting equation on the surface $\Sigma(t)$. We will not present the general result with respect to its complexity. It can be found in the above mentioned papers of P.Chadwick, P.K.Currie 1972 and M.F.McCarthy 1975.

It is worthy to show the particular but representative result for a plane wave. Assuming that the material is in the state of homogeneous deformations and uniform temperature ahead of the wave we arrive at the following equation for the magnitude of the amplitude a_k

$$\frac{da}{dn} + \sigma a - K a^2 = 0 \tag{5.42}$$

where

$$\begin{aligned} a &:= (a_k a_k)^{\frac{1}{2}}, \quad a_k := a l_k, \\ \sigma &:= -\frac{T}{2U_0 k} \frac{\partial^2 \tilde{F}}{\partial\theta \, \partial\chi_{k,L}} (\chi_{k,K})^{-1} N_K \chi_{j,L} l_j, \\ K &:= \frac{B_{(n)}}{2U_0^2} \frac{\partial^3 \tilde{F}}{\partial\chi_{k,K} \partial\chi_{l,L} \partial\chi_{m,M}} N_K N_L N_M l_k l_l l_m, \\ k &:= -J^{-1} B_{(n)}^2 \frac{\partial q_K}{\partial\theta_{,L}} N_K N_L, \end{aligned} \tag{5.43}$$

σ , K being constants (see: W.A.Green 1964 and M.F.McCarthy 1975, Sec.2.6). The differentiation in the equation (5.42) is carried out in the direction normal to the wavefront. The integration of this equation yields

$$a = \left[\left(1 - \frac{K a_0}{\sigma}\right) e^{\sigma n} + \frac{K a_0}{\sigma}\right]^{-1} a_0 . \tag{5.44}$$

In the case of $\left|\frac{Ka_0}{\sigma}\right| \ll 1$, we have

$$a \approx a_0 e^{-\sigma n} \tag{5.45}$$

which is the amplitude decay law of classical linear thermoelasticity. In general, the solution (5.44) admits the unbounded increase of the amplitude in the finite time

$$t_c = -\frac{\log\left(1-\frac{\sigma}{Ka_0}\right)}{\sigma U}. \tag{5.46}$$

The limit value of the amplitude is, of course, $\frac{\sigma}{K}$, for which the amplitude remains constant. This value of the amplitude is called critical. P.Chadwick, Pk.Currie 1972, have computed its value for copper in the temperature of about 300°K and have got the result

$$a_{cr} = 2.8 \times 10^7 \text{ megabars}\cdot s^{-1}.$$

Such a high rate of loading yields an instantaneous disruption of the material. It means that the possibility of the critical growth of the homothermal acoustic wave is highly unlikely.

5.1.3. The governing equations of acoustical waves in nonconductors

According to the definition (4.5), the nonconductor is characterized by the relation

$$q_K(C_{MN}, \theta, \theta_{,M}, \dot{\theta}) = 0 \tag{5.47}$$

for any temperature gradient $\theta_{,M}$. Consequently, the energy balance (5.35) takes the form

$$\rho_0 T\dot{\eta} = 0. \tag{5.48}$$

In particular, on the surface $\Sigma(t)$ we have

$$[\dot{\eta}] = -U_0 H = 0 \tag{5.49}$$

where H is the amplitude of the discontinuity of entropy rate. The relation (5.49) means that acoustic waves in nonconductors are necesserily homentropic, i.e. $H \equiv 0$.

Let us derive the propagation condition for such waves. Due to the constitutive relation for the entropy, we have

$$\dot{\eta} = \frac{\partial \tilde{\eta}}{\partial \chi_{k,K}} \dot{\chi}_{k,K} + \frac{\partial \tilde{\eta}}{\partial \theta} \dot{\theta}. \tag{5.50}$$

Making use of the compatibility conditions and the relation (5.49), we obtain

$$\frac{\partial \tilde{\eta}}{\partial \chi_{k,K}} a_k N_K + \frac{\partial \tilde{\eta}}{\partial \theta} \Theta = 0, \tag{5.51}$$

or, bearing in mind the formula $(5.3)_9$,

$$\Theta = -\left[\frac{\partial^2 \tilde{F}}{\partial \theta^2} - \left(\frac{dT}{d\theta}\right)^{-1} \frac{d^2 T}{d\theta^2} \frac{\partial \tilde{F}}{\partial \theta} \right]^{-1} L_{kK} N_K a_k. \tag{5.52}$$

The above formula allows to eliminate the temperature amplitude from the equation (5.27). Namely, we get

$$\left(\tilde{Q}_{kl} - \rho_0 U_0^2 \delta_{kl}\right) a_k = 0 \tag{5.53}$$

where

$$\tilde{Q}_{kl} := \left(A_{kLlM} - \rho_0 \left[\frac{\partial^2 \tilde{F}}{\partial \theta^2} - \left(\frac{dT}{d\theta}\right)^{-1} \frac{d^2 T}{d\theta^2} \frac{\partial \tilde{F}}{\partial \theta} \right]^{-1} L_{kL} L_{lM}\right) N_L N_M \tag{5.54}$$

is the acoustical tensor for homentropic waves. Due to the formal similarity of the propagation condition (5.53) and the corresponding condition (5.31) for homothermal waves, the analysis in both cases is also similar. In particular, it leads to the Fresnel-Hadamard theorem of the same form as previously.

Let us mention, finally, that in the case of normal and anomalous conductors the acoustical waves are usually neither homothermal nor homentropic. The investigation of these cases has been carried out by P.Chadwick and P.K.Currie[1972].

5.2. Shock Waves

The shock wave in nonlinear materials is the most difficult type of wave to analyze. The existence of such waves in elastic solids has been proved in many papers, starting from the early work of D.R.Bland[1962]. Some experimental evidence for viscoelastic materials has been reported by J.W.Nunziato at al.[1974]. The theoretical indication of the formation of shock in thermoelastic materials follows from the previous discussion of the growth of amplitude of the acoustical wave (see: the formula (5.44)).

Our considerations will concern only the basic properties of plane shock waves. We base the presentation on the papers of P.J.Chen, M.E.Gurtin[1970] and M.F.McCarthy[1975].

The confinement of considerations to plane waves eliminates the very important effect of curvature but due to the technical difficulties the progress in this field is very slow.

We start with some kinematical preliminaries. In the case of shock waves we assume the motion χ_k to be continuous but the velocity v_k and, consequently, the deformation gradient $\chi_{k,K}$ as well as the temperature suffer the discontinuities on the surface $\sigma(t)$. It is again convenient to use the material description of the wave. The wavefront $\Sigma(t)$ is the the image of $\sigma(t)$ in the reference configuration. Let us assume that the surface $\Sigma(t)$ moves in the $X_I \equiv X$ - direction. Then $X = Y(t)$ is the material point at which $\Sigma(t)$ intersects the X-axis at the instant of time t. The derivative

$$U_0 = \frac{dY}{dt}(t) \tag{5.55}$$

defines the intrinsic velocity of $\Sigma(t)$. We find the connection of this velocity with the previously introduced spatial velocities. The point of $\mathcal{G}(t)$, corresponding to the point Y(t) of the surface $\Sigma(t)$ is given by the relation

$$x = \chi(Y(t); t) \; ; \qquad x \equiv x_1 . \tag{5.56}$$

It means that the speed of displacement is described by the formula

$$c = \frac{d\chi}{dt}(Y(t); t). \tag{5.57}$$

Taking into account the discontinuity of the deformation gradient we have

$$c = \left(\frac{\partial\chi}{\partial X}\right)^{+} \dot{Y} + \left(\frac{\partial\chi}{\partial t}\right)^{+} = \left(\frac{\partial\chi}{\partial X}\right)^{-} \dot{Y} + \left(\frac{\partial\chi}{\partial t}\right)^{-}. \tag{5.58}$$

It is convenient to introduce the notation

$$e := \frac{\partial\chi}{\partial X}. \tag{5.59}$$

Then

$$c = e^{+} U_0 + v^{+} = e^{-} U_0 + v^{-}. \tag{5.60}$$

Now, bearing in mind the definition (2.69), we get

$$\begin{aligned} U^{+} &= c - v^{+} = e^{+} U_0 , \\ U^{-} &= c - v^{-} = e^{-} U_0 , \end{aligned} \tag{5.61}$$

and, consequently,

$$[\![U]\!] = U_0 [\![e]\!] \tag{5.62}$$

These relations describe the velocity fields, utilized in the earlier parts of this paper. Obviously, the relation (5.62) implies the following kinematical condition of compatibility

$$[\![e]\!] U_0 = -[\![v]\!] \tag{5.63}$$

and, according to the formula (5.17), we have

$$\frac{d[\![e]\!]}{dt} = [\![\dot{e}]\!] + U_0 [\![\frac{\partial e}{\partial X}]\!],$$
$$\frac{d[\![v]\!]}{dt} = [\![\dot{v}]\!] + U_0 [\![\dot{e}]\!]. \tag{5.64}$$

We confine our attention to the longitudinal shock waves. The problem of the propagation is described by the balance equations. The first group of these equations - the Rankine-Hugoniot's conditions- asserts that

$$\varrho_0 U_0 [\![v]\!] + [\![\sigma]\!] = 0, \qquad \sigma := t_{1I} N_I,$$
$$\varrho_0 U_0 [\![\varepsilon]\!] + \langle\sigma\rangle [\![v]\!] - [\![q]\!] = 0, \qquad q := q_I N_I. \tag{5.65}$$

Joining the equation $(5,65)_1$ and the relation (5,63), we obtain the following classical result for the speed of propagation

$$\varrho_0 U_0^2 = \frac{[\![\sigma]\!]}{[\![e]\!]}. \tag{5.66}$$

The above conditions should be supplemented by the jump equations following from the local balance of momentum and energy. In the considered case, the relations (5.23) and (5.36), describing these jumps, take the form

$$[\![\frac{\partial \sigma}{\partial X}]\!] = \varrho_0 [\![\dot{v}]\!],$$
$$\varrho_0 [\![T\dot{\eta}]\!] + [\![\frac{\partial q}{\partial X}]\!] = 0. \tag{5.67}$$

Due to the compatibility conditions (5.64), we can write the relation (5.67) as follows

$$[\![\frac{\partial \sigma}{\partial X}]\!] = \varrho_0 U^2 [\![\frac{\partial e}{\partial X}]\!] - 2\varrho_0 \sqrt{U_0} \frac{d}{dt}\left(\sqrt{U_0} [\![e]\!]\right). \tag{5.68}$$

Taking into account the constitutive relation $(5.3)_{10}$, we finally arrive at

$$\left[\frac{\partial^2\tilde{F}}{\partial e^2}\dot{e}+\frac{\partial^2\tilde{F}}{\partial e\,\partial\theta}\dot{\theta}\right]=U_0^2\left[\frac{\partial e}{\partial X}\right]-2\varrho_0\sqrt{U_0}\,\frac{d}{dt}\left(\sqrt{U_0}\,[e]\right). \tag{5.69}$$

On the other hand, the equation (5.67) together with the constitutive relations $(5.3)_7$, $(5.3)_8$ and $(5.3)_9$ yields

$$\begin{aligned}&\varrho_0\left[T\left(\frac{dT}{d\theta}\right)^{-1}\frac{\partial^2\tilde{F}}{\partial\theta\,\partial e}\dot{e}+T\frac{\partial}{\partial\theta}\left(\left(\frac{dT}{d\theta}\right)^{-1}\frac{\partial\tilde{F}}{\partial\theta}\right)\dot{\theta}\right]-\\&-\left[\frac{\partial\tilde{q}}{\partial e}\dot{e}+\frac{\partial\tilde{q}}{\partial\theta}\dot{\theta}+\frac{\partial\tilde{q}}{\partial\theta_{,X}}\cdot\frac{\partial^2\theta}{\partial t\,\partial X}\right]=0.\end{aligned} \tag{5.70}$$

The last two equations and the appropriate compatibility relations for the deformation gradient e and the temperature θ form the set describing the growth of jumps $[e]$ and $[\theta]$. This set is extremally complicated even in the considered simple case of the plane longitudinal shock wave. The general solution is unknown. Some results have been achieved for nonconductors, in which the wave propagates into the undisturbed region. These results as well as the analysis of basic properties of shocks of a small amplitude $[e]$ are reported in the paper of M.F.McCarthy[1975]. The growth equations for heat-conducting materials have not been investigated.

REFERENCES

1960 C.Truesdell, R.A.Toupin; The Classical Field Theories, Handbuch der Physik, III/1, Springer-Verlag.

1962 D.R.Bland; On Shock Waves in Hyperelastic Media, The Intern.Symp. on Second-Order Effects in Elasticity, Plasticity and Fluid Dynamics, Haifa, Pergamon Press Ltd., p-14.

1963 B.D.Coleman, V.J.Mizel; Thermodynamics and Departures from Fourier's Law of Heat Conduction, Arch.Rat.Mech. Anal., 13, 4, 245-261.

1964 B.D.Coleman, V.J.Mizel; Existence of Caloric Equations of State in Thermodynamics, J.Chem.Phys., 40,4,1116-1125.

R.Giles; Mathematical Foundations of Thermodynamics, Pergamon Press Ltd.

W.A.Green; The Growth of Plane Discontinuities Propagating into a Homogeneously Deformed Elastic Material, Arch.Rat.Mech.Anal., 16, 79-88.

1965 B.D.Coleman, M.E.Gurtin; Waves in Materials with Memory; III. Thermodynamic Influences on the Growth and Decay of Acceleration Waves, Arch.Rat.Mech.Anal., 19, 4,266-298.

B.D.Coleman, M.E.Gurtin; Waves in Materials with Memory; IV. Thermodynamics and the Velocity of General Acceleration Waves, Arch.Rat.Mech Anal., 19,5,317-338.

C.Truesdell, W.Noll; The Non-linear Field Theóries of Mechanics, Handbuch der Physik, III/3, Springer-Verlag.

1966 M.Marvan; Negative Absolute Temperatures, Iliffe Books ltd., London.

W.Noll; The foundations of Mechanics, CIME, Non-linear continuum Theories, ed. C.Truesdell and G.Grioli, Rome, 161-200.

C.Truesdell; The Elements of Continuum Mechanics, Springer-Verlag, New York Inc.

1967 J.L.B.Cooper; The Foundations of Thermodynamics, J.Math. Anal.App., 17, 172-193.

M.E.Gurtin, W.O.Williams; An Axiomatic Foundation for Continuum Thermodynamics, Arch.Rat.Mech.Anal., 26, 2, 83-117.

1968 G.M.C.Fisher, M.J.Leitman; On Continuum Thermodynamics with Surfaces, Arch.Rat.Mech.Anal., 30, 225-262.

M.E.Gurtin, V.J.Mizel, W.O.Williams; A Note on Cauchy's Stress Theorem, J.Math.Anal.App., 22,2, 398-401.

1969 C.Truesdell; Rational Thermodynamics, McGraw-Hill, New York.

1970 P.J.Chen, M.E.Gurtin; On the Growth of One-dimensional Shock Waves in Materials with Memory, Arch.Rat.Mech. Anal., 36,1, 33-46.

W.O.Williams; Thermodynamics of Rigid Continua, Arch. Rat.Mech.Anal., 36, 4, 270-284.

1971 I.Müller; Die Kältefunktion, eine universelle Funktion in der Thermodynamik viskoser Wærmeleitender Flüssigkeiten, Arch.Rat.Mech.Anal., 40, 1-36.

I.Müller; The Coldness, a Universal Function in Thermoelastic Bodies, Arch.Rat.Mech.Anal., 41, 319-332.
C.C.Wang; Field equations for Thermoelastic Bodies with Uniform Symmetry. Acceleration Waves in Isotropic Thermoelastic Bodies, CISM, # 112, Springer-Verlag.
W.O.Williams; Axioms for Work and Energy in General Continua, I.Smooth Velocity Fields, Arch.Rat.Mech. Anal., 42,2,93-114.

1972 P.Chadwick, P.K.Currie; The Propagation and Growth of Acceleration Waves in Heat-Conducting Elastic Materials, Arch.Rat.Mech.Anal.,49,2,137-158.
I-Shih Liu; Method of Lagrange Multipliers for Exploitation of the Entropy Principle, Arch.Rat.Mech.Anal., 46, 2, 13-148.
R.Stojanowič; Non-linear Thermoelasticity, CISM, #120, Springer-Verlag.
C.Truesdell; First Course in Rational Continuum Mechanics, The Johns Hopkins University, Baltimore.
W.O.Williams; Axioms for Work and Energy in General Continua, II.Surfaces of Discontinuity, Arch.Rat.Mech. Anal., 49, 3, 225-240.
K.Wilmanski; On Thermodynamics and Functions of States of Nonisolated Systems, Arch.Rat.Mech.Anal., 45, 4, 251-281.

1974 B.D.Coleman, D.R.Owen; A Mathematical Foundation for Thermodynamics, Arch.Rat.Mech.Anal., 54, 1, 1-104.
G.P.Moeckel; Thermodynamics of an Interface, Arch.Rat. Mech.Anal., 57, 3, 255-280.
J.W.Nunziato, E.K.Walsh, K.W.Schuler, L.M.Barker; Wave Propagation in Nonlinear Viscoelastic Solids, Handbuch der Physik, VIa/4, Springer-Verlag.
K.Wilmanski; Note on Clausius-Duhem Inequality for a Singular Surface, Bull.Acad.Polon.Sci., Ser.Sci.Techn., 22, 10, 493-500.

1975 M.F.McCarthy; Singular Surfaces and Waves, in: Continuum Physics, vol.II: Continuum Mechanics of Single Substance Bodies, ed.A.C.Eringen, 449-521, Academic Press.

K.Wilmanski; Thermodynamic Properties of Singular Surfaces in Continuous Media, Arch.Mech., 27, 3, 517-529.

1976 K.Wilmanski; Foundations of Neoclassical Thermodynamics, Metrization of Direct Thermodynamic Processes, in: Trends in Applications of Pure Mathematics to Mechanics ed.G.Fichera, 425-445, Pitman Publ.

F,J.Zeleznik; Thermodynamics, J.Math.Phys., 17, 8, 1579-1610.

1977 P.Chadwick; Restrictions on Heat Conduction Materials, Arch.Mech., 29, 5, 653-658.

K.Wilmanski; On the Galilean Invariance of Balance equation for a Singular Surface in Continuum, Arch.Mech. 29, 3, 459-475.

K.Wilmanski; On the Continuity of Fluxes in Axiomatic Thermodynamics, Lett.App.Eng.Sci., in: Int.J.Engn.Sci., 16, 5, 1978, 321-333.

1979 C.E.Beevers, J.Bree; A thermodynamic Theory of Isotropic Elastic-Plastic Materials, Arch.Mech., 31, 2,

K.Wilmanski; Localization Problem of Nonlocal Continuum Theories, Arch.Mech., 31,1,77-89.

CONTENTS

Preface
1. The Notion of a Thermodynamic Process in Continua
2. Balance Laws
    2.1. Continuity Assumptions
    2.2. Local Balance Laws
    2.3 Singulae Surfaces
    2.4. Second Law of Thermodynamics
    2.5. Summary
3. Thermoelastic Materials
    3.1. Constitutive Relations
        3.1.1. The structure of the constitutive functionals
        3.1.2. Constitutive functionals for singular surfaces
    3.2. Thermodynamically Admissible Constitutive Relations
    3.3. Elastic Heat Conductors
4. Heat Conduction in Thermoelastic Materials
    4.1. Classification of Elastic Heat Conductors
    4.2. Properties of the Heat Conduction Equation

5. Waves in Thermoelastic Materials
5.1. Acoustical Waves; Propagation Condition
5.1.1. Preliminaries
5.1.2. The governing equations of acoustical waves in a strong heat conductor
5.2. Shock Waves
References

THERMODYNAMICS OF DISSIPATIVE MATERIALS

Piotr PERZYNA
Institute of Fundamental Technological Research, Polish Academy of Sciences, Warsaw, Poland

PREFACE

The main objective of the present lecture is to show thermodynamic foundations of dissipative materials and in particular to discuss description of thermo-mechanical behaviour of elastic-viscoplastic and elastic-plastic materials.

In Section 1 the global formulation of thermodynamics of continua is given. After defining a continuous body, we assume a thermodynamic process as a fundamental concept. In a global sense, a thermodynamic process is understood as one parameter family of mappings which satisfy the laws of thermodynamics. The local formulation of a thermodynamic process is presented. It is shown that the laws of mechanics are appropriate implications of the first law of thermodynamics.

Section 2 presents a mathematical theory of dissipative materials. The intrinsic state of a particle X is defined as a pair - the actual deformation-temperature configuration of this particle X and its method of preparation. General principle of determinism is stated and the definition of the unique material structure is given.

In all thermodynamic considerations for dissipative materials the thermodynamic restrictions have been investigated based on a precisely selected method of preparation assumed.

We intend to investigate thermodynamic requirements for a general unique material structure without introducing particular realizations of the method of preparation space. We need only assume the topology for the method of preparation space and of course the smoothness requirements for processes and the response functions (functionals) considered.

The secondary purpose is to show some connection between rational and classical formulations of the principles of thermodynamics.

Particular attention is given to the discussion of the consequences of the dissipation principle assumed in the form of the Clausius-Duhem inequality. An attempt is made to examine the criterion of the selection of the response functions (functionals) and the accessibility criterion in the intrinsic state space and to study the principle of the increase of entropy. The importance of these principles for the evolution considered in the intrinsic state space is shown. The results obtained have fundamental implications for the thermodynamic theory of dissipative materials.

By different realizations of the method of preparation space we have obtained the material structure with internal state variables (Section 3) and the material structure of the rate type (Section 4). An essential feature of both structures considered is the form of the equation describing the evolution function. For every material structure the intrinsic state is determined in a different way. The intrinsic state and its evolution are substantial for the comparison of the material structures considered. An investigation of conditions under which these both material structure are isomorphic is given. An isomorphism of two unique material structures is understood as a similarity of these structures in the sense that these both structures describe the same material in a particle X of a body B (Section 5).

In Section 3 a particular discussion of coupling of dissipative effects for the internal friction and the thermally activated (or the high velocity dislocation damping) mechanisms is also given. Two groups of internal state variables are introduced. It has been shown that the group of internal state variables describing dissipative effects of the internal friction mechanism can be eliminated. In this case, of course, the evolution equations for a plastic flow mechanism are modified in such a way that the influence of internal friction effects are taken into account. The evolution equations for internal state variables describing the main mechanism take the form of integral or integro-differential equations.

Section 6 is devoted to the description of elastic-viscoplastic material for finite strains. After definition of elastic-viscoplastic response and plastic (or viscoplastic) deformation an analysis of dissipative mechanisms of viscoplastic flow is presented. It has been shown that two mechanisms, namely the thermal activation process and the damping of dislocation by phonon viscosity are the most important for proper explanation of the strain rate and temperature sensitivity of a material. Based on the physical motivation and available experimental results physical interpretation of internal state variables is given. One dimensional examples are considered. Experimental results for titanium, aluminium and copper are used to choose functions involved in the description. The method is developed to generalize the procedure proposed to three-dimensional case.

Section 7 is concerned with a thermoplastic behaviour of a material. A class of rate-type constitutive equations, which characterize the behaviour of elastic- plastic solids under large deformation, is developed. The relationship between the rate-type theory and the internal state variable theory of elastic-plastic solids is indicated. Plasticity as a limit case of viscoplasticity is also discussed.

CONTENTS

1. GLOBAL FORMULATION OF THERMODYNAMICS OF CONTINUA
 1.1. Material system and continuous body
 1.2. Motion and deformation
 1.3. Velocity field
 1.4. Thermodynamic process. Thermodynamic state. Temperature field
 1.5. Local thermodynamic process
2. GENERAL MATERIAL STRUCTURE
 2.1. Method of preparation. Intrinsic state
 2.2. General principle of determinism
 2.3. Evolution of intrinsic states
 2.4. Consequences of the dissipation principle
 2.5. Topological and smoothness assumptions
 2.6. Constitutive restrictions
 2.7. Accessibility criterion
 2.8. Principle of the increase of entropy

3. INTERNAL STATE VARIABLE MATERIAL STRUCTURE

3.1. Fundamental assumptions

3.2. Coupling of dissipative mechanisms

3.3. Particular example of coupling effects

4. RATE TYPE MATERIAL STRUCTURE

4.1. Definitions and fundamental assumptions

4.2. Comments on possible applications

5. ISOMORPHIC MATERIAL STRUCTURES (ISOTHERMAL PROCESSES)

5.1. General definition. Definition of a material

5.2. Discussion of a material isomorphism between internal state variable and rate type material structures (isothermal processes)

6. THERMO-VISCOPLASTICITY FOR FINITE STRAINS

6.1. Nature of an elastic-viscoplastic response

6.2. Definition of permanent (plastic or viscoplastic) deformation

6.3. Analysis of dissipation mechanisms

6.4. Physical foundations. Interpretation of internal state variables

6.5. Two approximations

6.6. General theory of viscoplasticity

6.7. Particular constitutive equations

6.8. On material isomorphism in description of viscoplasticity

7. THERMOPLASTICITY FOR FINITE DEFORMATIONS

7.1. Postulates of plasticity

7.2. General theory. Rate type formulations

7.3. Plasticity as a limit case of viscoplasticity

1. GLOBAL FORMULATION OF THERMODYNAMICS OF CONTINUA

1.1. Material system and continuous body(*)

Throughout this paper $\mathbb{R}$ denotes the set of all real numbers. The space $\mathbb{E}$, whose elements $x, y, \ldots$, we call spatial points, has the structure of a three-dimensional Euclidean point space. The translation space of $\mathbb{E}$ is denoted by $\mathbb{V}$, and $\mathbb{U}$ is the set of all skew linear transformations from $\mathbb{V}$ into $\mathbb{V}$.

A material system $\mathbb{B}$ is a set whose elements A,B,C,... are called bodies, and is endowed with a structure defined by the relations; $<$, $\vee$ and $\wedge$, i.e., by the inclusion, meet and join relations; $\mathcal{A}^e$ will designate the complement of $\mathcal{A} \in \mathbb{B}$; 0 will denote the null element in $\mathbb{B}$. Let us assume that $\mathcal{B}$ is the body whose complement is minimal.

In the following we will use the therminology: ele-

(*) We will follow here the approach presented by GURTIN, NOLL and WILLIAMS [33], (cf. also NOLL[61] and WILLIAMS [96,97], GURTIN and WILLIAMS [34,35], GREEN and RIVLIN [27] and GURTIN [38]).

ments $\mathcal{A} < \mathcal{B}$ are subbodies, $\mathcal{A}^e$ is the exterior of $\mathcal{A} \in \mathcal{B}$, two elements $\mathcal{A}, \mathcal{D} \in \mathbb{B}$ are separate if and only if $\mathcal{A} \in \mathcal{D}^e$, or equivalently if and only if $\mathcal{A} \wedge \mathcal{D} = 0$.

The requirement that $\mathcal{B}^e$ be minimal is simply that the only $\mathcal{A} \in \mathbb{B}$ with $\mathcal{A} < \mathcal{B}^e$ are $\mathcal{A} = 0$ or $\mathcal{A} = \mathcal{B}^e$.

We will denote by

$\mathbb{D}$ = the set of all subbodies

$$\mathrm{sep}\,(\mathbb{D} \times \mathbb{B}) = \{(\mathcal{A}, \mathcal{D}) \in \mathbb{D} \times \mathbb{B} \mid \mathcal{A} \wedge \mathcal{D} = 0\}$$

A body $\mathcal{B}$ is endowed with additional mathematical structure defining the properties of the body considered.

A body $\mathcal{B}$ is a set whose members X are called material points, and which is endowed with a structure defined as follows:

(i) $\mathcal{B}$ consists of a finite number of parts (called subbodies) which can be mapped onto regions in Euclidean space $\mathbb{E}$. Thus there exists a class $\mathbb{K}$ of mappings $\varkappa : \mathcal{B} \to \mathbb{E}$. The mappings $\varkappa \in \mathbb{K}$ are called the configurations of $\mathcal{B}$ (in the space $\mathbb{E}$). After NOLL[61] we shall assume that $\mathcal{B}$ is a continuous body of class C^p ($p \geq 1$) if the class $\mathbb{K}$ of configurations satisfies the following axioms:

a) Every $\varkappa \in \mathbb{K}$ is one-to-one and its range $\varkappa(\mathcal{B})$ is an open subset of $\mathbb{E}$, which is called the region occupied by $\mathcal{B}$ in the configuration $\varkappa$.

b) If $\gamma, \varkappa \in \mathbb{K}$ then the composite $\lambda = \gamma \circ \varkappa^{-1} : \varkappa(\mathcal{B}) \to \gamma(\mathcal{B})$ is a deformation of class C^p, which is called the deformation of $\mathcal{B}$ from the configuration $\varkappa$ into the configuration γ.

c) If $\varkappa \in \mathbb{K}$ and if $\lambda : \varkappa(\mathcal{B}) \to \mathbb{E}$ is a deformation of class C^p, then $\lambda \circ \varkappa \in \mathbb{K}$. The mapping $\lambda \circ \varkappa$ is called the configuration obtained from the configuration $\varkappa$ by the deformation λ.

The axioms a) - c) ensure that the class $\mathbb{K}$ endows the body $\mathcal{B}$ with the structure of a C^p - manifold modelled on $\mathbb{E}$.

(ii) $\mathcal{B}$ is (in its configuration) a measure space. The mass, which is a non-negative measure M, is defined once and for all over measurable subsets of the body and is also assumed to be an absolutely continuous function of volume in space i.e.,

$$M(\mathcal{R}) = \int_{\mathcal{R}} \varrho(x,t)\, dV(x) \qquad (1.1)$$

(iii) $\mathcal{B}$ has material structure defined by the system of constitutive equations.

Owing to the axioms stated we can identify a continuous body $\mathcal{B}$ with a region $\varkappa(\mathcal{B})$ in $\mathbb{E}$. The set $\mathbb{D}$ of subbodies is then a subclass of the set of all regular closed subregions of $\mathcal{B}$, $\mathcal{R}^e = \overline{\mathbb{E} - \mathcal{R}}$. The relations $<$, $\vee$ and $\wedge$ can be now understood as the set theory relations.

We denote the relative exterior of $\mathcal{R}$ in $\mathcal{B}$ by

$$\mathcal{R}^b = \overline{\mathcal{B} - \mathcal{R}} \tag{1.2}$$

and of course

$$\mathcal{R}^e = \mathcal{R}^b \cup \mathcal{B}^e \tag{1.3}$$

1.2. Motion and deformation(*)

The spatial point $\varkappa(X) \in \mathbb{E}$ is called the place of the material point $X \in \mathcal{B}$ in the configuration $\varkappa$. Let us denote the actual configuration by χ. A motion of a body $\mathcal{B}$ is a one parameter family of configurations $\chi_t \in \mathbb{K}$, where the real parameter t is called time. We can write

$$x = \chi_t(X) = \chi(X, t) \tag{1.4}$$

The point $x = \chi(X, t)$ is the spatial positions occupied by the material point X at time t, Fig. 1.

It is often convenient to identify the material point X with its position $\mathbf{X}$ in a fixed reference configuration $\varkappa$ and to write

$$x = \chi(\mathbf{X}, t) \tag{1.5}$$

(*) Cf. NOLL and TRUESDELL[92].

The gradient $\mathbf{F}$ of $\chi(\mathbf{X},t)$ with respect to $\mathbf{X}$, i.e.,

$$F(\mathbf{X},t)=\nabla\chi(\mathbf{X},t) \tag{1.6}$$

is the deformation gradient at $\mathbf{X}$. The deformation gradient describes all local properties of deformation at $\mathbf{X}$.

Owing to the axioms a) - c) for the class $\mathbb{K}$ the inverse F^{-1} of F exists.

The velocity $\dot{x}$ is the rate of change of position of a particle $\mathbf{X}$ and is defined by

$$\dot{x}=\left[\frac{\partial}{\partial t}\chi(X,t)\right]_{X=\text{const}} \tag{1.7}$$

Similarly we define the acceleration $\ddot{x}$ as

$$\ddot{x}=\left[\frac{\partial^2}{\partial t^2}\chi(X,t)\right]_{X=\text{const}} \tag{1.8}$$

In the following as fundamental local measure of deformation we will use the right CAUCHY-GREEN tensor $C(X,t)=F^TF$.

1.3. Velocity fields

We shall use the notion of a velocity field introduced by GURTIN, NOLL and WILLIAMS [33]. This notion is an abstraction of the usual concept for a continuum.

Definition 1. An instantaneous velocity space for $\mathcal{B}$ is a vector space $\mathbb{F}$ together with two mappings

$$\mathcal{T}: \mathbb{V} \longrightarrow \mathbb{F} \tag{1.9}$$

$$\mathcal{R}: \mathbb{E} \times \mathbb{U} \longrightarrow \mathbb{F} \tag{1.10}$$

such that

(i) $\mathcal{T}$ is linear;

(ii) for each $z \in \mathbb{E}$, $\mathcal{R}(z, \cdot)$ is linear;

(iii) for each $y, z \in \mathbb{E}$, $\Omega \in \mathbb{U}$

$$\mathcal{R}(y, \Omega) = \mathcal{T}\big(\Omega(z-y)\big) + \mathcal{R}(z, \Omega)$$

Elements $\upsilon \in \mathbb{F}$ are called velocity fields; $\mathcal{T}(c)$ is a translation with velocity c and $\mathcal{R}(y, \Omega)$ is a rigid rotation with spin Ω about y. A velocity field of the form $\mathcal{T}(c) + \mathcal{R}(y, \Omega)$ is called a rigid velocity field.

Let us define on F an equivalence relation by writing $u \sim \upsilon$ if and only if $u - \upsilon$ is a rigid velocity field. This equivalence relation defines a partition $\mathbb{F}$ into equivalence classes which we call intrinsic velocity fields.

For the continuous body $\mathcal{B}$ the vector space $\mathbb{F}$ may be chosen to be the set of all continuous vector fields on $\mathcal{B}$, $\mathcal{T}(c)$ is the constant vector field on $\mathcal{B}$ with value C, and $\mathcal{R}(y, \Omega)$ is the mapping defined on $\mathcal{B}$ by

$$X \longrightarrow \Omega\left[\chi(X) - y\right]$$

An intrinsic velocity is the set of all velocity fields which can be observed from various frames at a given time in a motion(*).

We associate with $\mathbb{F}$ an operation

$$\upsilon \longrightarrow \upsilon|_{\mathcal{R}}$$

defined for each $\mathcal{R} \in \mathbb{D}$ and $\upsilon \in \mathbb{F}$. The operation $(\upsilon, \mathcal{R}) \longrightarrow \upsilon|_{\mathcal{R}}$ is called the restriction of υ to $\mathcal{R}$.

We assume that all maps in $\mathbb{F}$ are of class C^2 on $\mathcal{B}$. The last restriction on $\mathbb{F}$ is a matter of convenience to exclude fields with discontinuities.

1.4. Thermodynamic process. Thermodynamic state. Temperature field

Definition 2. A process for $\mathcal{B}$ is a mapping that assigns to each time $t \in [t_o, t_1] \subset \mathbb{R}$ an ordered array

$$\{E_t, S_t, H_t, M_t, W_t\}$$

such that for each t the following axioms are satisfied:

(i) $E_t : \mathbb{D} \longrightarrow \mathbb{R}$, the derivative $\dot{E}_t(\mathcal{R}) = \frac{d}{dt} E_t(\mathcal{R})$ exists, and $\dot{E}_t(\mathcal{R})$ is an additive, absolutely continuous function of volume i.e.,

(*) A velocity field is determined uniquely only to within a rigid velocity field due to the necessity of making a choice of frame of reference.

$$\dot{E}_t(\mathcal{R}) = \int_{\mathcal{R}} \dot{e}(x,t)\, dV(x) \tag{1.11}$$

$E_t(\mathcal{R})$ is the internal energy of subbody $\mathcal{R}$ and $\dot{e}(x,t)$ is the rate of the specific internal energy.

(ii) $S_t : \mathbb{D} \to \mathbb{R}$, the derivative $\dot{S}_t(\mathcal{R}) = \frac{d}{dt} S_t(\mathcal{R})$ exists, and $\dot{S}_t(\mathcal{R})$ is an additive, absolutely continuous function of volume, i.e.,

$$\dot{S}_t(\mathcal{R}) = \int_{\mathcal{R}} \dot{s}(x,t)\, dV(x) \tag{1.12}$$

$S_t(\mathcal{R})$ is the internal entropy of subbody $\mathcal{R}$ and $\dot{s}(x,t)$ is the rate of the specific entropy.

(iii) $H_t : \mathrm{sep}(\mathbb{D} \times \mathbb{B}) \to \mathbb{R}$. The heating $H_t(\mathcal{R},\mathcal{D})$ represents the heat flux into $\mathcal{R}$ from $\mathcal{D}$. For each $\mathcal{D}$, $H_t(\cdot,\mathcal{D})$ admits the unique decomposition

$$H_t(\cdot,\mathcal{D}) = R_t(\cdot,\mathcal{D}) + Q_t(\cdot,\mathcal{D}) \tag{1.13}$$

where $R_t(\mathcal{R},\mathcal{D})$ is the body heating and $Q(\mathcal{R},\mathcal{D})$ is the contact heating; the former being an absolutely continuous function of volume, and the latter, an absolutely continuous function of surface area, i.e.,

$$H_t(\mathcal{R},\mathcal{R}^e) = R_t(\mathcal{R},\mathcal{R}^e) + Q(\mathcal{R},\mathcal{R}^e) \tag{1.14}$$

$$= \int_{\mathcal{R}} r_e(x,t)\, dV(x) + \int_{\partial\mathcal{R}} q(x,t,n)\, dA(x)$$

for every subbody $\mathcal{R}$; $r_e(x,t)$ denotes the volume density of heating supply and $q(x,t,n)$ the surface density of heating influx. R_t and Q_t are often said to describe radiation and conduction, respectively, but no such specific connotation need to be made in general

(iv) $M_t : \mathrm{sep}(\mathbb{D}\times\mathbb{B}) \to \mathbb{R}$. $M_t(\mathcal{R},\mathcal{D})$ is the entropy flux into $\mathcal{R}$ from $\mathcal{D}$. For any $\mathcal{D}$, $M_t(\cdot,\mathcal{D})$ admits the unique decomposition

$$M_t(\cdot,\mathcal{D}) = K_t(\ ,\mathcal{D}) + J_t(\cdot,\mathcal{D}) \qquad (1.15)$$

where $K_t(\cdot,\mathcal{D})$ is absolutely continuous with respect to the body heating $R_t(\cdot,\mathcal{D})$ and $J_t(\cdot,\mathcal{D})$ is absolutely continuous with respect to the contact heating $Q_t(\cdot,\mathcal{D})$, i.e.,

$$M_t(\mathcal{R},\mathcal{R}^e) = K_t(\mathcal{R},\mathcal{R}^e) + J_t(\mathcal{R},\mathcal{R}^e)$$

$$= \int_{\mathcal{R}} \frac{r_e(x,t)}{\varphi(x,t)}\, dV(x) + \int_{\partial\mathcal{R}} \frac{q(x,t,n)}{Q(x,t)}\, dA(x) \qquad (1.16)$$

for every subbody $\mathcal{R}$. The function $\varphi = \Theta = \vartheta(x,t)$ is called a positive - valued temperature field(*). K_t and J_t are often called the radiative entropy flux and the conductive entropy flux, respectively.

(v) $W_t : \mathrm{sep}(\mathbb{D}\times\mathbb{B})\times\mathbb{F} \longrightarrow \mathbb{R}$

For each $\upsilon\in\mathbb{F}$ the mapping $(\mathcal{A},\mathcal{D})\longrightarrow W_t(\mathcal{A},\mathcal{D},\upsilon)$ is an interaction, i.e.,

$\mathcal{A}\longrightarrow W_t(\mathcal{A},\mathcal{D},\upsilon)$ and $\mathcal{D}\longrightarrow W_t(\mathcal{A},\mathcal{D},\upsilon)$

are additive, whenever $\mathcal{A}$ and $\mathcal{D}$ are separate subbodies. For each $(\mathcal{A},\mathcal{D})\in\mathrm{sep}(\mathbb{D}\times\mathbb{B})$ the mapping

$\upsilon\longrightarrow W_t(\mathcal{A},\mathcal{D},\upsilon)$ is linear.

$W_t(\mathcal{A},\mathcal{D},\upsilon)$ is the power expended on $\mathcal{A}$ by $\mathcal{D}$ under the velocity field υ in a motion $\chi_t(\mathbb{B})$.

Definition 3. A thermodynamic process is a process that satisfies the following postulates:

1°. Balance of energy (the first law of thermodynamics)

$$\dot{E}_t(\mathcal{A}) = H_t(\mathcal{A},\mathcal{A}^e) + W_t(\mathcal{A},\mathcal{A}^e,\upsilon) \tag{1.17}$$

for every subbody $\mathcal{A}$.

2°. Growth of entropy (the second law of thermodynamics)

$$\dot{S}_t(\mathcal{A}) \geq M_t(\mathcal{A},\mathcal{A}^e) \tag{1.18}$$

for every subbody $\mathcal{A}$.

(*) GURTIN[37] has shown a global hypothesis involving the heat flux and the entropy flux which is equivalent to the existence of a single temperature in continuum thermodynamics.

The function

$$N_t(\mathcal{R}) = \dot{S}_t(\mathcal{R}) - M_t(\mathcal{R}, \mathcal{R}^e) \tag{1.19}$$

is called the rate of entropy production in the subbody $\mathcal{R}$. Thus, the second law of thermodynamics gives the condition for the rate of entropy production to be non negative, i.e.,

$N_t(\mathcal{R}) \geq 0$ for every subbody $\mathcal{R}$.

We understand the first law of thermodynamics as the fundamental restriction for phenomena and interactions we can consider. The second law, on the other hand, will give restrictions for constitutive relations assumed for a material.

Definition 4. A thermo-mechanical state of body $\mathbb{B}$ at time $t \in [t_0, t_1]$ is a collection of values which take the functions

$$\{E_t, S_t, H_t, M_t, W_t\}$$

for this particular time t.

Consider the map $c \longrightarrow \mathcal{T}(c) \in \mathbb{F}$ previously defined. This map is linear. From (v) it follows that

$$c \longrightarrow W_t(\mathcal{R}, \mathcal{D}, \mathcal{T}(c))$$

is a linear form on $\mathbb{V}$. Thus there exists a unique vector $f_t(\mathcal{R}, \mathcal{D})$ such that

$$W_t(\mathcal{R}, \mathcal{D}, \mathcal{T}(c)) = f_t(\mathcal{R}, \mathcal{D})\cdot c \tag{1.20}$$

for all $c \in \mathbb{V}$. We call $f_t(\mathcal{R}, \mathcal{D})$ the force exerted on $\mathcal{R}$ by $\mathcal{D}$ at time t .

Similarly given a point y in $\mathbb{E}$ there exists a unique skew transformation $M_{yt}(\mathcal{R}, \mathcal{D})$ such that

$$W_t\big(\mathcal{R}, \mathcal{D}, \mathcal{R}(y, \Omega)\big) = M_{yt}(\mathcal{R}, \mathcal{D})\,\Omega \tag{1.21}$$

We call $M_{yt}(\mathcal{R}, \mathcal{D})$ the moment about y exerted on $\mathcal{R}$ by $\mathcal{D}$ at time t .

The force $f_t(\mathcal{R}, \mathcal{D})$ and the moment $M_{yt}(\mathcal{R}, \mathcal{D})$ are interactions.

Let w denote the rigid displacement field $\mathcal{T}(c) + \mathcal{R}(y, \Omega)$. Then

$$W_t(\mathcal{R}, \mathcal{D}, w) = f_t(\mathcal{R}, \mathcal{D})\cdot c + M_{yt}(\mathcal{R}, \mathcal{D})\,\Omega \tag{1.22}$$

The first law of thermodynamics yields

$$\dot{E}_t(\mathcal{R}) = H_t(\mathcal{R}, \mathcal{R}^e) + W_t(\mathcal{R}, \mathcal{R}^e, v + w) \tag{1.23}$$

for every subbody $\mathcal{R}$ and any $v \in \mathbb{F}$ and any rigid displacement field w . Since W_t is linear in its third argument

$$\dot{E}_t(\mathcal{R}) = H_t(\mathcal{R}, \mathcal{R}^e) + W_t(\mathcal{R}, \mathcal{R}^e, v) + W_t(\mathcal{R}, \mathcal{R}^e, w) \tag{1.24}$$

and this relation can hold for every w only if

$$W_t(\mathcal{R}, \mathcal{R}^e, w) = 0 \tag{1.25}$$

Thus we have the following

Theorem 1. (Principle of virtual power). The power expended on a subbody by its exterior over any rigid velocity field vanishes.

As an immediate consequence of this theorem we have the corollary:

Balance of forces and moments. For every subbody $\mathcal{R}$ and every point $y \in \mathbb{E}$

$$f_t(\mathcal{R}, \mathcal{R}^e) = 0 \quad , \quad M_{yt}(\mathcal{R}, \mathcal{R}^e) = 0 \tag{1.26}$$

We shall call these equations <u>the Euler laws.</u>

It is necessary to recall that in the above structure the inertial force and inertial moment were not separated from other forces and moments. To do this separation let us write

$$\begin{aligned} f_t(\mathcal{R}, \mathcal{R}^e) &= f_t(\mathcal{R}, \mathcal{R}^b) + f_t(\mathcal{R}, \overline{\mathcal{R}^e - \mathcal{R}^b}) \,, \\ M_{yt}(\mathcal{R}, \mathcal{R}^e) &= M_{yt}(\mathcal{R}, \mathcal{R}^b) + M_{yt}(\mathcal{R}, \overline{\mathcal{R}^e - \mathcal{R}^b}) \end{aligned} \tag{1.27}$$

Following NOLL[59] and BEATTY[1], we set down

Postulate 1. The force exerted by $\overline{\mathcal{R}^e - \mathcal{R}^b}$ on subbody $\mathcal{R}$ is the negative of the rate of change of momentum of the body with respect to the preferred frame

$$f_t(\mathcal{R}, \overline{\mathcal{R}^e - \mathcal{R}^b}) = -\frac{d}{dt}\int_{\mathcal{R}} \varrho(x,t)\,\dot{x}(x,t)\,dV(x) \tag{1.28}$$

The moment exerted by $\overline{\mathcal{R}^e - \mathcal{R}^b}$ on subbody $\mathcal{R}$ is the negative of the rate of change of moment of momentum of the body relative to the preferred frame

$$M_{yt}(\mathcal{R}, \overline{\mathcal{R}^e - \mathcal{R}^b}) = -\frac{d}{dt}\int_{\mathcal{R}} [\mathbf{x} \times \dot{\mathbf{x}}(x,t)]\,\varrho(x,t)\,dV(x) \tag{1.29}$$

The force $f_t(\mathcal{R}, \mathcal{R}^b)$ is assumed to be the sum of two forces of special kinds, namely, the total body force f_{bt} and the total contact force f_{ct} ; the former is an absolutely continuous function of mass, and the latter, an absolutely continuous function of surface area:

$$\begin{aligned} f_t(\mathcal{R}, \mathcal{R}^b) &= f_{bt}(\mathcal{R}, \mathcal{R}^b) + f_{ct}(\mathcal{R}, \mathcal{R}^b) \\ &= \int_{\mathcal{R}} \varrho(x,t)\,b(x,t)\,dV(x) + \int_{\partial\mathcal{R}} \mathbf{t}(x,t,\mathbf{n})\,dA(x) \end{aligned} \tag{1.30}$$

The field $b(x,t)$ is the body force per unit mass; it represents forces acting at a distance. The traction $\mathbf{t}(x,t,\mathbf{n})$ represents the action of neighbouring parts of material upon one another, or any forces applied in any way to the boundary $\partial\mathcal{R}$ of the subbody $\mathcal{R}$ in its present configuration.

Suppose that $t(x,t,n)$ are continuous on $\mathcal{B}$ for each n, then according to the CAUCHY theorem there exists a continuous stress tensor field $T_c(x,t)$ such that

$$t(x,t,n) = T_c(x,t)\mathbf{n} \tag{1.31}$$

where n is the outer unit normal to $\partial\mathcal{R}$.

If we restrict attention to the case when all torques are moments of forces, then

$$M_{yt}(\mathcal{R},\mathcal{R}^b) = \int_{\mathcal{R}} \left[x \times b(x,t)\right] \varrho(x,t)\, dV(x) + \int_{\partial\mathcal{R}} \left[x \times \mathbf{t}(x,t,\mathbf{n})\right] dA(x) \tag{1.32}$$

Substituting of these results to the Euler laws and taking into account that the mass balances, i.e.,

$$\frac{d}{dt}(\varrho\, dV) = 0 \tag{1.33}$$

it is easy to derive a local equivalent to the principle of linear momentum for continua

$$\operatorname{div} T_c + \varrho b = \varrho \ddot{x} \tag{1.34}$$

called Cauchy's first law of motion, and an assertion that the stress tensor T_c is symmetric

$$T_c = T_c^T \tag{1.35}$$

which is Cauchy's second law of motion.

The power expended on $\mathcal{R}$ by its exterior under the velocity field is given by

$$W_t(\mathcal{R}, \mathcal{R}^e, v) = \int_{\mathcal{R}} \operatorname{tr}\left[T_c(x,t)\,\operatorname{grad}\dot{x}(x,t)\right] dV(x). \tag{1.36}$$

The scalar

$$w(x,t) = \operatorname{tr}\left[T_c(x,t)\ \operatorname{grad}\dot{x}\ (x,t)\right] \tag{1.37}$$

is the net working per unit volume; it is often called the density of the stress power.

As the result of the equation for the balance of energy we obtained the equation for the balance of mass and the equations expressing the Cauchy`s laws of motion. Additionally, we should satisfy the following differential equation

$$\dot{e}(x,t) = \operatorname{div}\mathbf{q}(x,t) + r_e(x,t) + \operatorname{tr}\left[T_c\ \operatorname{grad}\dot{x}\right] \tag{1.38}$$

where $\qquad q(x,t,\mathbf{n}) = \mathbf{q}(x,t)\cdot\mathbf{n}$

This equation is sometimes regarded as a local statement of the first law of thermodynamics.

The second law of thermodynamics gives a local inequality as follows (Clausius - Duhem inequality)

$$\dot{s}(x,t) \geqslant \operatorname{div}\left[\frac{q(x,t)}{\vartheta(x,t)}\right] + \frac{r_e(x,t)}{\vartheta(x,t)} . \tag{1.39}$$

It will be convenient to introduce the new variables

$$\varrho\dot{\varepsilon} = \dot{e} \quad , \quad \varrho\dot{\eta} = \dot{s} \qquad \text{and} \quad \varrho r = r_e \tag{1.40}$$

where $\dot{\varepsilon}$ denotes the rate of the specific internal energy per unit mass, $\dot{\eta}$ the rate of the specific entropy per unit mass and r is the heat supply per unit time and unit mass.

The rate of entropy production in the subbody $\mathcal{R}$ can be now written in the form

$$N_t(\mathcal{R}) = \frac{d}{dt}\int_{\mathcal{R}} \varrho\eta \, dV - \int_{\mathcal{R}} \frac{\varrho r}{\vartheta} dV - \int_{\partial\mathcal{R}} \frac{q \cdot n}{\vartheta} dA \tag{1.41}$$

Under suitable smoothness assumption we have

$$N_t(\mathcal{R}) = \int_{\mathcal{R}} \xi \varrho \, dV \tag{1.42}$$

where

$$\xi = \dot{\eta} - \frac{r}{\vartheta} - \frac{1}{\varrho\vartheta} \operatorname{div} q + \frac{1}{\varrho\vartheta^2} q \operatorname{grad} \vartheta \tag{1.43}$$

is the specific rate of entropy production per unit mass.

We now introduce the specific free energy per unit mass (the Helmholtz free energy) $\psi(x,t)$, which we define by

$$\psi = \varepsilon - \vartheta\eta . \tag{1.44}$$

In the following we shall use the symbol ∇ to indicate a gradient in the reference configuration $\varkappa$, i.e., a gradient computed taking X as the independent variable, whereas grad is used when the position x in the present configuration is taken as the independent variable. For the temperature field ϑ we have

$$\nabla\vartheta = F^T \operatorname{grad}\vartheta . \tag{1.45}$$

We shall also introduce the second Piola-Kirchhoff stress tensor

$$T = \det F \, F^{-1} T_c (F^{-1})^T \tag{1.46}$$

and similarly, the heat flux vector per unit surface in the reference configuration

$$q_\varkappa = -\det F \, F^{-1} q . \tag{1.47}$$

Using these new variables we can write the Cauchy's laws of motion in the form

$$\operatorname{Div}(FT) + \rho_\varkappa b = \rho_\varkappa \ddot{x} , \tag{1.48}$$

$$T = T^T \tag{1.49}$$

a local form of the first law of thermodynamics

$$\frac{1}{2}\operatorname{tr}(T\dot{C}) - \operatorname{Div} q_\varkappa - \varrho_\varkappa(\dot{\psi} + \vartheta\dot{\eta} + \dot{\vartheta}\eta) + \varrho_\varkappa r = 0 \tag{1.50}$$

and the Clausius-Duhem inequality as follows

$$-\dot{\psi} - \dot{\vartheta}\eta + \frac{1}{2\varrho_\varkappa}\operatorname{tr}(T\dot{C}) - \frac{1}{\varrho_\varkappa\vartheta} q_\varkappa \cdot \nabla\vartheta \geqslant 0 \tag{1.51}$$

where the operator Div is computed with respect to the material coordinates X and $\varrho_\varkappa$ is the mass density in the reference configuration $\varkappa$, the dot denotes the material differentiation with respect to time t.

1.5. Local thermodynamic process

The consideration of a global thermodynamic process for a body $\mathcal{B}$ gives the following equations describing a local thermodynamic process in a particle X (cf. Refs.[71,73]):

$$\operatorname{Div}\left[F(t), T(t)\right] + \varrho_\varkappa b(t) = \varrho_\varkappa \ddot{x}(t),$$

$$T(t) = T(t)^T, \tag{1.52}$$

$$\frac{1}{2}\operatorname{tr}\left[T(t)\dot{C}(t)\right] - \operatorname{Div} q_\varkappa(t) - \varrho_\varkappa\left[\dot{\psi}(t) + \vartheta(t)\eta(t)\right.$$

$$+\dot{\vartheta}(t)\eta(t)\Big]+\varrho_{\varkappa}r(t)=0\ ,$$

$$\dot{\psi}(t)-\dot{\vartheta}(t)\eta(t)+\frac{1}{2\varrho_{\varkappa}}\mathrm{tr}\big[T(t)\dot{C}(t)\big]-\frac{1}{\varrho_{\varkappa}\vartheta(t)}q_{\varkappa}(t)\cdot\nabla\vartheta(t)\geq 0\ ,$$

where $F(t)$ denotes the value of the deformation gradient of a particle X at time t and is determined by the function of motion χ by the relation (1.6) if ∇ denotes gradient with respect to the material coordinates $\mathbf{X}$, the particle X is identified with its position $\mathbf{X}$ in a fixed reference configuration $\varkappa$, see Fig. 1, $T(t)$ is the value of the Piola-Kirchhoff stress tensor of X at t, $b(t)$ the value of the density of the body force of X at t, $\varrho_{\varkappa}$ denotes the mass density in the reference configuration $\varkappa$, $C(t)$ the value of the right Cauchy-Green deformation tensor in X at t, $q_{\varkappa}(t)$ is the value of the heat flux vector per unit surface in the reference configuration in X at t, $\psi(t)$ denotes the value of the specific free energy per unit mass in X at t, $\eta(t)$ the value of the specific entropy per unit mass in X at t, $\vartheta(t)$ is the value of absolute temperature in X at t, $r(t)$ the value of the heat supply per unit mass and unit time in X at t, dot denotes differentiation with respect to time and the operator Div is computed with respect to the material coordinates $\mathbf{X}$.

The equations $(1.52)_1$ - $(1.52)_2$ are called Cauchy's laws of motion, the Eq.$(1.52)_3$ represents the first local law of thermodynamics and the inequality $(1.52)_4$ represents the second law of thermodynamics and is called the Clausius-Duhem inequality

Definition 5. The three values

$$\mathfrak{g} = \left(C(t), \vartheta(t), \nabla\vartheta(t) \right) \tag{1.53}$$

computed in a particle X at the instant of time $t \in [0, d_p]$ we shall call the local deformation-temperature configuration of X at time t .

A set of all possible local configurations of a particle X will be denoted by $\mathfrak{G}$ and will be called the configuration space (the deformation-temperature configuration space).

Definition 6. The four values

$$\mathfrak{s} = \left(\Psi(t), \eta(t), T(t), q_n(t) \right) \tag{1.54}$$

given in a particle X at time $t \in [0, d_z]$ we shall call the local response of X at time t .

A set of all possible local responses of a particle X will be denoted by $\mathfrak{S}$ and will be called the response space.

We shall consider processes in the configuration space $\mathfrak{G}$ and processes in the response space $\mathfrak{S}$.

A process(*)

$$P \equiv (C, \vartheta, \nabla\vartheta) : [0, d_P] \longrightarrow \mathcal{G} \tag{1.55}$$

will determine the change of the deformation-temperature configuration of a particle X in the interval of time $[0, d_P]$. A number d_p will be called the duration of the process P, and $P^i = P(.0)$ and $P^f = P(d_p)$ the initial and final values of the process P, respectively.

A process

$$Z \equiv (\Psi, \eta, T, q_\varkappa) : [0, d_Z] \longrightarrow \mathcal{S} \tag{1.56}$$

will determine the change of the response of a particle X in the interval of time $[0, d_Z]$, i.e., the change of the free energy, the entropy, the Piola-Kirchhoff stress, and the heat flux.

It is important to note that if the deformation-temperature configuration g and the response s of a particle X at time t are known and we have the function of motion χ then we can determine the value of the body force b(t) from the first Cauchy's law of motion $(1.52)_1$ and the value of the heat supply per unit mass and unit time r(t) from the first local law of thermodynamics $(1.52)_3$.

(*) For a thorough discussion of properties of a process and for the definitions of a segment of the given process P and the continuation of the process P_1 with P_2 see NOLL[60].

Let us denote by

$$\pi \equiv \left\{ P \mid P : [0, d_P] \longrightarrow \mathcal{G} \right\} \qquad (1.57)$$

a set of all deformation temperature configuration processes, and by

$$\mathcal{Z} \equiv \left\{ Z \mid Z : [0, d_Z] \longrightarrow \mathfrak{Z} \right\} \qquad (1.58)$$

a set of all response processes for a particle X.

Definition 7. Every pair $(P, Z) \in \pi \times \mathcal{Z}$ such that $\mathrm{Dom}\, P = \mathrm{Dom}\, Z$ and for every instant of time $t \in [0, d_P]$ the dissipation principle in the form of the Clausius-Duhem inequality $(1.52)_4$ is satisfied will be called a local thermodynamic process.

2. GENERAL MATERIAL STRUCTURE

2.1. Method of preparation. Intrinsic state

In a class of local thermodynamic processes we shall consider a subset which will be compatible with the internal constitutive assumptions describing the internal physical constitution of a body B, i.e. compatible with a material(*) of a body B. Such a subset of a local thermodynam-

(*) A material as defined by NOLL[60] is an equivalence class of material structures, the equivalence being material isomorphy, cf. also Ref.[80].

ic process space, will be called admissible for the constitutive assumptions in question.

To discuss the general relation between processes $P \in \pi$ and $Z \in \mathcal{Z}$ which defines a material structure of a body B let us introduce a space $\mathcal{K}$ connected with the configuration space $\mathcal{G}$ in such a way that elements of the space $\mathcal{K}$, which will be denoted by $k \in \mathcal{K}$, are the method of preparation of the corresponding configurations g from $\mathcal{G}$. The space $\mathcal{K}$ will be called the method of preparation space(*).

A main objective of thermodynamics of continuous media is to predict the response of a particle X of a body B, of which physical properties are known, at the end of a deformation-temperature process. We can give an answer to this question if and only if we have full information about particle X before the test, i.e. before a deformation-temperature process. This information, which is needed for unique prediction of a future response of a particle X for every deformation-temperature process, is called the method of preparation of the actual deformation-temperature

(*) For a notion of the method of preparation see Refs.[70,71,72], cf. also BRIDGMAN[6] and GILES[25]. The precise definition of the method of preparation space for a pure mechanical process was first given in Ref.[75]. We generalize here the presentation from Ref.[75] to a thermodynamic process.

configuration. In other words the method of preparation should give the additional information required to define uniquely the internal state of particle X of a body B during the local thermodynamic process.

It will be shown that a method of preparation of the deformation-temperature configuration of a particle X is needed to describe the internal dissipation of a material. This is a very important feature of the notion of the method of preparation.

Several different methods of preparation may correspond to one configuration, but it is very important that for a given method of preparation of the initial configuration only one response process corresponds to one deformation-temperature process beginning at this configuration.

Definition 8. A non-empty set $\mathcal{K}$ will be called the method of preparation space for a particle X if

$$\bigvee_{\Sigma \subset \mathcal{G} \times \mathcal{K}} \; \bigvee_{\mathfrak{R} : (\Sigma \times \pi)^{*} \to \mathcal{Z}} \; \bigwedge_{g \in \mathcal{G}} \; \bigwedge_{P \in \pi_g} \; \bigvee_{\mathcal{K}_g \subset \mathcal{K}} \mathfrak{R}(g, \cdot, P) : \mathcal{K}_g \to \mathcal{Z}_P, \tag{2.1}$$

is bijection,

where

$$(\Sigma \times \pi)^{*} \equiv \left\{ (\sigma, P) \in \Sigma \times \pi \;\middle|\; \bigvee_{\mathcal{K}_{pi} \in \mathcal{K}} \sigma \in \{P^{i}\} \times \mathcal{K}_{pi} \right\} \tag{2.2}$$

$$\pi_g \equiv \{P \in \pi \mid P^i = g\} , \tag{2.3}$$

and $\mathcal{X}_P$ is a subset of $\mathcal{X}$ corresponding to the process P.

Definition 9. A set

$$\Sigma \equiv \bigcup_{g \in S} \{g\} \times \mathcal{K}_g , \quad \mathcal{K}_g \subset \mathcal{K} \tag{2.4}$$

(constructed by the Definition 8) is called the intrinsic state space(*) of a particle X.

The element $\sigma \in \Sigma$ is a pair of the deformation-temperature configuration and the method of preparation, i.e.

(*) The intrinsic state space Σ is due to Ref. [75]. It plays similar role in the theory as the state space introduced by NOLL [60]. The difference between these two notions of state is in the conception of the method of preparation. The elements of the intrinsic state space are pairs. Every pair consists of the local configuration and its method of preparation. There is no notion of the method of preparation in NOLL's conception of state. The idea of splitting every element of the intrinsic state space into the local configuration and its method of preparation allows us to characterize precisely the intrinsic state of a particle X and is of great importance for the development of thermodynamics of dissipative material structure.

$$\sigma \equiv (P(t), A(t)) = (g, k), \; g \in \mathcal{G}, \; k \in \mathcal{K}_g \qquad (2.5)$$

where by A we denote a process in the method of preparation space $\mathcal{K}$, i.e., $A : [0, d_p] \to \mathcal{K}$.

We define two mappings as follows(*)

$$\hat{G} \equiv pr_{\mathcal{G}} : \Sigma \to \mathcal{G} \qquad (2.6)$$

$$\hat{K} \equiv pr_{\mathcal{K}} : \Sigma \to \mathcal{K} \qquad (2.7)$$

which determine the projections from the intrinsic state space on the configuration space $\mathcal{G}$ and on the method of preparation space $\mathcal{K}$, respectively.

2.2. General principle of determinism

The notion of a method of preparation is connected with a general principle of determinism in mechanics of continua. A principle of determinism can be stated as follows. Between an initial deformation-temperature configuration, its method of preparation, a deformation-temperature

(*) The mappings $\hat{G}$ and $\hat{K}$ were first introduced for a purely mechanical case, the former by NOLL[60] and the latter in Ref.[75].

process beginning at this configuration and a response process of a particle X there exists a functional relationship. The functional relation will describe a unique material structure in a particle X of a body B .

According to the Definition 8 there exists a mapping

$$\mathcal{R} : (\Sigma \times \pi)^* \longrightarrow \mathcal{Z} \tag{2.8}$$

The mapping $\mathcal{R}$ is called the constitutive mapping.

The constitutive mapping $\mathcal{R} : (\Sigma \times \pi)^* \longrightarrow \mathcal{Z}$ · has the property as follows

$$\bigwedge_{\sigma_1, \sigma_2 \in \Sigma} \bigwedge_{P \in \pi_g} \left\{ (\sigma_i, P) \in (\Sigma \times \pi)^*, \ i = 1, 2 \ \text{and} \atop \hat{G}(\sigma_1) = \hat{G}(\sigma_2) \right. \tag{2.9}$$

$$\left. \mathcal{R}(\sigma_1, P) = \mathcal{R}(\sigma_2, P) \right\} \Longrightarrow \hat{K}(\sigma_1) = \hat{K}(\sigma_2) .$$

Definition 10. A system ($\mathcal{G}, \pi, \Sigma, \mathcal{R}$) is called a unique material structure in a particle X of a body B .

The constitutive mapping $\mathcal{R}$ with the property (2.9) expresses a general principle of determinism for dissipative continuum body.

A general principle of determinism: A unique response process $Z \in \mathcal{Z}$ corresponds to every deformation-temperature process $P \in \pi$ beginning at the given intrinsic state $\sigma \in \Sigma$.

This statement of a principle of determinism is very

general. It concerns thermodynamic processes and is valid for the arbitrary method of preparation space introduced.

2.3. Evolution of intrinsic states

Let us assume that a unique material structure is given. If we have the initial intrinsic state and the deformation-temperature process beginning at this intrinsic state we are interested then in the intrinsic state at the end of the process. The problem will be solved if the mapping between the intrinsic state at the end of the deformation-temperature process and the initial intrinsic state will be given.

Definition 11. It is said that the mapping

$$\hat{e} : (\Sigma \times \pi)^{*} \longrightarrow \Sigma \tag{2.10}$$

is the evolution function, if for every pair the equation

$$\mathfrak{R}\left(\hat{e}(\sigma, P),\ P^{f}_{(0)}\right) = \left[\mathfrak{R}(\sigma, P)\right]^{f} \tag{2.11}$$

is satisfied, where $[\mathfrak{R}(\sigma, P)]^{f}$ denotes the final value of the response process $Z = \mathfrak{R}(\sigma, P)$ and $P^{f}_{(0)}$ is the deformation-temperature process of duration zero.

In practical applications it will be convenient to have mapping from the intrinsic state space Σ into the response space $\mathcal{S}$. So, it is useful to define a new mapping

$$\hat{S} : \Sigma \longrightarrow \mathcal{S} \tag{2.12}$$

by the expression

$$\hat{S}(\sigma) = \mathcal{R}\left(\sigma, \hat{G}(\sigma)_{(o)}\right). \tag{2.13}$$

The principle of determinism can be expressed by the relation(*)

$$s = Z(t) = \hat{S}\left(\hat{e}(\sigma_o, P_{[o,t]})\right) = \hat{S}(\sigma) \tag{2.14}$$

for every $(\sigma_o, P) \in (\Sigma \times \pi)^*$, see Fig. 2.

The principle of determinism can be stated as follows:

A unique value of the response $s \in \mathcal{S}$ i.e. unique values of the free energy $\psi(t)$, the entropy $\eta(t)$, the Piola-Kirchhoff stress tensor $T(t)$ and the heat flux vector $q_x(t)$ corresponds to every intrinsic state $\sigma \in \Sigma$.

The mapping $\hat{S}$ is called the response function.

The system $(\mathcal{G}, \pi, \Sigma, \hat{S}, \hat{e})$ is also a unique material structure in a particle X.

The response function $\hat{S}$ represents the free energy response function $\hat{\Psi}$, the entropy response function $\hat{N}$,

(*) It is worth to note that both mappings $\hat{e}$ and $\hat{S}$ are similar to those introduced by NOLL[60] in a purely mechanical consideration but in the present theory as in Ref.[75], the mappings $\hat{e}$ and $\hat{S}$ are generated by the constitutive mapping $\mathcal{R}$.

the stress response function $\hat{T}$ and the heat flux response function $\hat{Q}$, i.e.

$$\hat{S} \equiv \left\{ \hat{\Psi}, \hat{N}, \hat{T}, \hat{Q} \right\} . \tag{2.15}$$

Definition 12. A local thermodynamic process compatible with a unique material structure will be called an admissible process.

2.4. Consequences of the dissipation principle

According to the Definition 3 every local thermodynamic process has to satisfy the dissipation principle in the form of the Clausius-Duhem inequality $(1.52)_4$.

Using the Definition 8 of an admissible process we may state now the main problem of the thermodynamics of materials[(x)]: In an assigned class of process and within an assigned class of response functions (functionals) $\hat{S} \equiv \left\{ \hat{\Psi}, \hat{N}, \hat{T}, \hat{Q} \right\}$ to determine those that satisfy the Clausius-Duhem inequality $(1.52)_4$.

Thus, it can be said that the main problem of the thermodynamics of materials is to determine an admissible thermodynamic process.

(x) Cf. COLEMAN and NOLL[13] and TRUESDELL[93].

It is noteworthy that the answer to the main problem of the thermodynamics of materials depends on the topology assumed for the method of preparation space.

It will be proved that the dissipation principle will imply two fundamental criteria in the theory of materials:

1°. The criterion of the selection of response functions (functionals) $\hat{S} \equiv \{\hat{\Psi}, \hat{N}, \hat{T}, \hat{Q}\}$ (*)

2°. The criterion of the accessibility of an intrinsic state σ from the given initial intrinsic state σ_0 (**).

Both these criteria are main consequences of the dissipation principle.

2.5. Topological and smoothness assumptions

To investigate restrictions placed on a local thermodynamic process by the dissipation principle we assume:

1°. The space of the method of preparation $\mathcal{K}$ is complete metrizable topological space(***).

(*) It seems that ECKART [22] was the first who understood properly the consequences of the Clausius-Duhem inequality for constitutive assumptions. Further development of his idea was done by COLEMAN and NOLL [13], COLEMAN [14], COLEMAN and GURTIN [15], COLEMAN and MIZEL [16] and COLEMAN and OWEN [17].

(**) The exploration of similar criterion to this has been recently taken, in another connection, by COLEMAN and OWEN [18].

(***) For the exact meaning of the mathematical terms introduced please consult the book by KELLY [43].

2°. Processes $P \in \pi$ considered in the configuration space $\mathcal{G}$ are continuously differentiable, i.e. for every time $t \in [0, d_p]$ exists the derivative $\frac{d}{d\tau} P(\tau)\Big|_{\tau=t} = \dot{P}(t)$. The derivative $\dot{P}(t)$ determines the rate of change of the process P at t .

3°. Processes A considered in the method of preparation space $\mathcal{K}$ are continuously differentiable, i.e., for every time $t \in [0, d_p]$ exists the derivative $\frac{d}{d\tau} P(\tau)\Big|_{\tau=t} = \dot{A}(t)$. The derivative $\dot{A}(t)$ determines the rate of change of the process A at t .

4°. The rate $\dot{A}(t)$ for $t \in [0, d_p]$ is independent of the rate $\dot{P}(t)$, i.e. we may assume that

$$\dot{A}(t) = \hat{\alpha}(\sigma) = \hat{\alpha}\big(P(t), A(t)\big), \quad t \in [0, d_p] \tag{2.16}$$

Under this assumption the rate of change of the process A in the method of preparation space $\mathcal{K}$ is completely determined by the actual intrinsic state.

Two interesting cases can be considered:

$$\begin{aligned} &(i) \quad A(0) = \kappa_0 \in \mathcal{K} \\ &(ii) \quad A(-\infty) = \kappa_{(-\infty)} \in \mathcal{K} \end{aligned} \tag{2.17}$$

The differential equation (2.16) is called the evolution equation in the method of preparation space $\mathcal{K}$ and together with appropriate initial value (i) or (ii) for the given deformation-temperature process $P:[0,d_P]\longrightarrow \mathcal{S}$ completely determines the evolution of internal states, i.e. the evolution function $\hat{e}:(\Sigma\times\pi)^*\longrightarrow\Sigma$.

The differential equation (2.16) with the initial value (2.17) (i) leads to the unique material structure with internal state variables, and the initial value problem (2.16) and (2.17) (ii) is isomorphic with the unique material structure with memory(*).

5°. A real value free energy response function $\hat{\psi}$ defined on $\Sigma\in\mathcal{S}\times\mathcal{K}$ is continuously differentiable on Σ with respect to the topology in Σ, i.e. the gradients $\partial_{P(t)}\hat{\psi}$ and $\partial_{A(t)}\hat{\psi}$ exist and are continuous functions on Σ. This property is called a chain rule property.

The deformation-temperature configuration space $\mathcal{S}$ has properties of a thirteen-dimensional vector space $\mathcal{V}_{13}$, hence it is a complete normed space (a topology of $\mathcal{S}$ is implied by a natural norm).

The intrinsic state space Σ with a topology implied by the topology assumed for $\mathcal{K}$ and the natural topology of $\mathcal{S}$ will be denoted by $\mathfrak{D}$.

For future applications it will be useful to assume

(*) Cf. KOSIŃSKI and WOJNO [44] and LEITMAN and MIZEL [53].

stronger topology for $\mathcal{K}$, namely that $\mathcal{K}$ is a complete normed space (a Banach space). Since $\mathcal{G}$ is also a complete normed space, hence Σ will have properties of a complete normed space denoted by $\mathcal{D}^*$.

In the following we shall assume that the response functions (functionals) $\hat{S} \equiv \{\hat{\Psi}, \hat{N}, \hat{T}, \hat{Q}\}$ are defined on a set $\mathcal{D}$ (or $\mathcal{D}^*$).

To state precisely, a chain rule property for the free energy response function (functional) $\hat{\Psi}$ is assumed with respect to the topology of $\mathcal{D}$ (or $\mathcal{D}^*$).

It follows from the above assumptions that in each admissible local thermodynamic process

$$\begin{aligned} \dot{\psi}(t) &= \frac{d}{d\tau}\hat{\Psi}\left(P(\tau), A(\tau)\right)\Big|_{\tau=t} \\ &= \partial_{P(t)}\hat{\Psi}(\cdot) * \dot{P}(t) + \partial_{A(t)}\hat{\Psi}(\cdot) \,\square\, \dot{A}(t), \end{aligned} \tag{2.18}$$

where

$$\begin{aligned} \partial_{P(t)}\hat{\Psi}(\cdot) * \dot{P}(t) &\equiv \mathrm{tr}\left[\partial_{C(t)}\hat{\Psi}(\cdot)\,\dot{C}(t)\right] \\ &+ \partial_{\vartheta(t)}\hat{\Psi}(\cdot)\,\dot{\vartheta}(t) + \partial_{\nabla\vartheta(t)}\hat{\Psi}(\cdot)\cdot\overline{\nabla\vartheta(t)}^{\,\cdot}, \end{aligned} \tag{2.19}$$

and $\partial_{A(t)}\hat{\Psi}(\cdot) \,\square\, A(t)$ denotes a linear function (functional) with respect to $\dot{A}(t)$. The result of $\partial_{A(t)}\hat{\Psi}(\cdot) \,\square\, \dot{A}(t)$ does depend on the realization of the method of preparation space $\mathcal{K}$ and on the topology induced in $\mathcal{K}$.

To make clear our reasoning let us consider two examples.

1°. If $\mathcal{K}$ is the past history function space then $\partial_{A(t)}\hat{\psi}(\cdot) \,\Box\, \dot{A}(t) = \delta\hat{\psi}(\cdot\,|\,\dot{A}(t))$ denotes the Frechet derivative (in a Banach function space), i.e. the linear functional with respect to $\dot{A}(t)$. This realization of the method of preparation space leads to the unique material structure with memory (cf. Ref. [79]).

2°. If $\mathcal{K}$ is a finite-dimensional vector space then $\partial_{A(t)}\hat{\psi}(\cdot) \,\Box\, \dot{A}(t) = \partial_{A(t)}\hat{\psi}(\cdot)\cdot\dot{A}(t)$ denotes the scalar product in $\mathcal{K}$. This realization of the method of preparation space leads to the unique material structure with internal state variables (cf. Ref. [79]).

2.6. Constitutive restrictions

The dissipation principle requires that $(1.52)_4$ hold at every time $t\in[0,d_p]$. We may use the Eqs. (2.14), (2.15) and (2.18) to write $(1.52)_4$ in the form

$$\frac{1}{2\varrho_\varkappa}\mathrm{tr}\left\{\left[T(t)-2\varrho_\varkappa\partial_{C(t)}\hat{\psi}\right]\dot{C}(t)\right\}-\left[\eta(t)+\partial_{\vartheta(t)}\hat{\psi}\right]\dot{\vartheta}(t)$$
$$-\partial_{\nabla\vartheta(t)}\hat{\psi}\cdot\overline{\nabla\vartheta(t)}^{\,\bullet}-\partial_{A(t)}\hat{\psi}\,\Box\,A(t)-\frac{1}{\varrho_\varkappa\vartheta(t)}q_\varkappa(t)\cdot\nabla\vartheta(t)\geq 0. \tag{2.20}$$

Since $\dot{C}(t)$, $\dot{\vartheta}(t)$ and $\overline{\nabla\vartheta(t)}^{\,\bullet}$ can be selected independently and may be arbitrarily chosen$^{(*)}$ hence the inequality (2.20) yields the results as follows

$$
\begin{aligned}
&\partial_{_\kappa\vartheta(t)}\hat{\psi} = 0 ,\\
&T(t) = 2\varrho_\kappa \, \partial_{C(t)}\hat{\psi}(\cdot) ,\\
&\eta(t) = -\partial_{\vartheta(t)}\hat{\psi}(\cdot) ,\\
-&\partial_{A(t)}\hat{\psi}(\cdot) \square \dot{A}(t) - \frac{1}{\varrho_\kappa\vartheta(t)}\hat{Q}(\sigma)\cdot\nabla\vartheta(t) = 0 ,
\end{aligned}
\tag{2.21}
$$

satisfied at every time $t \in [0, d_p]$

Let us introduce the following notations

$$
\begin{aligned}
&\hat{d}(\sigma) = -\partial_{A(t)}\hat{\psi}(\cdot) \square \dot{A}(t) - \frac{1}{\varrho_\kappa\vartheta(t)}\hat{Q}(\sigma)\cdot\nabla\vartheta(t) ,\\
&\hat{\iota}(\sigma) = -\frac{1}{\vartheta(t)}\partial_{A(t)}\hat{\psi}(\cdot) \square \dot{A}(t) .
\end{aligned}
\tag{2.22}
$$

The mapping $\hat{d} : \Sigma \longrightarrow \mathbb{R}^+$ (where $\mathbb{R}^+$ denotes the set of non-negative real numbers) is called the general dissipation function, and $\hat{d}(\sigma)$ denotes the value of the general dissipation function at the intrinsic state $\sigma \in \Sigma$.

The mapping $\hat{\iota} : \Sigma \longrightarrow \mathbb{R}$ (the set of real numbers) is called the internal disspation function, and $\hat{\iota}(\sigma)$ is its

(*) To prove this statement we can use a similar procedure as COLEMAN and GURTIN [15] and VALANIS [94] for a material with internal state variables or as WANG and BOWEN [95] for a non-linear material with quasi-elastic response.

value at the intrinsic state $\sigma \in \Sigma$.

The inequality $(2.21)_4$ is called the general dissipation inequality and using the Eqs. (2.22) it can be written in the form

$$\hat{d}(\sigma) = \vartheta(t)\hat{\iota}(\sigma) - \frac{1}{\varrho_\varkappa \vartheta(t)} \hat{Q}(\sigma) \cdot \nabla\vartheta(t) \geqslant 0 . \tag{2.23}$$

The four results (2.21) express the criterion of the selection of the response functions (functionals) $\hat{S} \equiv \{\hat{\Psi}, \hat{N}, \hat{T}, \hat{Q}\}$, which can be stated as follows:

Choosing the free energy response function $\hat{\Psi}$ which is independent of the actual temperature gradient $\nabla\vartheta(t)$ and the heat flux response function $\hat{Q}$ such that the general dissipation inequality (2.23) is satisfied at every instant of time $t \in [0, d_p]$ or for every intrinsic state σ determined by the relation $\sigma = \hat{e}(\sigma_0, P_{[0,t]})$, where σ_0 denotes the initial intrinsic state and $P_{[0,t]}$ the segment of the given deformation-temperature process P , the response stress function $\hat{T}$ and the response entropy function $\hat{N}$ are uniquely determined by the relations $(2.21)_2$ and $(2.21)_3$.

It is worth to note that for the case $q_\varkappa(t) \equiv 0$ or $\nabla\vartheta(t) \equiv 0$ for $t \in [0, d_p]$ the general dissipation inequality (2.23) takes the particular form

$$\hat{\iota}(\sigma) = -\frac{1}{\vartheta(t)}\,\partial_{A(t)}\hat{\psi}(\cdot) \,\square\, \dot{A}(t) \geq 0\,,\quad t\in[0,d_p] \quad (2.24)$$

which is called the internal dissipation inequality.

We can return now to the discussion of the notion of the method of preparation as such information which is required for the description of the internal dissipation of an inelastic material. The expression $(2.22)_2$ which defines the value of the internal dissipation at the intrinsic state shows that full information given in the method of preparation, i.e. $A(t)=k$ and the evolution equation $\dot{A}(t)=\hat{\alpha}(\sigma)$, essentially determines the internal dissipation for this intrinsic state.

If there is no need to introduce any information in the method of preparation this case corresponds to an ideal material without internal dissipation - this ideal material is called a perfectly elastic material.

2.7. Accessibility criterion

Let us assume that the initial intrinsic state $\sigma_0\in\Sigma$ is known, and let us choose an arbitrary intrinsic state $\sigma^*\in\Sigma$, see Fig. 3. The question arises whether the intrinsic state σ^* is accessible from the initial intrinsic state σ_0 , or in other words what is the condition of accessibility of σ^* from σ_0 .

If the intrinsic state σ^* is accessible from the initial intrinsic state σ_0 then the deformation-temperature process P has to exist which generates the process in the method of preparation space $A:[0,d_p]\longrightarrow\mathcal{K}$ such that

$$\sigma_0=\big(P(0),A(0)\big)=(g_0,\kappa_0) \text{ and } \sigma^*=\big(P(d_p),A(d_p)\big)=(g^*,\kappa^*), \quad (2.25)$$

and for every instant of time $t\in[0,d_p]$ the dissipation principle is satisfied.

The response of a material corresponding to the intrinsic state σ^* is determined by the constitutive relation

$$Z(d_p)=\hat{S}\big(\hat{e}(\sigma_0,P)\big)=\hat{S}(\sigma^*). \quad (2.26)$$

Since the deformation-temperature process P is assumed to be known and this process generates a process A in the method of preparation space $\mathcal{K}$ hence the dissipation principle will give fundamental restriction on the process $A:[0,d_p]\longrightarrow\mathcal{K}$.

The dissipation principle requires that for a given process P such a process $A:[0,d_p]\longrightarrow\mathcal{K}$ is chosen that the general dissipation inequality

$$\hat{d}(\sigma)\geqslant 0 \quad (2.27)$$

for every instant of time $t \in [0, d_p]$ must be satisfied.

This is the second fundamental criterion obtained as the result of the dissipation principle

<u>Accessibility criterion</u>: An arbitrary intrinsic state $\sigma^* \in \Sigma$ is accessible from the initial intrinsic state $\sigma_o \in \Sigma$ if there exists a pair of processes $(P, A) : [0, d_p] \longrightarrow \mathcal{G} \times \mathcal{K}$ such that the following two conditions are satisfied:

$$1^\circ. \quad \big(P(0), A(0)\big) = (g_o, k_o) = \sigma_o \,, \big(P(d_p), A(d_p)\big) = (g^*, k^*) = \sigma^* ,$$

(2.28)

$$2^\circ. \quad \hat{d}\big(P(t), A(t)\big) \geqslant 0$$

for every instant of time $t \in [0, d_p]$.

It is noteworthy that the accessibility criterion places some restrictions on the evolution function $\hat{e} : (\Sigma \times \pi)^* \longrightarrow \Sigma$ Indeed, if we assume the deformation-temperature process P such that the intrinsic state $\sigma^* = \hat{e}(\sigma_o, P)$ is accessible from the initial intrinsic state σ_o by this process P then the condition 2° of (2.28) represents the restriction on the evolution function $\hat{e}$.

This conclusion is obvious if the condition 2° of (2.28) is written in the form

$$\hat{d}\left(\hat{e}(\sigma_o, P_{[0,t]})\right) \geqslant 0 \qquad \text{for} \quad t \in [0, d_p] \qquad (2.29)$$

2.8. Principle of the increase of entropy

Let us consider the intrinsic state space Σ . In this space we choose the initial intrinsic state σ_0 and an arbitrary intrinsic state σ . Let us assume that there exist a pair of processes $(P,A):[0,d_p] \longrightarrow \mathcal{G} \times \mathcal{K}$ such that

$$\big(P(0),A(0)\big)=(g_0,k_0)=\sigma_0 \ , \ \big(P(t),A(t)\big)=(g,k)=\sigma \tag{2.30}$$

The pair of the processes (P,A) is represented in the intrinsic state space Σ by the curve α , see Fig. 4.

We define the curvelinear integral along the curve α which due to natural time parametrization can be written in the form

$$\begin{aligned} \mathcal{J}(\sigma_0,\sigma) &= \int_0^t \hat{d}\big(P(\tau),A(\tau)\big)\,d\tau \\ &= \int_0^t \hat{d}\big(\hat{e}(\sigma_0,P_{[0,\tau]})\big)\,d\tau \ , \quad t\in[0,d_p] \ . \end{aligned} \tag{2.31}$$

In similar way we can define the integral

$$\begin{aligned} I(\sigma_0,\sigma) &= \int_0^t \hat{i}\big(P(\tau),A(\tau)\big)\,d\tau \\ &= \int_0^t \hat{i}\big(\hat{e}(\sigma_0,P_{[0,\tau]})\big)\,d\tau \ , \quad t\in[0,d_p] \ . \end{aligned} \tag{2.32}$$

The integrals $\mathcal{J}(\sigma_0,\sigma)$ and $I(\sigma_0,\sigma)$ are called the general dissipation integral and the internal dissipation integral, respectively.

Let us consider in the internal state space Σ two states σ_a and σ_b which lie on the curve $\mathcal{L}$, see Fig. 5. The state σ_a corresponds to the instant of time $t_a \in [0, d_p]$ and the state σ_b to the instant $t_b \in [0, d_p]$, and of course $t_b > t_a$.

The dissipation principle requires that

$$\mathfrak{z}(\sigma_o, \sigma_b) - \mathfrak{z}(\sigma_o, \sigma_a) \geq 0 . \qquad (2.33)$$

Principle of the increase of entropy: For all $\sigma_a, \sigma_b \in \Sigma$, σ is accessible from σ_a if and only if the general dissipation integral of σ_b is not less than that of σ_a.

It is important to note that the internal dissipation integral (2.32) is a measure of irreversibility of a local thermodynamic process for the case when $q_\varkappa(t) \equiv 0$ or $\nabla\vartheta(t) \equiv 0$ for $t \in [0, d_p]$ and may be interpreted as the empirical entropy$^{(*)}$ or as the irreversibility function$^{(**)}$.

If we assume this interpretation and the condition $q_\varkappa(t) \equiv 0$ for $t \in [0, d_p]$, we can state the principle of

(*) The clear meaning of the empirical entropy can be found in the papers by BUCHDAHL [7,8], cf. also BUCHDAHL and GREVE [9], COOPER [19], RASTALL [87] and BOYLING [3,4].

(**) The concept of the irreversibility function was introduced by GILES [25].

the increase of entropy in the form as follows (cf. RASTALL [87]): For all $\sigma_a, \sigma_b \in \Sigma$, σ_b is adiabatically accessible from σ_a if and only if the empirical entropy of σ_b is not less than that of σ_a.

The principle of the increase of entropy was first formulated by PLANCK [85,86].

It is worth pointing out some features of the thermodynamics of materials presented. We started from the dissipation principle in the form of the Clausius-Duhem inequality and we deduced two fundamental criteria for the thermodynamics of materials, namely the criterion of the selection of the response functions (functionals) occuring in the mathematical statement of the general principle of determinism and the accessibility criterion in the intrinsic state space Σ.

The accessibility criterion is connected with the CARATHEODORY formulation of the second law of thermodynamics(*).

As a consequence of the dissipation principle we also

(*) Cf. CARATHEODORY [12]. Caratheodory's formulation of classical thermodynamics has been developed further by BORN [5], BUCHDAHL [10], FALK and JUNG [23], BERNSTEIN [2], BUCHDAHL and GREVE [9], LANDSBERG [48,49], COOPER [19], RASTALL [87] and BOYLING [3,4].

deduced the principle of the increase of entropy. Appropriate interpretation of the general dissipation integral (or the internal dissipation integral) led to the very old statement of the second law of thermodynamics first presented by PLANCK(*).

It is very important to stress that all considerations concerned a general unique material structure before a particular realization of the method of preparation was given.

3. INTERNAL STATE VARIABLE MATERIAL STRUCTURE(**)

3.1. Fundamental assumptions

Let us assume that a class of deformation-temperature processes contains only piecewise continuously differentiable processes.

PROPOSITION 1. The method of preparation space is a finite dimensional vector space $\mathcal{V}_n$ i.e.

$$\mathcal{K} \equiv \mathcal{V}_n \tag{3.1}$$

(*) Cf. PLANCK [85,86]. Extension of Planck's idea has been done by BUCHDAHL [7,8], RASTALL [87] and BOYLING [3,4].

(**) Cf. Refs. [60, 45].

the intrinsic state space Σ is the set

$$\Sigma \equiv \{ (g, \varkappa) \mid g \in \mathcal{G}, \varkappa \in \vartheta_{ng} \} \tag{3.2}$$

and

$$(\Sigma \times \pi)^{*} \equiv \{ (\sigma, P) \in \Sigma \times \pi \mid \hat{G}(\sigma) = P^{i} \} . \tag{3.3}$$

PROPOSITION 2. (i) There exists a mapping

$$\hat{\alpha} : \Sigma \longrightarrow \vartheta_{n} \tag{3.4}$$

such that for every $P \in \pi$ and $\varkappa_{o} \in \vartheta_{n_{P^i}}$ the initial value problem

$$\frac{d}{d\tau} A(\tau) = \hat{\alpha}\big(P(\tau), A(\tau)\big), \; A(0) = \varkappa_{o} \tag{3.5}$$

has a unique solution $A : [0, d_{P}] \longrightarrow \mathcal{K} \equiv \vartheta_{n}$.

(ii) The constitutive mapping $\mathcal{R} : (\Sigma \times \pi)^{*} \longrightarrow \mathcal{Z}$ satisfying for every pair $(\sigma_{o}, P) \in (\Sigma \times \pi)^{*}$ the condition

$$\mathcal{R}\big(\hat{G}(\sigma), \cdot, P\big) : \vartheta_{n_{P^i}} \longrightarrow \mathcal{Z}_{P} \text{ must be bijection} \tag{3.6}$$

is such that the evolution function $\hat{e}$ is of the form

$$\hat{e}(\sigma_{o}, P) = \big(pf, \mathcal{F}(P, \varkappa_{o})\big) , \tag{3.7}$$

where $\mathcal{F}$ denotes the solution functional of the initial value problem (3.5).

Definition 13. The unique material structure $(\mathcal{G}, \pi, \Sigma, \mathcal{R})$ satisfying the Propositions 1 and 2 is called the material structure with internal state variables.

The principle of determinism for the material structure with internal state variables is expressed by the constitutive equation

$$Z(t) = \hat{S}\big(P(t), A(t)\big). \tag{3.8}$$

By using the evolution function (3.7) the constitutive equation (3.8) can be written in the form

$$Z(t) = \hat{S}\big(\hat{e}(\sigma_0, P_{[0,t]})\big). \tag{3.9}$$

The function $\hat{S}$ represents tne constitutive functions $\hat{\Psi}, \hat{N}, \hat{T}, \hat{Q}$. The domain of the constitutive functions is a subset of the intrinsic state $\Sigma = \mathcal{G} \times \mathcal{V}_n$. Since the method of preparation space $\mathcal{K} \equiv \mathcal{V}_n$ is a finite-dimensional vector space (n is finite) and the deformation-temperature configuration space $\mathcal{G}$ is a thirteen-dimensional vector space V_{13}, hence the intrinsic state space Σ is a finite-dimensional vector space (cf. discussion in Sec.2.5).

We can define the domain of the constitutive functions for the material structure with internal state variables as a complete normed space

$$\mathcal{D}^{\#} \equiv \mathcal{V}_{13} \times \mathcal{V}_n \tag{3.10}$$

We have satisfied 1° - 4° conditions discussed in Sec.2.5. Let us assume that the condition 5° of that section is also satisfied. Hence all results obtained in Secs. 2.6 - 2.8 are valid for the material structure with internal state variables.

The general dissipation inequality (2.23) takes the form

$$\hat{d}(\sigma) = -\partial_{A(t)}\hat{\psi}(\cdot)\cdot\dot{A}(t) - \frac{1}{\varrho_{\varkappa}\vartheta(t)}\hat{Q}(\sigma)\cdot\nabla\vartheta(t) \geqslant 0. \tag{3.11}$$

The internal dissipation function $\hat{\imath}:\Sigma\to\mathbb{R}$ for the material structure with internal state variables is defined as follows

$$\hat{\imath}(\sigma) = -\frac{1}{\vartheta(t)}\partial_{A(t)}\hat{\psi}(\cdot)\cdot A(t) = -\frac{1}{\vartheta(t)}\partial_{A(t)}\hat{\psi}(\cdot)\cdot\hat{a}(\sigma). \tag{3.12}$$

These results show that the internal state variables and the rates of the internal state variables determine the value of the internal dissipation function $\hat{\imath}:\Sigma\to\mathbb{R}$ at the intrinsic state σ .

3.2. Coupling of dissipative mechanisms

In the case of an inelastic material there are two main reasons for internal dissipation. First, the pure viscous effects and second the visco-plastic flow.

To make our reasoning clear let us consider two main physical mechanisms responsible for inelastic effects, namely the internal friction mechanism and the thermo-activated mechanism (or the damping mechanism).

To describe both inelastic effects simultaneously two groups of internal state variables are introduced(*), i.e.

$$A(t) = (\alpha(t), \omega(t)) \qquad (3.13)$$

for which the following initial value problem

$$\begin{aligned} \dot{\alpha}(t) &= \mathfrak{a}(\sigma) = \mathfrak{a}(P(t), \alpha(t), \omega(t)), \quad \alpha(0) = \alpha_0, \\ \dot{\omega}(t) &= \Omega(\sigma) = \Omega(P(t), \alpha(t), \omega(t)), \quad \omega(0) = \omega_0, \end{aligned} \qquad (3.14)$$

is postulated.

It is assumed that the first group of internal state variables $\alpha(t)$ describes the dissipation effects of the internal friction mechanism, and the second $\omega(t)$ the effects of the thermo-activated mechanism (or damping mechanism).

It is noteworthy that two groups of internal state variables $\alpha(t)$ and $\omega(t)$ introduced by Eqs,(3.13) and (3.14) can described coupling between dissipative mechanisms.

(*) This idea was first introduced by the author in the paper [69].

3.3. Particular example of coupling effects

In many cases of practical interest we can find a formalism in which the internal state variables describing effects of the first mechanism are eventually eliminated. We can achieve this in a rather simple manner if the first mechanism is described by the linear evolution equations(*).

If we again consider the internal friction mechanism and the thermo-activated mechanism (or the damping mechanism) as two main dissipative mechanisms then we can show a very good example of such a procedure(**).

It is very well known fact that the internal friction mechanism leads to visco-elastic effects and is described by the linear evolution equations. On the other hand the thermoactivated mechanism (or the damping mechanism) is responsible for elastic-viscoplastic effects. The internal state variables introduced to describe these effects are governed by the non-linear evolution equations.

Let us assume that the evolution equations (3.2) have the form

$$\dot{\alpha}(t) = \mathfrak{a}_1[\alpha(t)] - \mathfrak{a}_2(P(t), \omega(t)), \tag{3.15}$$

(x) For statistical justification of this procedure see ZWANZIG [99] and HAKEN [39].

(**) This procedure has been developed in the previous paper of the author (cf. Ref. [83]).

$$\dot{\omega}(t) = \Omega_1\big(P(t), \omega(t)\big) + \Omega_2\big(P(t), \omega(t)\big)\big[\alpha(t)\big],$$

where $\mathfrak{A}_1$ is a constant matrix, $\mathfrak{A}_2$, Ω_1 and Ω_2 may depend in an arbitrary manner on $P(t)$ and $\omega(t)$.

Equation $(3.15)_1$ possesses the formal solution

$$\alpha(t) = \exp(\mathfrak{A}_1 t)\,\alpha_0 - \int_0^t \exp[\mathfrak{A}_1(t-\tau)]\,\mathfrak{A}_2\big(P(\tau), \omega(\tau)\big)\,d\tau \tag{3.16}$$

which may be transformed by partial integration into

$$\begin{aligned}\alpha(t) = {} & \mathfrak{A}_1^{-1}\mathfrak{A}_2\big(P(t), \omega(t)\big) + \exp(\mathfrak{A}_1 t)\big[\alpha_0 - \mathfrak{A}_1^{-1}\mathfrak{A}_2\big(P(0), \omega(0)\big)\big] \\ & - \int_0^t \exp\big[\mathfrak{A}_1(t-\tau)\,\mathfrak{A}_1^{-1}\big\{\partial_{P(\tau)}\mathfrak{A}_2\big(P(\tau), \omega(\tau)\big)\cdot\dot{P}(\tau) \\ & + \partial_{\omega(\tau)}\mathfrak{A}_2\big(P(\tau), \omega(\tau)\big)\cdot\dot{\omega}(\tau)\big\}\,d\tau\end{aligned} \tag{3.17}$$

Inserting the result (3.17) into $(3.15)_2$ yields the desired evolution equation

$$\begin{aligned}\dot{\omega}(t) = {} & \Omega_1\big(P(t), \omega(t)\big) + \Omega_2\big(P(t), \omega(t)\big)\,\mathfrak{A}_1^{-1}\mathfrak{A}_2\big(P(t), \omega(t)\big) \\ & - \int_0^t \Omega_2\big(P(t), \omega(t)\big)\exp\big[\mathfrak{A}_1(t-\tau)\big]\,\mathfrak{A}_1^{-1}\big\{\partial_{P(\tau)}\mathfrak{A}_2\big(P(\tau), \omega(\tau)\big)\cdot\dot{P}(\tau) \\ & + \partial_{\omega(\tau)}\mathfrak{A}_2\big(P(\tau), \omega(\tau)\big)\cdot\dot{\omega}(\tau)\big\}\,d\tau \\ & + \Omega_2\big(P(t), \omega(t)\big)\exp(\mathfrak{A}_1 t)\big[\alpha_0 - \mathfrak{A}_1^{-1}\mathfrak{A}_2\big(P(0), \omega(0)\big)\big].\end{aligned} \tag{3.18}$$

The internal state variables of the first group enter into the integro-differential equation (3.18) only through their initial values α_0 .

We can obtain equivalent result by inserting the formal solution (3.16) directly into the evolution equation $(3.15)_2$. The result is as follows

$$\dot{\omega}(t) = \Omega_1\big(P(t), \omega(t)\big) + \Omega_2\big(P(t), \omega(t)\big)\Big[\exp(\mathfrak{a}_1 t)\,\alpha_0 - \int_0^t \exp\big[\mathfrak{a}_1(t-\tau)\big]\,\mathfrak{a}_2\big(P(\tau), \omega(\tau)\big)\,d\tau\Big]\ , \quad \omega(0) = \omega_0 . \tag{3.19}$$

The initial value problem (3.19) is equivalent to the Volterra integral equation

$$\begin{aligned}\omega(t) = \omega_0 &+ \int_0^t \Omega_1\big(P(\zeta), \omega(\zeta)\big)\,d\zeta \\ &+ \int_0^t \Omega_2\big(P(\zeta), \omega(\zeta)\big)\Big\{\exp(\mathfrak{a}_1\zeta)\,\alpha_0 \\ &- \int_0^t \exp\big[\mathfrak{a}_1(\zeta-\tau)\big]\,\mathfrak{a}_2\big(P(\tau), \omega(\tau)\big)\,d\tau\Big\}\,d\zeta\ .\end{aligned} \tag{3.20}$$

The result in the form of the integro-differential equation (3.18) has more direct interpretation than the evolution equation (3.19) or than the Volterra integral equation (3.20).

All terms, except first, on the right hand side in the integro-differential equation (3.18) reflect influences of the first mechanism on the evolution of the internal state variables of the second group. The second term describes an instantaneous interaction, the integral describes a general retarded interaction and the last term may be interpreted as a fluctuation interaction. All these terms represent the coupling effects between two dissipative mechanisms considered.

The result obtained in this section is of great importance to us. To describe interactions of both mechanisms considered it is sufficient to introduce only one group of the internal state variables, namely ω , and to use the modified evolution equation (3.18) or (3.20). So, in what follows we can assume that the intrinsic state is given by

$$\sigma^* = \left(P(t), \omega(t)\right). \tag{3.21}$$

4. RATE TYPE MATERIAL STRUCTURE(*)

4.1. Definitions and fundamental assumptions

We assume that the deformations processes $P \in \pi$ are piecewise continuously differentiable, i.e. for every time $t \in [0, d_P]$ exist left and right derivatives $\frac{d}{d\tau} P(\tau)\Big|_{\tau=t} = \dot{P}(t)$ which are equal for all but a finite number of $t \in [0, d_P]$

For such a class of processes π we define the set

$$\vartheta_g \equiv \left\{ \frac{d}{d\tau} P(\tau) \,\middle|\, P \in \pi, \ \tau \in [0, d_P], \ P(\tau) = g \in \mathcal{G} \right\}, \tag{4.1}$$

(*) For the rate type material structure see NOLL [60] .

i.e. the class of all possible derivatives of deformation process at a time at which they have the value g .

PROPOSITION 3. (i) The method of preparation space in the stress space $\mathcal{T}$, i.e.

$$\mathcal{K} \equiv \mathcal{T} \tag{4.2}$$

the intrinsic state space Σ is the set

$$\Sigma \equiv \left\{ (g, \mathfrak{t}) \,\middle|\, g \in \mathcal{G} \, , \; \mathfrak{t} \in \mathcal{T}_g \equiv \mathcal{K}_g \right\} \, , \; \mathfrak{t} \equiv T(t) \, , \tag{4.3}$$

and

$$(\Sigma \times \pi)^* = \left\{ (\sigma, P) \in \Sigma \times \pi \, , \; \hat{G}(\sigma) = P^i \right\} . \tag{4.4}$$

(ii) There exists a mapping

$$\hat{\beta} : \bar{\mathcal{D}} \longrightarrow \mathcal{T} \, , \; \bar{\mathcal{D}} \equiv \left\{ (g, \dot{g}, \mathfrak{t}) \,\middle|\, g \in \mathcal{G} \, , \; \dot{g} \in \mathcal{V}_g \, , \mathfrak{t} \in \mathcal{T}_g \right\} \tag{4.5}$$

such that for every $P \in \pi$ and $\mathfrak{t}_0 \in \mathcal{T}_{p^i}$ the initial value problem

$$\frac{dT(\tau)}{d\tau} = \hat{\beta}\left(P(\tau) \, , \dot{P}(\tau) \, , T(\tau) \right) \, , \; T(0) = \mathfrak{t}_0 \tag{4.6}$$

has a unique solution $T : [0, d_p] \longrightarrow \mathcal{T}$.

(iii) The constitutive mapping $\mathfrak{R} : (\Sigma \times \pi)^* \longrightarrow \mathcal{X}$ satisfying for every pair $(\sigma_0 , P) \in (\Sigma \times \pi)^*$ the condition

$$\mathcal{R}\left(\hat{G}(\sigma_0), \cdot, P\right) : \mathcal{T}_{P^i} \longrightarrow \mathcal{I}_P \qquad \text{must be bijection} \quad (4.7)$$

in such that the evolution function $\hat{e}$ is of the form

$$\hat{e}(\sigma_0, P) = \left(P^f, \mathfrak{z}(P, \mathfrak{k}_0)\right) \qquad (4.8)$$

where $\mathfrak{z}(P, \mathfrak{k}_0)$ denotes the solution functional of the initial value problem (4.6).

We have constitutive equations

$$Z(t) = \hat{S}(\sigma) = \hat{S}\left(\hat{e}(\sigma_0, P_{[0,t]})\right) = \hat{S}(g, \mathfrak{k}) \qquad (4.9)$$

with condition for the stress response function

$$\hat{T}(\sigma) = \mathfrak{z}(P_{[0,t]}, \mathfrak{k}_0) \qquad (4.10)$$

Definition 14. The unique material structure $(\mathcal{G}, \pi, \Sigma, \mathcal{R})$ satisfying the Proposition 3 is called the material structure of a rate type.

4.2. Comments on possible applications

The rate-type material structure has been used to describe elastic-plastic response of a material. The theory of plastic flow has been developed within the framework of the rate type material structure.

This is why the study of the rate type structure is of great importance for us.

The next section will show conditions under which the internal state variable and the rate type material structure are equivalent.

5. ISOMORPHIC MATERIAL STRUCTURES (ISOTHERMAL PROCESSES)(*)

5.1. General definition. Definition of a material

Suppose that we have defined, in one and the same particle X two material structures $(\mathcal{G}, \pi, \Sigma_1, \hat{S}_1, \hat{e}_1)$ and $(\mathcal{G}, \pi, \Sigma_2, S_2, e_2)$.

Definition 15. Two material structures $(\mathcal{G}, \pi, \Sigma_1, S_1, e_1)$ and $(\mathcal{G}, \pi, \Sigma_2, S_2, e_2)$ are materially isomorphic if there exists a bijection $\iota : \Sigma_1 \longrightarrow \Sigma_2$ with the properties

1) $\hat{S}_2(\iota(\sigma_1)) = \hat{S}_1(\sigma_1)$,

2) $\hat{G}_2(\iota(\sigma_1)) = \hat{G}_1(\sigma_1)$,

3) $\hat{e}_2(\iota(\sigma_1), P) = \iota(\hat{e}_1(\sigma_1, P))$.

This definition is a special case of the definition proposed by NOLL [60].

(*) Cf. Ref. [80].

If two material structures defined in the same particle X are materially isomorphic, we also say that they describe at X the same material. So, <u>a material</u> is an equivalence class of material structures, the equivalence being material isomorphy (given by the Definition 15).

An isomorphism is thus a similarity. More important property of an isomorphism is that, if the unique material structure $(\mathcal{S}, \pi, \Sigma_1, S_1, e_1)$ is isomorphic with the unique material structure $(\mathcal{S}, \pi, \Sigma_2, S_2, e_2)$ and if some features expressed by terms Σ_1, S_1, e_2 are valid for the structure $(\mathcal{S}, \pi, \Sigma_1, S_1, e_1)$ then the same features expressed by terms Σ_2, S_2, e_2 are preserved for the structure $(\mathcal{S}, \pi, \Sigma_2, S_2, e_2)$.

An isomorphism of two unique material structures is thus a similarity of these structures, under which the actions $\hat{S}_1$ and $\hat{e}_1$ defined on the set Σ_1 correspond to the actions $\hat{S}_2$ and $\hat{e}_2$ defined on the set Σ_2.

This is the main reason for which the study of isomorphism for different material structures is of great practical importance.

5.2. Discussion of a material isomorphism between internal state variable and rate type material structures (isothermal processes)(*)

Let us write the main equations for the internal state

(*) For an isothermal proces $\mathfrak{Z} = \mathfrak{T}$, i.e. $Z(t) = T(t) = \mathfrak{z} = \mathfrak{t}$

variable material structure in the form

$$\sigma_1 = \big(P(t), A(t)\big) = (g, k) \in \Sigma_1,\ g \in \mathcal{G},\ k \in \vartheta_{ng}, \tag{5.1}$$

$$Z(t) = \hat{S}_1\big(P(t), A(t)\big), \tag{5.2}$$

$$\frac{dA(\tau)}{d\tau} = \hat{\alpha}\big(P(\tau), A(\tau)\big),\ A(0) = \hat{K}(\sigma_{1_0}) = k_0 . \tag{5.3}$$

We assume additionally that the Eq.(5.2) is such that $A(t)$ can be expressed as a function of $P(t)$ and $Z(t)$ for any time $t \in [0, d_P]$, i.e.

$$A(t) = \hat{M}\big(P(t), Z(t)\big) \tag{5.4}$$

The pair $\big(P(t), Z(t)\big)$ determines the intrinsic state σ_2 at time t, i.e.

$$\sigma_2 = \big(P(t), Z(t)\big) = (g, t) \in \Sigma_2,\ g \in \mathcal{G},\ t \in \mathcal{T}_g . \tag{5.5}$$

The bijection $\iota : \Sigma_1 \longrightarrow \Sigma_2$ is determined by the mapping $\hat{M} : \Sigma_2 \longrightarrow \mathcal{K}_1 \equiv \vartheta_n$.

Let us assume that the deformation processes $P \in \pi$ are continuous and piecewise differentiable with respect to time in the interval $[0, d_P]$ and that the constitutive function $\hat{S}_1$ is differentiable with respect to both variables $P(\tau)$ and $A(\tau)$, i.e. that the gradients $\partial_{P(\tau)}\hat{S}_1$ and $\partial_{A(\tau)}\hat{S}_1$ exist.

By differentiating the constitutive equation (5.2)

with respect to time, we obtain the following evolution equation for the rate type structure

$$\frac{dZ(\tau)}{d\tau} = \hat{\beta}_0(P(\tau), Z(\tau)) + \hat{\beta}_1(P(\tau), Z(\tau))\,[\dot{P}(\tau)], \tag{5.6}$$

where

$$\begin{aligned} \hat{\beta}_0 &= \partial_{A(\tau)} \hat{S}_1(P(\tau), M(P(\tau), Z(\tau)))\,\hat{\alpha}\{P(\tau), \hat{M}(P, Z)), \\ \hat{\beta}_1 &= \partial_{P(\tau)} \hat{S}_1(F(\tau), M(P(\tau), Z(\tau))), \end{aligned} \tag{5.7}$$

and the initial value Z(0) is determined by the relation

$$Z(0) = \hat{S}_1(P(0), A(0)) = \mathfrak{k}_0 . \tag{5.8}$$

If the functions $\hat{\beta}_0$ and $\hat{\beta}_1$ are such that the unique solution of the initial value problem (5.6) - (5.8) exists, then we can write

$$Z(t) = \mathfrak{z}(P_{[0,t]}, t_0) = \hat{S}_2(\sigma_2) = \hat{K}(\sigma_2) = \mathfrak{k}. \tag{5.9}$$

The connections between different mappings introduced are explained in Fig. 6.

Under the conditions introduced, there is simple isomorphism between internal state variable material structure and the rate type material structure.

Under above conditions both these material structures describe the same material at a particle X of a body B .

6. THERMO-VISCOPLASTICITY FOR FINITE STRAINS

6.1. Nature of an elastic-viscoplastic response

It is well known that in many practical problems the actual behaviour of a material is governed by plastic as well as by rheologic effects. It can even be said that for many important structural materials rheologic effects are more pronounced after the plastic state has been reached.

Recent research concerning the description of dynamic properties of materials has shown that the application of the theory of plasticity, in which the fundamental assumption is that of time independence of the equations of state, leads to too large discrepancies between the theoretical and experimental results.

Thus there is no need to point out the advantages that can be gained by simultaneous description of rheologic and plastic effects, and the general problem is that of viscoplasticity. However, the difficulties of combined treatment of rheologic and plastic phenomena are enormous. The viscous properties of the material introduce, a time dependence of equations of state and the plastic properties, on the other hand, make these equations depend on the loading path.

Thus, as a result of simultaneous introduction of viscous and plastic properties, we obtain a dependence on the load history and on the time. A description of strain in viscoplasticity will therefore involve the history of the specimen, expressed in the type of the loading process, and the time. Different results will be obtained for different loading paths and different duration of the process.

6.2. Definition of permanent (plastic or viscoplastic) deformation

The definition of permanent (plastic) deformation tensor for finite deformation has been introduced by GREEN and NAGHDI [28-30].

To show all detail concerning this definition let us consider the neighbourhood $\mathcal{N}$ of a particle X in the reference configuration $\varkappa(\mathcal{B})$ of a body $\mathcal{B}$, Fig. 7. In actual deformed configuration a particle X has a position $x=\chi(X,t)$ and its neighbourhood is $\chi(\mathcal{N})$. After LEE and LIU [51] let us introduce for an element of the body $\mathcal{N}$ the stress free configuration $\vartheta(\mathcal{N})$. It is assumed that in the stress free configuration an element $\vartheta(\mathcal{N})$ has uniform reference temperature ϑ_0. If this is done for each element of the body, then the stress free elements no longer fit together to form a continuum which can be mapped continuously back to the reference configuration of the body (cf. GREEN and NAGHDI [30]). However, the line element dy in the stree-free configuration (see Fig. 7) and in the neighbourhood of the point corresponding to x is related to the line elements at x and X by

$$dx = F_e dy \quad , \quad dy = F_p dX \tag{6.1}$$

such that

$$dx = F dX \quad \text{where} \quad F = F_e F_p \tag{6.2}$$

Let ds_o, ds, ds_y denote the lengths of corresponding line elements in the initial reference configuration, the deformed configuration at time t and the stress-free configuration (also at time t), respectively. Then, from (6.1) we have

$$ds^2 = dy^T C_e dy \quad , \quad ds_y^2 = dX^T C_p dX \quad , \tag{6.3}$$

where

$$C_e = F_e^T F_e \quad , \quad C_p = F_p^T F_p \tag{6.4}$$

LEE [52] has adopted C_e as a measure of "elastic" strain. He does not appear to use C_p as a measure of plastic strain. Instead, he represents plastic deformation by the tensor F_p . The tensors C_e and F_p depend on the orientation of the stress-free configuration whose line element is dy . To show this let us use transformation for superposed rigid body motions

$$x^* = c_o + Q(x - x_o) \ ,$$
$$y^* = \bar{c}_o + \bar{Q}(y - y_o) \ , \qquad (6.5)$$

where c_o , x_o , $\bar{c}_o$ and y_o are vector-valued functions of t ; Q and $\bar{Q}$ are proper orthogonal tensor valued functions of t . We have

$$C_e^* = \bar{Q}\, C_e \bar{Q}^T \ , \quad F_p^* = \bar{Q}\, F_p \ . \qquad (6.6)$$

To introduce tensors as measure of elastic and plastic deformations invariant under superposed rigid body motions let us observed that

$$ds^2 = dX^T C\, dX \ ,$$
$$ds^2 - ds_y^2 = dX^T (C - C_p)\, dX \ . \qquad (6.7)$$

After GREEN and NAGHDI [30] we can identify C_p as a measure of plastic strain with

$$C'' = C_p \qquad (6.8)$$

and we can interpret

$$E' = \frac{1}{2}C' = \frac{1}{2}(C - C'') = E - E'' = E_e \qquad (6.9)$$

as "elastic" strain. For the tensor E' we have required property

$$E'^{*} = E' . \tag{6.10}$$

As a result we also have

$$E = E' + E'' \qquad \text{or} \qquad C = C' + C_p . \tag{6.11}$$

6.3. Analysis of dissipation mechanisms

We aim to construct a reasonable and physically justified theory of viscoplasticity within the framework of a material structure with internal state variables.

In order to give a proper physical interpretation of internal state variables introduced so far we should analyse mechanisms which govern the internal dissipation of a material.

Let us assume that the cause of rheological dissipation is the internal friction which may arise by means of a number of mechanisms depending on many sources. In specific situations and for particular materials we can determine the most probable mechanism of internal friction experimentally. In this case, we can specify the functions $\mathfrak{A}_1$ and $\mathfrak{A}$ in Eq. $(3.15)_1$.

The second kind of dissipation caused by internal changes of a material during plastic deformation can be explained by different mechanisms.

The basic result of the microscopic investigations is

that the elementary process of plastic deformation is the motion of a line-shaped crystal defect called dislocation. It is generally agreed that the thermal obstacles impeding the motion of dislocations through a crystal are responsible for the dynamic aspects of plastic deformation.

Let us consider mild steel as an example. ROSENFIELD and HAHN [88] have shown that in the temperature-strain rate spectrum of plain carbon steel we can consider four regions which reflect different mechanisms of plastic deformation, Fig. 8. It is convenient to discuss the material behaviour in relation to each of the regions.

Region I is characterized by a yield stress relatively insensitive both to strain rate and temperature. The plastic flow is governed by athermal mechanisms.

The distinctive features of Regions II lie not only in the fact that the yield stress is more markedly temperature and rate sensitive but also that the semilogarithmic rate sensitivity is independent of temperature, cf. CAMPBELL and FERGUSON [11], Fig. 9.

It is generally accepted that at low temperatures the plastic strain rate in c.p.h., f.c.c. and b.c.c. metals is controlled by the thermal activation of dislocation motion, the activation energy being a function of the applied stress and temperature.

Common thermal obstacles or mechanisms in pure metals

are the Peierls-Nabarro stress, forest dislocations, the motion of jogs in screw dislocations, cross-slip of screw dislocations, and climb of edge dislocations (cf. Fig. 10).

Region III is characterized by a lower rate and temperature dependence of the yield stress. The experimental results do not shade light on the rate sensitivity in Region III.

Region IV encompasses very high strain rates, 10^3 to 10^6 s^{-1}. The yield stress of mild steel and other metals in this region seems to become extremely strain-rate sensitive. Thermal activation is still operative in Region IV, but the increase in rate sensitivity in this region must be attributed to an additional mechanism opposing the motion of dislocations. An alternative interpretation would be that the total stress is the sum of two parts, one determined by the short-range barriers and the other - by additional dissipative mechanisms.

CAMPBELL and FERGUSON [11] have shown that the increase of the rate sensitivity in Region IV is due to viscous resistance to dislocation motion and the observed macroscopic viscosity is consistent with that expected from the theory of the damping of dislocation motion by phonon viscosity.

6.4. Physical foundations. Interpretation of internal state variables

The effects of temperature and strain rate on the plastic flow behaviour of metals have been successfully rationalized in terms of the dynamics of dislocations.

To make our reasoning clear let us consider the movement of a dislocation through the rows of barriers. The moving dislocation is dissipated energy mainly due to two sources. The first is connected with the overcoming of obstacles which involves thermally activated process and the second with the interactions of dislocation with lattice thermal vibrations (phonon drag) or with electrons (electron viscosity) or are based on relaxation effects in the dislocation core (glide-plane viscosity).

For that reason the theoretical mechanisms governing the velocity of dislocation in relatively pure materials can be divided into two groups, i.e., mechanisms which do and do not involve a thermal activation process.

According to that explanation it is convenient to divide the strain rate sensitivity behaviour of metals into two regions (cf. Fig. 11):

a) thermally activated region (Region II in Fig. 11) and

b) the phonon damping region (Region IV in Fig. 11).

We use here denotation of the dissipative regions as

proposed by ROSENFIELD and HAHN (cf. Fig. 8).

The flow stress τ can be represented by the equation

$$\tau = \tau_A + \tau^* + \tau_D \tag{6.12}$$

where τ_A is the athermal component of stress, τ^* is the thermally activated component of stress, and τ_D is the stress attributed to phonon damping.

If a dislocation is moving through the rows of barriers then its velocity can be determined by the expression(*)

$$v = AL^{-1}/(t_s + t_B) \tag{6.13}$$

where AL^{-1} is the average distance of dislocation movement after each thermal activation, t_S is the time a dislocation spent at the obstacle and t_B is the time of travelling between the barriers.

The plastic strain rate $\dot{E}_p$ is given by the OROWAN relation

$$\dot{E}_p = \rho_M b v , \tag{6.14}$$

where ρ_M is the mobil dislocation density, and b denotes the Burger`s vector and v is the average dislocation velocity.

(*) See a theoretical analysis presented by KUMAR and KUMBLE [47] and also very recent paper by TEODOSIU and SIDOROFF [91] .

The average velocity of a dislocation damped by the phonons between thermally activated events computed by the relation (6.13) takes the form of the expression

$$v = AL^{-1}\left[\vartheta^{-1}\exp\left(U/k\vartheta\right) + ABL^{-1}/(T - T_B)b\right]^{-1}, \qquad (6.15)$$

where ν is the vibration frequency of the dislocation segment, $U(T^*)$ is the activation energy, $k\vartheta$ is the Boltzmann constant times the absolute temperature, B denotes the dislocation drag coefficient, and T_B is attributed to the stress needed to overcome the forest dislocation barriers to the dislocation motion (cf. Fig. 11).

After inserting the result (6.15) into Eq.(6.14) we obtain the desired equation for the plastic strain rate

$$\dot{E}_p = \rho_M b AL^{-1}\left[\nu^{-1}\exp\left(U(T^*)/k\vartheta\right) + ABL^{-1}/(T - T_B)b\right]^{-1} \qquad (6.16)$$

Equation (6.16) is the evolution equation for the plastic strain E_p in the case of simultaneous interaction of the thermally activated mechanism and the phonon damping mechanism.

It will be useful to consider two limit cases as follows.

a) If the time t_B taken by the dislocation to travel between the barriers in a viscous phonon medium is negligible when compared with the time t_S spent at the obstacle then we can focus our attention on the analysis of the thermally activated process(*) (Region II in Fig. 11). The dislocation velocity (6.13) can be approximated by the expression

$$v = \frac{AL^{-1}}{t_s}. \tag{6.17}$$

The evolution equation for the plastic strain E_p takes now the form

$$\dot{E}_p = \varrho_M L^{-1} A b \nu \exp\left[-U(T^*)/k\vartheta\right]. \tag{6.18}$$

Since the flow stress T for this case is given by the relation

$$T = T_A + T^* \tag{6.19}$$

the activation energy U can be assumed as the nonlinear function of the overstress $\frac{T}{T_A} - 1$, i.e.

$$U(T^*) = \varphi\left[a\left(\frac{T}{T_A} - 1\right)\right], \tag{6.20}$$

where a is a constant.

If we additionally introduce the denotation

$$\gamma_1 = \varrho_M L^{-1} b A \nu, \tag{6.21}$$

(*) The mechanism for overcoming the dislocation forest which may appear in crystals of different metals in the various temperature ranges has been developed theoretically by SEEGER [89,90].

the evolution equation (6.18) takes the very well known form

$$\dot{E}_p = \gamma_1 \exp\left\{ -\varphi\left[a\left(\frac{T}{T_A}-1\right)\right] / k\vartheta \right\} . \tag{6.22}$$

In the description proposed we have three parameters E_p, γ_1 and T_A which can be treated as the internal state variables (cf. here the results of the paper by PERZYNA and WOJNO [84]).

b) With increasing dislocation velocities, the ratio t_B/t_S increases and at high enough stress or in a perfect crystal the velocity is only governed by the phonon damping mechanism(*) (cf. Region IV in Fig. 11). At very high strain rates the applied stress is high enough to overcome instantaneously the dislocation barriers without any aid from thermal fluctuations. This is true for the flow stress $T > T_B$, where T_B is attributed to the stress needed to overcome the forest dislocation barriers to the dislocation motion and is called the back stress.

In this region the dislocation velocity (6.13) can be approximated by the expression

(*) The phonon viscosity theory has been developed by MASON [55]. For a thorough theoretical discussion of damping mechanisms see NABARRO [56] and GORMAN, WOOD and VREELAND [26].

$$\upsilon = \frac{AL^{-1}}{t_B} \tag{6.23}$$

and the evolution equation for the plastic strain takes the form(*)

$$\dot{E}_p = \frac{\rho_M b^2}{B}(T - T_B) \,. \tag{6.24}$$

The flow stress T consists of two terms

$$T = T_B + T_D \qquad \text{where} \quad T_B = T_A + T_o \,. \tag{6.25}$$

If we introduce the denotation

$$\gamma_2 = \frac{\rho_M b^2 T_B}{B} \,, \tag{6.26}$$

then the evolution equation (6.26) has more simple form

$$\dot{E}_p = \gamma_2 \left(\frac{T}{T_B} - 1 \right) . \tag{6.27}$$

In the description here discussed we have similarly as in the previous case three parameters, namely E_p, γ_2 and

(*) The dislocation drag coefficient B can be interpreted as generalized damping constant for phonon viscosity and phonon scattering mechanisms, i.e. $B = B_{pv} + B_{ps}$, cf. Ref.[77]. The ratio B/ρ_M can be obtained from the slop of the linear portion of the stress against strain rate curve, and from this ratio the value of B can be calculated if a value of ρ_M is assumed.

T_B which can be treated as the internal state variables.

When a crystal is set into vibration, the vibrations decay even if there is no loss of energy from the crystal to its surroundings. The cause of the decay is called internal friction. Internal friction may arise by a number of mechanisms$^{(*)}$. In the linear approximation the theory of each of these mechanisms leads to the known Boltzmann constitutive equations of viscoelasticity$^{(**)}$.

As it has been suggested by NABARRO [56] the presence of dislocations in the crystal may increase the internal friction in three general ways. Firstly, the dislocations act as sources of internal stresses, and contribute to the thermo-elastic damping. Secondly, the dislocations may move under the action of an applied stress, thereby producing a modulus defect. If the motion of the dislocations is resisted by one of the mechanisms discussed in this section their displacement is not in phase with the applied stress, and the modulus defect has an imaginary component which corresponds to an internal friction. Thirdly, the motion of

(*) Comprehensive reviews of the internal friction mechanisms can be found in the treatise by NABARRO [56].

(**) Statistical mechanics of viscoelasticity based on the internal friction mechanisms has been developed by DeVAULT and McLENNAN [20] and by MARIS [54].

the dislocations may be hindered by localized obstacles. A dislocation acquires energy in order to overcome an obstacle, and this energy is dissipated when the obstacle has been passed by a mechanism which contribute to the mechanisms of dislocation damping.

There is also interaction between dislocations and point defects. This interaction is of great importance to description of fracture phenomena. Point defects influence the motion of dislocation and as a result they affect the dissipative mechanisms of plastic flow. To rationalize internal state variable description of inelastic material it will be reasonable to introduce additional scalar internal state variable ξ . This state variable is interpreted as scalar measure of the concentration of point defects.

It will be shown that the internal state variable ξ will be useful in the description of flow and fracture phenomena of inelastic materials. It could supply additional information needed to predict failure modes of a material.

Basing on the results obtained for both limit cases considered and taking into account interaction between dislocations and point defects we propose the internal state variables as follows

$$\omega(t) = \left[E_p(t), \gamma(t), \varkappa(t), \xi(t)\right], \tag{6.28}$$

where for the one-dimensional case $E_p(t)$ is scalar value

interpreted as the inelastic strain, $\gamma(t)$ is the viscosity parameter, $\varkappa(t)$ is the work-hardening parameter and $\xi(t)$ denotes the concentration of point defects.

6.5. Two approximations

It will be useful for future applications to consider approximations of the description previously presented. We shall show the nature of these approximations for one-dimensional case(*).

a) For some metals (e.g. iron, mild steel and titanium) for medium strain rates the thermally-activated mechanism becomes dominant. In this particular case we can extend Region II to be valid for the entire range of strain rate and temperature changes, Fig. 12.

Let us assume the internal state variables in the form introduced by (6.28). For the thermally-activated approximation we interprete $\gamma(t)$ as the viscosity parameter defined by Eq. (6.21), and $\varkappa(t)$ as the athermal stress $T_A(t)$.

We postulate the evolution equations for the internal state variables (6.28) as follows

$$\dot{E}_p(t) = \gamma(t)\left\langle \Phi\left[\frac{T(t)}{\varkappa(t)} - 1\right]\right\rangle,$$

$$\dot{\gamma}(t) = \hat{\Gamma}(\sigma^*)\dot{E}_p(t), \qquad (6.29)$$

(*) An idea of these two approximations is modelled after Ref. [83].

$$\dot{\varkappa}(t) = \hat{K}(\sigma^*)\,\dot{E}_p(t),$$

$$\dot{\xi}(t) = \hat{\Xi}_1(\sigma^*)\dot{E}_p(t) + \hat{\Xi}_2(\sigma^*),$$

where Φ is a dimensionless function of overstress and $\hat{\Gamma}$, $\hat{K}$, $\hat{\Xi}_1$ and $\hat{\Xi}_2$ are functions of the intrinsic state

$$\sigma^* = \big(E(t), \vartheta(t), \omega(t)\big). \tag{6.30}$$

It is intended that all these functions may be chosen to represent the results of tests on the dynamic behaviour of metals.

In the evolution equation $(6.29)_4$ we took advantage of the assumption that the main sources of the generation or the annihilation of point defects are the plastic deformation and the temperature dependent annealing process(*).

For practical application it will be useful to reduce number of the internal state variables. To be more precise we shall consider the example of experimental data for pure titanium. LAWSON and NICHOLAS [50] have performed series of experiments on pure titanium in which they obtained the linear dependence of shear flow stress on log (strain rate) for a very large range of strain rate changes (from 10^{-4}

(*) In the previous papers of the author, [74,81,82] the production of point defects by the neutron irradiation has been taken into consideration. Thorough discussion of generation of point defects has been presented by SEEGER [90].

to $10^4 s^{-1}$). These data are shown in Fig. 13 and 14.

To represent these experimental data we assume the evolution equations for the internal state variables E_p and ξ to be as follows

$$\dot{E}_p(t) = \gamma(t)\left\{\exp\left[\frac{I(t)}{\varkappa(t)} - 1\right] - 1\right\},$$
$$\dot{\xi}(t) = \gamma(t)\,\hat{\Xi}_1(\sigma^*)\left\{\exp\left[\frac{I(t)}{\varkappa(t)} - 1\right] - 1\right\} + \hat{\Xi}_2(\sigma^*), \tag{6.31}$$

in which $\Phi(\cdot) = \exp(\cdot) - 1$ and $\gamma(t)$ and $\varkappa(t)$ are postulated in the form

$$\gamma(t) = \hat{\gamma}\left(\vartheta(t), E_p(t), \xi(t)\right),$$
$$\varkappa(t) = \hat{\varkappa}\left(\vartheta(t), E_p(t), \xi(t)\right). \tag{6.32}$$

The function $\hat{\varkappa}$ represents a description of the static curve for titanium. For an isothermal case (ϑ = const) and for given ξ = const this function can be taken as the approximation of the curve obtained for $\dot{E}_p = \text{const} = 10^{-4} s^{-1}$ (cf. Fig. 13).

To describe the coupling effects between the internal friction and thermally-activated mechanisms let us assume evolution equations for the internal state variables of the second group, i.e. for $\omega(t) = \left[E_p(t), \xi(t)\right]$ in the following form

$$\dot{\omega}(t) = \Omega_1(\sigma^*) + \Omega_2(\sigma^*)[\alpha(t)] \tag{6.33}$$

with the initial value $\omega(0) = \omega_0$, where

$$\Omega_1(\sigma^*) = \left[\gamma(t)\Phi\left(\frac{T(t)}{\kappa(t)} - 1\right), \gamma(t)\hat{\Xi}_1(\sigma^*)\Phi\left(\frac{T(t)}{\kappa(t)} - 1\right) + \hat{\Xi}_2(\sigma^*)\right],$$
$$\Omega_2(\sigma^*) = [\hat{h}(\sigma^*), \hat{m}(\sigma^*)]. \tag{6.34}$$

The functions $\hat{h}(\sigma^*)$ and $\hat{m}(\sigma^*)$ describe the influence of the internal friction mechanism on the change of the in - elastic strain E_P and of the concentration of the point defects ξ .

If the evolution equations for the first group of the internal state variables (i.e. for $\alpha = [\alpha_1, \dots \alpha_n]$, n is finite) are postulated in the form of Eq.(3.15)$_1$ then to describe coupling effects it is sufficient to work with the evolution equation (3.18) or (3.19) with the functions Ω_1 and Ω_2 given by Eqs.(6.34).

b) Some recent investigations on metals have shown that at very high strain rates the phonon damping mechanism becomes most influential. For some metals (e.g. aluminium and copper) the damping mechanism can approximate the plastic flow phenomena in the entire range of strain rates changes. The situation is schematically shown in Fig. 15. This approximation has been suggested for aluminium single crystals by FERGUSON, KUMAR and DORN [24] , KUMAR, HAUSER and

DORN [46] and GORMAN, WOOD and VREELAND [26], for polycrystalline aluminium by HAUSER, SIMMONS and DORN [40], KARNES and RIPPERGER [42] and KUMAR, HAUSER and DORN [46] and for copper by KUMAR and KUMBLE [47].

The flow stress of aluminium crystals at different strain rates and strains at temperature 10 K, 77 K, 300 K and 500 K obtained by KUMAR, HAUSER and DORN [46] is shown in Fig. 16. It is apparent that the linear relationship between stress and strain rate does continue to prevail at high strain rates without any deviation even at 20 percent strain. The linear portion of the stress versus strain rate curve when extrapolated intersects the stress axis at T_B. This has been defined as a back stress. The results plotted in Fig. 16 clearly show that the back stress T_B for aluminium single crystals is a function of strain or dislocation density and is independent of temperature within the accuracy of the measurements.

Data from the paper by HAUSER, SIMMONS and DORN [40] for polycrystalline aluminium which were originally plotted on a semilogarithmic scale are shown in Fig. 17 on a linear scale. Again it is observed that the magnitudeof T_B increases with an increase in strain. However, for the polycrystalline aluminium the back stress T_B does depend on temperature, namely decreases with an increase in temperature. At slower strain rates the damping term is negligible and the strain rate sensitivity in Eq.(5.1) arises from the thermally activated component of the stress τ^*. At very high strain rate there is no assistance from the thermally fluctuations and the strain rate sensitivity is attributed to the third term in Eq.(5.1) which is a dislocation damping term τ_D.

The mechanical behaviour of copper at strain rates from 10^{-3} to $10^3 s^{-1}$ at temperature 300 K, 420 K and 590 K was investigated by KUMAR and KUMBLE [47]. Data obtained by these authors have been plotted in a linear scale in Figs. 18, 19 and 20. In these plots it is observed that the strain rate behaviour of copper can be divided into two regions. Below 10 s^{-1} the mobile dislocations are thermally activated over the forest dislocation barriers and above $10^2 s^{-1}$ the dislocation motion is viscous drag limited. The flow stress is linear function of strain rate at high strain rates for all the states at 300, 420 and 590 K and can be represented by the Eq.(6.27). The magnitude of τ_B for copper and its variation with strain is wholly consistent with the concept that τ_B is the stress needed to overcome the usual barriers to the dislocation motion. The authors of the paper concluded that in the case of copper these barriers are forest dislocations. The magnitude of τ_B in Fig.20 decreases with increasing temperature for a strain-hardened state. This has been attributed to the recovery of the dislocation structure.

KUMAR and KUMBLE [47] observed also that the mobile dislocation density in the viscous phonon damping region is independent of strain up to 6% and decrease slightly with a further increase in strain. The mobile dislocation density increases with increasing temperature.

This last observation is of great importance to our phenomenological description based on internal state variables because it concerns the viscosity parameter γ_2 defined by the relation (6.26).

A discussion of the experimental data for aluminium single crystals, polycrystalline aluminium and copper clearly shows that the second approximation proposed is justified

To describe this simplified case let us introduce the internal state variables as postulated by Eq.(6.28) and interpret $\gamma(t)=\gamma_2(t)$ and $\varkappa(t)=T_B(t)$.

Basing on the experimental results discussed and the physical propositions given in Sect. 6.4 b we postulate the evolution equations for the internal state variables as follows

$$
\begin{aligned}
&\dot{E}_P(t) = \gamma(t)\left[\frac{T(t)}{\varkappa(t)} - 1\right] ,\\
&\dot{\gamma}(t) = \bar{\Gamma}(\sigma^*)\dot{E}_P(t) , \qquad (6.35)\\
&\dot{\varkappa}(t) = \bar{K}(\sigma^*)\dot{E}_P(t) ,\\
&\dot{\xi}(t) = \Xi_1(\sigma^*)\dot{E}_P(t) + \Xi_2(\sigma^*) .
\end{aligned}
$$

where $\bar{\Gamma}$, $\bar{K}$, $\bar{\Xi}_1$ and $\bar{\Xi}_2$ are new functions of the intrinsic state σ^*.

Our description will be simpler and more useful if, basing on the experimental data discussed, we can determine directly the material functions

$$\gamma(t) = \bar{\gamma}\left(\vartheta(t), E_p(t), \xi(t)\right),$$
$$\varkappa(t) = \bar{\varkappa}\left(\vartheta(t), E_p(t), \xi(t)\right). \tag{6.36}$$

If that is the case the phenomenological theory is straightforward and involves only two internal state variables $E_p(t)$ and $\xi(t)$ with the evolution equations in the form

$$\dot{E}_p(t) = \bar{\gamma}(\cdot)\left[\frac{T(t)}{\bar{\varkappa}(\cdot)} - 1\right],$$
$$\dot{\xi}(t) = \bar{\gamma}(\cdot)\bar{\Xi}_1(\sigma^*)\left[\frac{T(t)}{\bar{\varkappa}(\cdot)} - 1\right] + \bar{\Xi}_2(\sigma^*). \tag{6.37}$$

The coupling between the internal friction and the phonon damping mechanisms is described by the integral equation (3.20) in which the functions Ω_1 and Ω_2 are defined as follows

$$\Omega_1(\sigma^*) = \left[\bar{\gamma}(\cdot)\left[\frac{T(t)}{\bar{\varkappa}(\cdot)} - 1\right], \bar{\gamma}(\cdot)\bar{\Xi}_1(\sigma^*)\left[\frac{T(t)}{\bar{\varkappa}(\cdot)} - 1\right] + \bar{\Xi}_2(\sigma^*)\right],$$
$$\Omega_2(\sigma^*) = \left[\bar{h}(\sigma^*), \bar{m}(\sigma^*)\right], \tag{6.38}$$

where $\bar{h}(\sigma^*)$ and $\bar{m}(\sigma^*)$ are new influence functions.

The crucial problem for the theory describing the influence of the internal friction mechanism on the plastic flow mechanisms is determination of the functions $a_2(\sigma^*)$, $\hat{h}(\sigma^*)$ and $\hat{m}(\sigma^*)$ (or $\bar{h}(\sigma^*)$ and $\bar{m}(\sigma^*)$). It is noteworthy that the determination of above functions is very difficult. The reason for that is the lack of experimental tests investigating the coupling effects discussed.

In the case when material under consideration has two very distinct regions of viscoplastic flow we can apply an idea presented in Refs. [78] and [84].

6.6. General theory of viscoplasticity

To generalize the description of the coupling effects between the internal friction and plastic flow meehanisms for the case of combined states of stress and strain we postulate the internal state variables as follows

$$\omega(t)=[C_p(t),\ \gamma(t),\ \varkappa(t),\ \xi(t)] \tag{6.39}$$

where $C_p(t)$ denotes now the inelastic deformation tensor, $\gamma(t)$ is the viscosity parameter, $\varkappa(t)$ is interpreted as the work-hardening parameter and $\xi(t)$ as the concentration of defects.

As it was shown in the previous sections we can approximate the rate sensitive plastic flow by the thermally activated mechanism or by the phonon damping mechanism.

In both cases the description is given by the constititive equations (cf. Eqs.(3.8,)

$$Z(t) = \hat{S}(\sigma^*) = \hat{S}\big(P(t), \omega(t)\big), \quad \hat{S} \equiv \{\hat{\Psi}, \hat{N}, \hat{T}, \hat{Q}\} \tag{6.40}$$

with the restrictions

$$\begin{gathered} \partial_{\nabla\vartheta(t)}\hat{\Psi}(\cdot) = 0, \; T(t) = 2\varrho_\varkappa \partial_{C(t)}\hat{\Psi}(\cdot), \; \eta(t) = -\partial_{\vartheta(t)}\hat{\Psi}(\cdot) \\ -\partial_{\omega(t)}\hat{\Psi}(\cdot)\cdot\dot{\omega}(t) - \frac{1}{\varrho_\varkappa\vartheta(t)}\hat{Q}(\sigma^*)\cdot\nabla\vartheta(t) \geqslant 0, \\ t \in [0, d_p], \end{gathered} \tag{6.41}$$

and by the integro-differential equation (3.18) (or by the Volterra integral equation (3.20)) in which the functions Ω_1 and Ω_2 take the form

$$\begin{gathered} \Omega_1(\sigma^*) = \gamma(t)\left\langle\Phi\left(\frac{f(t)}{\varkappa(t)} - 1\right)\right\rangle \Big[\partial_{T(t)}f, \; \mathrm{tr}\big(\partial_{T(t)}f\,\hat{\Gamma}(\sigma^*)\big), \\ \mathrm{tr}\big(\partial_{T(t)}f\,\hat{K}(\sigma^*)\big), \; \mathrm{tr}\big(\partial_{T(t)}f\,\hat{\Xi}_1(\sigma^*)\big) + \frac{\hat{\Xi}_2(\sigma^*)}{\gamma(t)\langle\Phi(\cdot)\rangle}\Big] \\ \Omega_2(\sigma^*) = \big[\hat{h}_1(\sigma^*), \; \hat{h}_2(\sigma^*), \; 0, \; \hat{m}(\sigma^*)\big] \end{gathered} \tag{6.42}$$

for thermally activated plastic flow, and

$$\begin{gathered} \Omega_1(\sigma^*) = \gamma(t)\left\langle\frac{f(t)}{\varkappa(t)} - 1\right\rangle \Big[\partial_{T(t)}f, \; \mathrm{tr}\big(\partial_{T(t)}f\,\bar{\Gamma}(\sigma^*)\big), \\ \mathrm{tr}\big(\partial_{T(t)}f\,\bar{K}(\sigma^*)\big), \; \mathrm{tr}\big(\partial_{T(t)}f\,\Xi_1(\sigma^*)\big) + \frac{\Xi_2(\sigma^*)}{\gamma(t)\left\langle\frac{f(t)}{\varkappa(t)} - 1\right\rangle}\Big] \\ \Omega_2(\sigma^*) = \big[\bar{h}_1(\sigma^*), \; \bar{h}_2(\sigma^*), \; 0, \; \bar{m}(\sigma^*)\big] \end{gathered} \tag{6.43}$$

for phonon damping plastic flow, respectively.

In this description we took advantage of a notion of the quasi-static yield criterion for an elastic-viscoplastic material defined as follows(*)

$$\mathcal{F}(\sigma^*) = \frac{f(t)}{\varkappa(t)} - 1 , \tag{6.44}$$

where

$$f(t) = f\big(T(t), C_p(t), \vartheta(t), \xi(t)\big) , \tag{6.45}$$

and we introduce the symbol $\langle[\ ]\rangle$ according to the definition

$$\langle[\ ]\rangle = \begin{cases} 0 & \text{if} \quad f(t) \leqslant \varkappa(t) \\ [\ ] & \text{if} \quad f(t) > \varkappa(t) \end{cases} \tag{6.46}$$

The material functions $\hat{\Gamma}, \hat{K}, \hat{\Xi}_1, \hat{\Xi}_2, \hat{h}_1, \hat{h}_2, \hat{m}$ and $\bar{\Gamma}, \bar{K}, \bar{\Xi}_1, \bar{\Xi}_2, \hat{h}_1, \hat{h}_2, \bar{m}$ are understood as a simple generalization for the three-dimensional case of the respective functions defined and interpreted in the one-dimensional consideration.

6.7. Particular constitutive equations

To show particular constitutive equations of viscoplastic response of a material let us assume that the internal dissipation is governed by a single mechanism, namely the thermally activated process. We postulate the intrinsic state σ in the form

(*) The existence of the quasi-static yield criterion is one of the fundamental assumptions of viscoplasticity theory, cf. Refs. [63-65] .

$$\sigma = \left(C(t), \vartheta(t), \nabla\vartheta(t), C_p(t), \varkappa(t), \gamma(t) \right). \tag{6.47}$$

We assume also that the inelastic deformation rate tensor is a function of the excess of the applied state of stress over the quasi-static yield condition, the static yield function corresponding to that of the Huber-Mises yield criterion.

Thus the excess is assumed to be in the form

$$\mathcal{F}(\sigma) = \frac{\sqrt{II_{Td}}}{\varkappa(t)} - 1 , \tag{6.48}$$

where II_{Td} denotes the second invariant of the stress deviation T_d .

Under the above assumption and by the analogy to Eqs.(6.42) we postulate that the internal state variables C_p , $\varkappa$, γ are determined by the following initial value problem

$$\begin{aligned} \dot{C}_p(t) &= \gamma(t) \left\langle \Phi\left(\frac{\sqrt{II_{Td}}}{\varkappa(t)} - 1 \right) \right\rangle \frac{T_d}{\sqrt{II_{T_d}}} , & C_p(0) &= C_p^0 , \\ \dot{\varkappa}(t) &= \operatorname{tr}\left(\hat{K}(\sigma) \dot{C}_p(t) \right) , & \varkappa(0) &= \varkappa_0 , \\ \dot{\gamma}(t) &= \operatorname{tr}\left(\hat{\Gamma}(\sigma) \dot{C}_p(t) \right) , & \gamma(0) &= \gamma_0 , \end{aligned} \tag{6.49}$$

where $\Phi(0)=0$, and the symbol $\langle \Phi(\mathcal{F}) \rangle$ is defined as follows

$$\langle \Phi(\mathcal{F}) \rangle = \begin{cases} 0 & \text{if} \quad \sqrt{II_{T_d}} \leq \varkappa(t) , \\ \Phi(\mathcal{F}) & \text{if} \quad \sqrt{II_{T_d}} > \varkappa(t) . \end{cases} \tag{6.50}$$

Here $\hat{K}$ and $\hat{\Gamma}$ are second order tensor functions. The functions Φ , $\hat{K}$ and $\hat{\Gamma}$ must be chosen to represent results of tests on the dynamic behaviour of particular material.

The internal dissipation is given by

$$\hat{\imath}(\sigma) = -\frac{1}{\vartheta(t)}\frac{\gamma(t)}{\sqrt{II_{T_d}}}\left\langle \Phi\left(\frac{\sqrt{II_{T_d}}}{\varkappa(t)}-1\right)\right\rangle\Big\{ \operatorname{tr}\Big[\Big(\partial_{C_p(t)}\hat{\psi}(\cdot) + \partial_{\varkappa(t)}\hat{\psi}(\cdot)\,\hat{K}(\sigma) + \partial_{\gamma(t)}\hat{\psi}(\cdot)\,\hat{\Gamma}(\sigma)\Big)\,T_d\Big]\Big\}. \tag{6.51}$$

If the internal state variables $\varkappa$ and γ are assumed to be constant during the thermodynamic process for $t\in[0,d_p]$ then we have the evolution equation for C_p in the form(*)

$$\dot{C}_p(t) = \gamma_0\left\langle \Phi\left(\frac{\sqrt{II_{T_d}}}{\varkappa_0}-1\right)\right\rangle\frac{T_d}{\sqrt{II_{T_d}}} \tag{6.52}$$

with the initial value $C_p(0)=C_p^{\circ}$.

The internal dissipation for this particular case is given by

$$\hat{\imath}(\sigma) = -\frac{\gamma_0}{\vartheta(t)}\frac{1}{\sqrt{II_{T_d}}}\left\langle \Phi\left(\frac{\sqrt{II_{T_d}}}{\varkappa_0}-1\right)\right\rangle \operatorname{tr}\left[\partial_{C_p(t)}\hat{\psi}(\cdot)\,T_d\right]. \tag{6.53}$$

(*) Cf. Refs. [63,65] .

6.8. On material isomorphism in description of viscoplasticity.

Example 1. Elastic-viscoplastic material (medium strain rates, isothermal process) (*)

Let us postulate that

$$A(t)=C_p(t) \tag{6.54}$$

i.e. the inelastic deformation tensor $C_p(t)$ is assumed as the internal state variable. So, the intrinsic state is determined by a pair

$$\sigma_1=\left(C(t),C_p(t)\right)\in\Sigma_1 \tag{6.55}$$

The constitutive equation has the form

$$T(t)=\hat{S}_1\left(C(t),C_p(t)\right). \tag{6.56}$$

The evolution equation is assumed as follows

$$\frac{dC_p(\tau)}{d\tau}=\gamma\left\langle\Phi\left(\mathcal{F}(\tau)\right)\right\rangle\partial_{T(\tau)}f \tag{6.57}$$

with the initial value

$$C_p(0)=C_p^{o}, \tag{6.58}$$

(*) A through discussion of the internal state variable description of dynamic plasticity is given in Ref. [68], cf. also [76].

where γ is a viscosity constant, $\mathcal{F}(t)$ is the static yield condition and is assumed in the form

$$\mathcal{F}(t) = \frac{f(T(t))}{\varkappa_0} - 1 , \tag{6.59}$$

where $\varkappa_0$ is a yield constant, the dimensionless function $\Phi(\mathcal{F})$ may be chosen to represent results of test on the dynamic behaviour of materials and the symbol $\langle\Phi(\mathcal{F})\rangle$ is defined as follows

$$\langle\Phi(\mathcal{F}(t))\rangle = \begin{cases} 0 & \text{for} \quad \mathcal{F}(t) \leq 0 , \\ \Phi(\mathcal{F}(t)) & \text{for} \quad \mathcal{F}(t) > 0 . \end{cases} \tag{6.60}$$

A material described by the relations (6.54) - (6.59) is called an elastic-perfectly viscoplastic material. The relations (14.1) - (14.6) defined the internal state variable structure of an elastic-perfectly viscoplastic material.

If the constitutive equation (6.56) can be written in the form

$$C_p(t) = \hat{M}(C(t), T(t)) , \tag{6.61}$$

then the internal state σ_2 is determined by the expression

$$\sigma_2 = (C(t), T(t)) \in \Sigma_2 . \tag{6.62}$$

Let us assume that all conditions explained in the previous section are satisfied for an elastic-perfectly viscoplastic material.

Differentiating the constitutive equation (6.56) with respect to time gives the evolution equation for the stress $T(t)$ in the form

$$\frac{dT(\tau)}{d\tau} = \hat{\beta}_0\big(C(\tau), T(\tau)\big) + \hat{\beta}_1\big(C(\tau), T(\tau)\big)\,[\dot{C}(\tau)] \tag{6.63}$$

with the initial condition

$$T(0) = \hat{S}_1\big(C(0), C_p(0)\big), \tag{6.64}$$

where

$$\hat{\beta}_0 = \gamma\langle\Phi(\mathcal{F}(\tau))\rangle\, \partial_{C_{p}(\tau)}\hat{S}_1[\partial_{T(\tau)} f], \qquad \hat{\beta}_1 = \partial_{C(\tau)}\hat{S}_1 . \tag{6.65}$$

The equations (6.61) - (6.65) define the rate type structure for an elastic-perfectly plastic material.

Example 2. Elastic-viscoplastic material (description in the entire range of strain rates, isothermal process) (*)

Let us now postulate

(*) Cf. Ref. 80 .

$$A(t)=\left(C_p(t),\gamma(t),\varkappa(t)\right), \tag{6.66}$$

where $\varkappa(t)$ is the work-hardening scalar parameter, $\gamma(t)$ is the viscosity scalar parameter and $C_p(t)$ is the inelastic deformation tensor. The intrinsic state is determined by

$$\sigma_1=\left(C(t),\varkappa(t),\gamma(t),C_p(t),\right)\in\Sigma_1 . \tag{6.67}$$

We assume the constitutive equation in the form

$$T(t)=\hat{S}_1(\sigma_1), \tag{6.68}$$

and postulate the static yield condition

$$\mathcal{F}(t)=\frac{f\left(T(t),C_p(t)\right)}{\varkappa(t)}-1 . \tag{6.69}$$

The evolution equations for the internal state variables $C_p(t),\gamma(t),\varkappa(t)$ are postulated in the form

$$\begin{aligned}
\dot{C}_p(\tau)&=\gamma(\tau)\left\langle\Phi\left(\mathcal{F}(\tau)\right)\right\rangle\partial_{T(\tau)}f ,\\
\dot{\gamma}(\tau)&=\mathrm{tr}\left[\Gamma_0(\sigma_1)\dot{C}_p(\tau)\right],\\
\dot{\varkappa}(\tau)&=\mathrm{tr}\left[K_0(\sigma_1)\dot{C}_p(\tau)\right],
\end{aligned} \tag{6.70}$$

with the initial conditions

$$C_p(0)=C_p^{\circ}, \quad \gamma(0)=\gamma_0, \quad \varkappa(0)=\varkappa_0 . \tag{6.71}$$

It is easy to show that for this case there is no

simple transition to rate type structure of an elastic-viscoplastic material. But we shall prove that the internal state variable structure of an elastic-viscoplastic material determined by the Eqs.(6.66) - (6.71) is isomorphic with some mixed material structure.

To show this let us assume that the constitutive equation (6.68) can be written in the form

$$C_p(t)=\hat{M}\big(C(t),\,\varkappa(t),\,\gamma(t),\,T(t)\big). \tag{6.72}$$

Let us postulate that the intrinsic state σ_2 is now determined by the expression

$$\sigma_2=\big(C(t),\,\varkappa(t),\,\gamma(t),\,T(t)\big)\in\Sigma_2\,. \tag{6.73}$$

We have assumed that the method of preparation space $\mathcal{K}$ is the Cartesian product of the two-dimensional vector space $\mathcal{V}_2$ and the stress space $\mathcal{T}$, i.e.

$$\mathcal{K}\equiv\mathcal{V}_2\times\mathcal{T}. \tag{6.74}$$

In a similar way as in the previous section we can obtain the evolution equation for the stress tensor $T(t)$ in the form

$$\frac{dT(\tau)}{d\tau}=\hat{\hat{\beta}}_0(\sigma_2)+\hat{\hat{\beta}}_1(\sigma_2)\,[\dot{C}(\tau)]\,, \tag{6.75}$$

where

$$\hat{\hat{\beta}}_0 = \gamma(\tau)\langle\Phi(\mathcal{F}(\tau))\rangle\Big\{\partial_{\varkappa(t)}\hat{S}_1\,\mathrm{tr}\big[K_0\,\partial_{T(t)}f\big]$$

$$+\partial_{\gamma(\tau)}\hat{S}_1\,\mathrm{tr}\big[\Gamma_0\,\partial_{T(\tau)}f\big]+\partial_{C_p(\tau)}\hat{S}_1\big[\partial_{T(\tau)}f\big]\Big\}, \qquad (6.76)$$

$$\hat{\hat{\beta}}_1 = \partial_{C(\tau)}\hat{S}_1,$$

with the initial value

$$T(0) = \hat{S}_1\big(C(0),\ \varkappa(0),\ \gamma(0),\ C_p(0)\big). \qquad (6.77)$$

The evolution equation for $\varkappa(t)$ and $\gamma(t)$ have the same form as postulated by (6.70), i.e.

$$\dot{\gamma}(\tau) = \gamma(\tau)\langle\Phi(\mathcal{F}(\tau))\rangle\ \mathrm{tr}\big[\Gamma_0(\sigma_2)\,\partial_{T(\tau)}f\big],$$

$$\dot{\varkappa}(\tau) = \varkappa(\tau)\langle\Phi(\mathcal{F}(\tau))\rangle\ \mathrm{tr}\big[K_0(\sigma_2)\partial_{T(\tau)}f\big], \qquad (6.78)$$

with

$$\gamma(0) = \gamma_0, \quad \varkappa(0) = \varkappa_0. \qquad (6.79)$$

The equation (6.72) - (6.79) defined the mixed material structure which is isomorphic with the internal state variable structure of an elastic-viscoplastic material (6.66) - (6.71).

7. THERMOPLASTICITY FOR FINITE DEFORMATIONS

7.1. Postulates of plasticity

It is desirable to present here a brief heuristic discussion concerning main features of elastic-plastic response of a material.

The first important property which distinguishes the behaviour of an elastic-plastic continuum from that of a nonlinear elastic continuum is connected with permanent deformations. This feature of a material that it can deform permanently is the result of different paths for loading and unloading processes. Unloading process from an elastic-plastic state follows a path in stress space different from that of loading process.

The second feature of plastic response is its time independency. So, the constitutive equations for an elastic-plastic material have to be invariant under the time scale.

These both features, namely the occurrence of permanent deformations and time independent behaviour of a material are characteristic for plastic response.

7.2. General theory. Rate type formulations

We intend to discuss the constitutive equations of an elastic-plastic material within the framework of the rate type material structure developed in Section 4.

We follow here the theory of GREEN and NAGHDI [31] which characterizes the behaviour of an elastic-plastic material under large deformations. The theory is constructed without a priori introduction of the concept of plastic strain.

In this theory the intrinsic state σ of a particle X is assumed in the form

$$\sigma = \big(P(t), A(t)\big) = \big(E(t), \vartheta(t), \nabla\vartheta(t), T(t), \varkappa(t)\big). \qquad (7.1)$$

Thus the method of preparation is represented by the Piola-Kirchhoff stress tensor $T(t)$ and by the work-hardening parameter $\varkappa(t)$.

The constitutive assumption is as follows

$$\big(\psi(t), \eta(t), q(t)\big) = \big(\hat{\Psi}(\sigma), \hat{N}(\sigma), \hat{Q}(\sigma)\big). \qquad (7.2)$$

Essential assumption of the theory concerns the evolution equations for the Piola-Kirchhoff stress tensor $T(t)$ and for the work-hardening parameter $\varkappa(t)$.

The evolution equations are postulated in the general form

$$\hat{L}_1(\sigma)[\dot{E}(t)] + \hat{L}_2(\sigma)[\dot{T}(t)] + \hat{L}_3(\sigma)[\dot{\vartheta}(t)] + \hat{L}_4(\sigma)\left[\overline{\nabla\vartheta(t)}^{\boldsymbol{\cdot}}\right] = 0,$$

$$\dot{\varkappa}(t) = \mathrm{tr}\left[\hat{K}_1(\sigma)\dot{E}(t)\right] + \mathrm{tr}\left[\hat{K}_2(\sigma)\dot{T}(t)\right] + \hat{K}_3(\sigma)\dot{\vartheta}(t) \qquad (7.3)$$

$$+ \hat{K}_4(\sigma)\cdot\overline{\nabla\vartheta(t)}^{\boldsymbol{\cdot}},$$

where $\hat{L}_1,\ldots,\hat{L}_4,\hat{K}_1,\ldots,\hat{K}_4$ are tensor functions of the intrinsic state $\mathfrak{S}$. All these tensor functions are unaltered by superposed rigid body motion. They are also assumed to have some symmetry properties to satisfy the requirements of symmetry for the tensors E and T.

On the evolution equations (7.3) we should pose further restrictions to obtain features associated with rate-independent plastic deformations. To this end, GREEN and NAGHDI [31] have assumed the existence of a region $\mathcal{P}$ in the space ε^{16} (of variables $E, \vartheta, \nabla\vartheta, T$), bounded by a closed surface $\partial\mathcal{P}$, throughout which the equations $(7.3)_1$ are integrable in the form:

$$H(\mathfrak{S})=\text{const} \qquad \left(H_{\alpha\beta}(\mathfrak{S})=\text{const}\right), \tag{7.4}$$

whenever $\varkappa$ has a constant value. In Eq.(7.4) H is a symmetric tensor function and is invariant with respect to superposed rigid body motion. The region $\mathcal{P}$ depends on the time in the sense that when $\varkappa$ changes from one constant value to another, then $\mathcal{P}$ changes also.

After differentiating (7.4) and comparison of the result with $(7.3)_1$, we obtain

$$\begin{aligned} \hat{L}_{1\alpha\beta\gamma\mu} &= B_{\alpha\beta\varrho\varkappa}\frac{\partial H}{\partial E_{\gamma\mu}}, \\ \hat{L}_{2\alpha\beta\gamma\mu} &= B_{\alpha\beta\varrho\varkappa}\frac{\partial H_{\varrho\varkappa}}{\partial T_{\gamma\mu}}, \end{aligned} \tag{7.5}$$

$$\hat{L}_{3\alpha\beta} = B_{\alpha\beta\varrho\varkappa} \frac{\partial H_{\varrho\varkappa}}{\partial \vartheta} ,$$

$$\hat{L}_{4\alpha\beta\gamma} = B_{\alpha\beta\varrho\varkappa} \frac{\partial H_{\varrho\varkappa}}{\partial \vartheta_{,\gamma}} ,$$

which hold in the region $\mathscr{P}$ and where $B_{\alpha\beta\varrho\varkappa}$ is a tensor function of σ. With the use of (7.5) the evolution equations $(7.3)_1$ can be replaced by the more restricted system as follows

$$B_{\alpha\beta\gamma\mu} H_{\gamma\mu} = 0 , \tag{7.6}$$

where

$$H_{\alpha\beta} = H_{\beta\alpha} , \quad B_{\alpha\beta\gamma\mu} = B_{\beta\alpha\gamma\mu} = B_{\alpha\beta\mu\gamma} . \tag{7.7}$$

Let us assume that the 6 x 6 matrix of a system of six linear equations (7.6) is a continuous function of the intrinsic state σ. We denote this matrix by $\mathbf{M}$ and its determinant by

$$f(\sigma) = \det \mathbf{M} \tag{7.8}$$

For a given constant value of $\varkappa = \bar{\varkappa}$, the variables $(P(t), T(t)) = (E(t), \vartheta(t), \nabla\vartheta(t), T(t))$ satisfying the condition $f(P(t), T(t), \bar{\varkappa}) < 0$ form a set in a region $\mathscr{P}$ of the space ε^{16}. This region is bounded by the surface $f(P(t), T(t), \bar{\varkappa}) = 0$.

In the following we shall restrict our considerations to the situation in which $\mathcal{P}$ specified by

$$f\left(P(t), T(t), \bar{\varkappa}\right) < 0 \tag{7.9}$$

is a simply connected region of ε^{16} bounded by a single closed surface $\partial\mathcal{P}$ whose equation is $f\left(P(t), T(t), \bar{\varkappa}\right) = 0$ As $\bar{\varkappa}$ continuously assumes a series of different constant values the region $\mathcal{P}$ and its boundary $\partial\mathcal{P}$ change continuously but maintain the same topological character.

When $\varkappa = \bar{\varkappa}$ and the set $\left(E(t), \vartheta(t), \nabla\vartheta(t), T(t)\right)$ satisfy the condition (7.9), the only solution of the system of equations (7.6) is the zero solution

$$\dot{H} = 0 \tag{7.10}$$

Since $\dot{\varkappa}$ vanishes throughout $\mathcal{P}$ when $\varkappa = \bar{\varkappa}$, we replace $(7.3)_2$ by

$$\dot{\varkappa}(t) = \mathrm{tr}\left[\mathcal{h}(\sigma)\dot{H}(t)\right] , \qquad \mathcal{h} = \mathcal{h}^T , \tag{7.11}$$

where $\mathcal{h}$ is a symmetric function of the intrinsic state σ and is unaltered by superposed rigid body motion.

When the right-hand side of (7.8) vanishes, i.e. when

$$\mathcal{f}(\sigma) = 0 \tag{7.12}$$

the matrix $\mathbb{M}$ is of rank $\leqslant$ 5.

GREEN and NAGHDI [31] have restricted their considerations to the case in which the matrix $\mathbb{M}$ is of rank 5 when (7.12) holds. For this case the evolution equations (7.6) are linearly dependent and equivalent to a set of five linearly independent equations. The sixth equation is given by (7.12) and, alternatively, this may be expressed in the rate form

$$\dot{f}(t)=\hat{f}(t)+\partial_{\varkappa(t)}f\,\dot{\varkappa}(t)=0 \quad , \tag{7.13}$$

where

$$\begin{aligned}\hat{f}(t) &= \mathrm{tr}\left[\partial_{E(t)}f\,\dot{E}(t)\right]+\mathrm{tr}\left[\partial_{T(t)}f\,\dot{T}(t)\right]\\ &+\partial_{\vartheta(t)}f\,\dot{\vartheta}(t)+\partial_{\nabla\vartheta(t)}f\cdot\overline{\nabla\vartheta(t)}^{\,\cdot}\end{aligned} \tag{7.14}$$

Under the above conditions, the general solution of equations (7.6) is as follows

$$\dot{H}_{\alpha\beta}(t)=\Lambda G_{\alpha\beta}(\mathfrak{S}) \quad , \tag{7.15}$$

where Λ is an arbitrary scalar and $G_{\alpha\beta}$ is a tensor function of the intrinsic state $\mathfrak{S}$ such that

$$B_{\alpha\beta\gamma\mu}G_{\gamma\mu}=0 \; . \tag{7.16}$$

Let us now consider the case when

$$f\left(P(t),T(t),\bar{\varkappa}\right)=0 \quad ; \tag{7.17}$$

when $\varkappa=0$, it follows that

$$\Lambda = 0 \qquad \text{whenever} \quad \hat{f}(t) = 0. \tag{7.18}$$

Hence

$$\Lambda = \lambda \hat{f}(t), \tag{7.19}$$

where λ is a new scalar function. Thus we have

$$\dot{H}_{\alpha\beta}(t) = \lambda \hat{f}(t) G_{\alpha\beta}(\sigma). \tag{7.20}$$

When (7.20) and (7.13) are satisfied, $\dot{\varkappa} \neq 0$ and hence $\hat{f} = 0$. It follows that

$$\left(1 + \lambda h_{\alpha\beta}(\sigma) G_{\alpha\beta}(\sigma) \partial_{\varkappa(t)} f\right) \hat{f}(t) = 0 \tag{7.21}$$

which determines λ as a function of σ.

The above considerations provide background for the concept of a yield or loading surface and the associated flow rules.

It is also natural to identify the function H with the plastic strain tensor E''.

7.3. Plasticity as a limit case of viscoplasticity(*)

Let us assume that

(*) Cf. Refs. 68,69,72,73.

$$\alpha \equiv 0 \qquad (7.22)$$

and additionally that $\omega(t)$ is determined by the rate independent evolution equation.

An elastic-viscoplastic material as described in Sec.6 loses its strain rate sensitivity if, and only if, the viscosity parameter $\gamma(t) \to \infty$. In this case the statical yield condition (cf. with (6.64))

$$\mathcal{F} = \frac{f(t)}{\varkappa(t)} - 1 = 0 \qquad (7.23)$$

is satisfied, the material loses its viscosity properties and behaves as an elastic-plastic.

The second group of the internal state variables is as follows (*)

$$\omega(t) = \left[C_p(t), \varkappa(t), \xi(t)\right]. \qquad (7.24)$$

From the definition of the symbol $\langle \Phi(\mathcal{F}) \rangle$ we see that the differential equation determining the plastic deformation tensor $C_p(t)$ takes the following form

$$\dot{C}_p(t) = \Lambda^* \partial_{T(t)} f(t) \qquad (7.25)$$

where the parameter $\Lambda^* = \gamma \langle \Phi(\mathcal{F}) \rangle$ may be determined from

(*) The intrinsic state $\sigma = \left(C(t), \vartheta(t), \omega(t)\right)$.

the condition that the point representing in temperature-stress space the actual state of temperature and stress lies on the yield surface (7.23).

From the yield condition (7.23) and the general assumption that the rate of internal state variables $\dot{\varkappa}$ and $\dot{\xi}$ vanish when $\dot{C}_p(t)=0$, we can deduce the following criterion of loading

$$f(t)=\varkappa(t) \quad \text{and} \quad \operatorname{tr}\left(\partial_{T(t)} f\, \dot{T}(t)\right)+\partial_{\vartheta(t)} f\, \dot{\vartheta}(t)>0. \tag{7.26}$$

Similarly, the criteria

$$f(t)=\varkappa(t) \quad \text{and} \quad \operatorname{tr}\left(\partial_{T(t)} f\, \dot{T}(t)\right)+\partial_{\vartheta(t)} f\, \dot{\vartheta}(t) \leqslant 0 \tag{7.27}$$

define the unloading and neutral state, respectively.

To satisfy the condition that the point representing the actual state of loading and temperature lies on the yield surface, it is sufficient to fulfill $\dot{f}(t)=\dot{\varkappa}(t)$, i.e.,

$$\begin{aligned} &\operatorname{tr}\left(\partial_{T(t)} f\, \dot{T}(t)\right)+\partial_{\vartheta(t)} f\, \dot{\vartheta}(t)+\operatorname{tr}\left(\partial_{C_p(t)} f\, \dot{C}_p(t)\right) \\ &\quad +\partial_{\xi(t)} f\, \dot{\xi}(t)=\operatorname{tr}\left[\hat{K}(\sigma)\, \dot{C}_p(t)\right]. \end{aligned} \tag{7.28}$$

Using (7.25) and assuming

$$\dot{\xi}(t)=\operatorname{tr}\left[\hat{\Xi}_1(\sigma)\, \dot{C}_p(t)\right], \tag{7.29}$$

we have

$$\Lambda^{*} = \lambda^{*}\left[\operatorname{tr}\left(\partial_{T(t)} f\, \dot{T}(t)\right) + \partial_{\vartheta(t)} f\, \dot{\vartheta}(t)\right] \quad , \tag{7.30}$$

where

$$\lambda^{*} = \left\{ \operatorname{tr}\left[\left(\hat{K}(\mathbf{6}) - \partial_{C_p(t)} f - \partial_{\xi(t)} f\right) \partial_{T(t)} f\right]\right\} > 0 \tag{7.31}$$

The relation for Λ^{*} and the criteria of loading, unloading and neutral state have shown that the differential equation determining the internal state tensor $C_p(t)$ for an elastic-plastic material can be written as follows

$$\dot{C}_p(t) = \lambda^{*}\left\langle \operatorname{tr}\left[\partial_{T(t)} f\, \dot{T}(t)\right] + \partial_{\vartheta(t)} f\, \dot{\vartheta}(t)\right\rangle \partial_{T(t)} f \quad , \tag{7.32}$$

where

$$\langle\{\ \}\rangle = \begin{cases} \{\ \} & \text{if } f = \varkappa \text{ and } \{\ \} > 0 \\ 0 & \text{if } f = \varkappa \text{ and } \{\ \} \leqq 0 \text{ or if } f < \varkappa \end{cases} \tag{7.33}$$

The full system of constitutive equations together with the evolution equations describing the behaviour of an elastic-plastic material is as follows

$$\psi(t) = \hat{\Psi}(\mathbf{6}) \,, \qquad T(t) = 2\varrho_{\varkappa} \partial_{C(t)} \hat{\Psi}(\mathbf{6}) \quad ,$$

$$\begin{aligned}
&\eta(t) = -\partial_{\vartheta(t)}\hat{\psi}(\sigma), \\
&q(t) = \hat{Q}\big(\sigma, \nabla\vartheta(t)\big), \\
&\dot{C}_p(t) = \lambda^* \langle\{\ \}\rangle \partial_{T(t)} f, \\
&\dot{\varkappa}(t) = \operatorname{tr}\big[\hat{K}(\sigma)\dot{C}_p(t)\big], \quad \dot{\xi}(t) = \operatorname{tr}\big[\hat{\Xi}_1(\sigma)\dot{C}_p(t)\big]
\end{aligned} \tag{7.34}$$

To satisfy the thermodynamic postulate by an elastic-plastic material, the equations (7.34) should fulfill the general dissipation inequality

$$i(\sigma) - \frac{1}{\varrho_{\varkappa}\vartheta^2(t)} \hat{Q}\big(\sigma, \nabla\vartheta(t)\big)\cdot\nabla\vartheta(t) \geqslant 0 , \tag{7.35}$$

where i denotes the internal dissipation function for an elastic-plastic material, i.e.,

$$\begin{aligned}
\hat{i}(\sigma) = -\frac{\lambda^*}{\vartheta(t)} \langle\{\ \}\rangle\Big(&\operatorname{tr}\big\{\partial_{C_p(t)}\hat{\psi}(\sigma) + \\
&+ \partial_{\varkappa(t)}\hat{\psi}\,\hat{K}(\sigma) + \partial_{\xi(t)}\hat{\psi}(\sigma)\,\hat{\Xi}_1(\sigma)\big\}\partial_{T(t)} f\Big)
\end{aligned} \tag{7.36}$$

REFERENCES

1. BEATTY, M.E., On the foundation principles of general classical mechanics, Arch. Rat. Mech. Anal., 24, 264-273, 1967.
2. BERNSTEIN, B., Proof of Caratheodory's local theorem and its global application to thermostatics, J. Math. Phys., 1, 222-224, 1960.
3. BOYLING, J.B., Caratheodory's principle and the existence of global integrating factors, Commun. Math. Phys., 10, 52-68, 1968.
4. BOYLING, J.B., An axiomatic approach to classical thermodynamics, Proc. Royal Soc. London, A329, 35-70, 1972.
5. BORN, M., Kritische Betrachtungen zur traditionellen Darstellung der Thermodynamik, Physik. Zeitschr., 22, 218-224, 249-254, 282-286, 1921.
6. BRIDGMAN, P.W., The thermodynamics of plastic deformation and generalized entropy, Rev. Modern Phys., 22, 56-63, 1950.
7. BUCHDAHL, H.A., A formal treatment of the consequence of the second law of thermodynamics in Caratheodory's formulation, Zeitschrift Phys., 152, 425-439, 1958.
8. BUCHDAHL, H.A., Entropy concept and ordering of states. I, Zeitschrift Phys., 168, 316-321, 1962.

9. BUCHDAHL, H.A. and GREVE, W., Entropy concept and ordering of states. II, Zeitschrift Phys., 168, 386-391, 1962.

10. BUCHDAHL, H.A., The concepts of classical thermodynamics, Cambridge, University Press, 1966.

11. CAMPBELL, J.D. and FERGUSON, W.G., The temperature and strain rate dependence of the shear strength of mild steel, Phil. Mag. 21, 63-82, 1970.

12. CARATHEODORY, C., Untersuchungen über die Grundlagen der Thermodynamik, Math. Annalen, 67, 355-386, 1909.

13. COLEMAN, B.D. and NOLL, W., The thermodynamics of elastic materials with heat conduction and viscosity, Arch. Rat. Mech. Anal., 13, 167-178, 1963.

14. COLEMAN, B.D., Thermodynamics of materials with memory, Arch. Rat. Mech. Anal., 17, 1-46, 1964.

15. COLEMAN, B.D. and GURTIN, M.E., Thermodynamics with internal state variables, J. Chem. Phys., 47, 597-613, 1967.

16. COLEMAN B.D. and MIZEL, V.J., A general theory of dissipation in materials with memory, Arch. Rat. Mech. Anal., 27, 255-274, 1968.

17. COLEMAN, B.D. and OWEN, D.R., On the thermodynamics of materials with memory, Arch. Rat. Mech. Anal., 36, 245-269, 1970.

18. COLEMAN, B.D. and OWEN, D.R., A mathematical foundation for thermodynamics, Arch. Rat. Mech. Anal., 54, 1-104, 1974.

19. COOPER, J.L.B., The foundations of thermodynamics, J. Math. Anal. Appls., 17, 172-193, 1967.

20. De VAULT, G.P. and McLENNAN, J.A., Statistical mechanics of visco-elasticity, Phys. Rev., 137, A724-A730, 1965.

21. DHARAN, C.K.H. and HAUSER, F.E., High-velocity dislocation damping in aluminium, J. Appl. Phys., 44, 1468-1474, 1973.

22. ECKART, C., The thermodynamics of irreversible processes, Phys. Reviews, 58, 267-269, 269-275, 919-924, 1940; 73, 373-382, 1948.

23. FALK, G. and JUNG, H., Axiomatic der Thermodynamik, Handbuch der Physik, III/2, Springer 1959, pp.119-175.

24. FERGUSON, W.G., KUMAR, A. and DORN, J.E., Dislocation damping in aluminium at high strain rates, J. Appl. Phys., 38, 1863-1869, 1967.

25. GILES, R., Mathematical Foundation of Thermodynamic, Pergamon Press, Oxford 1964.

26. GORMAN, J.A., WOOD, D.S. and VREELAND, T., Mobility of dislocations in aluminium, J. Appl. Phys., 40, 833-841, 1969.

27. GREEN, A.E. and RIVLIN, R.S., On Cauchy's equations of motion, ZAMP, 15, 290292, 1964.

28. GREEN, A.E. and NAGHDI, P.M., A general theory of an elastic-plastic continuum, Arch. Rat. Mech. Anal., 18, 251-281, 1965.

29. GREEN, A.E. and NAGHDI, P.M., A thermodynamic development of elastic-plastic continua, Proc. IUTAM Symp., Vienna, June 22-28, 1966, Eds. H. Parkus and L.I. Sedov, Springer 1967, pp.117-131.

30. GREEN, A.E. and NAGHDI, P.M., Some remarks on elastic-plastic deformation at finite strain, Int. J. Engng. Sci., 9, 1219-1229, 1971.

31. GREEN, A.E. and NAGHDI, P.M., Rate-type constitutive equations and elastic-plastic materials, Int. J. Engng. Sci., 11, 725-734, 1973.

32. GREEN, A.E. and NAGHDI, P.M., On thermodynamics and the nature of the second law, Proc. Roy. Soc. London, A.357, 253-270, 1977.

33. GURTIN, M.E., NOLL, W. and WILLIAMS, W.O., On the foundations of thermodynamics, Carnegic-Mellow University Report, 42/68 1968.

34. GURTIN, M.E. and WILLIAMS, W.O., An axiomatic foundations for continuum thermodynamics, Arch. Rat. Mech. Anal., 26, 83-117, 1967.

35. GURTIN, M.E. and WILLIAMS, W.O., On the first law of thermodynamics, Arch. Rat. Mech. Anal., 42, 77-92,1971.

36. GURTIN, M.E., Modern continuum thermodynamics, Mechanics Today, vol. 1, 1972, 168-213.

37. GURTIN, M.E., On the existence of a single temperature in continuum thermodynamics, J. Appl. Math. Physics (ZAMP), 27, 776-779, 1976.

38. GURTIN, M.E., Thermodynamics and stability, Arch. Rat. Mech. Anal., 59, 63-96, 1975.

39. HAKEN, H., Cooperative phenomena in systems far from thermal equilibrium and in nonphysical systems, Rev. Mod. Phys., 47, 67-121, 1975.

40. HAUSER, F.E., SIMMONS, J.A. and DORN, J.E., Strain rate effects in plastic wave propagation, Response of Metals to High Velocity Deformation, Wiley (Interscience) New York, 1961, 93-114.

41. HUTTER, K., The functions of thermodynamics, its basic postulates and implications. A review of modern thermodynamics, Acta Mechanica, 27, 1-54, 1977.

42. KARNES, C.H. and RIPPERGER, E.A., Strain rate effects in cold worked high-purity aluminium, J. Mech. Phys. Solids, 14, 75-88, 1966.

43. KELLY, J.L., General Topology, Van NOSTRAND, New York 1955.

44. KOSIŃSKI, W. and WOJNO, W., Remarks on internal variable and history descriptions of material, Arch. Mech., 25, 709-713, 1973.

45. KOSIŃSKI, W. and PERZYNA, P., The unique material structure, Bull. Acad. Polon. Sci., Ser. Sci. Techn., 21, 655-662, 1973.

46. KUMAR, A., HAUSER, F.E. and DORN, J.E., Viscous drag on dislocations in aluminium at high strain rates, Acta Metall., 16, 1189-1197, 1968.

47. KUMAR, A. and KUMBLE, R.G., Viscous drag on dislocations at high strain rates in copper, J. Appl. Phys., 40, 3475-3480, 1969.

48. LANDSBERG, P.T., Foundations of thermodynamics, Rev. Modern Phys., 28, 363-392, 1956.

49. LANDSBERG, P.T., Main ideas in the axiomatics of thermodynamics, Pure and Applied Chemistry, 22, 215-227, 1970.

50. LAWSON, J.E. and NICHOLAS, T., The dynamic mechanical behaviour of titanium in shear, J. Mech. Phys. Solids, 20, 65-76, 1972.

51. LEE, E.H. and LIU, D.T., Finite-strain elastic-plastic theory with application to plane-wave analysis, J. Appl. Phys., 38, 19-27, 1967.

52. LEE, E.H., Elastic-plastic deformation at finite strains, J. Appl. Mech., 36, 1-6, 1969.

53. LEITMAN, M.J. and MIZEL, V.J., On fading memory spaces and hereditary integral equations, Arch. Rat. Mech. Anal., 48, 1-50, 1972.

54. MARIS, H.J., Phonon viscosity, Phys. Rev., 188, 1303-1307, 1969.

55. MASON, W.P., Phonon viscosity and its effect on acoustic wave attenuation and dislocation motion, J. Acoustical Soc. Amer., 32, 458-472, 1960.

56. NABARRO, F.R.N., Theory of crystal dislocation, Oxford, 1967.

57. NAGHDI, P.M. and TRAPP, J.A., Restrictions on constitutive equations of finitely deformed elastic-plastic materials, Quart. J. Mech. Appl. Math., 28, 25-46, 1975.

58. NAGHDI, P.M. and TRAPP, J.A., On the nature of normality of plastic strain rate and convexity of yield surfaces in plasticity, J. Appl. Mech., 42, 61-66, 1975.

59. NOLL, W., The foundations of classical mechanics in the light of recent advances in continuum mechanics, Proc. Int. Symp. Axiomatic Method Spec. Ref. Geom. Phys., 1957-1958, pp. 226-281.

60. NOLL, W., A new mathematical theory of simple materials, Arch. Rat. Mech. Anal., 48, 1-50, 1972.

61. NOLL, W., Lectures on the foundations of continuum

mechanics and thermodynamics, Arch. Rat. Mech. Anal., 52, 62-92, 1973.

62. OLSZAK, W. and PERZYNA, P., Stationary and nonstationary viscoplasticity, in Inelastic Behavior of Solids, Battelle Inst. Material Science Colloquia, Columbus, Sept. 15-19, 1969, McGraw-Hill 1970, pp. 53-75.

63. PERZYNA, P., The constitutive equations for rate sensitive plastic materials, Quart. Appl. Maths., 20, 321-332, 1963.

64. PERZYNA, P., The constitutive equations for work-hardening and rate sensitive plastic materials, Proc. Vibration Problems, 4, 281-290, 1963.

65. PERZYNA, P., Fundamental problems of viscoplasticity, Advances in Appl. Mech., 9, 243-377, 1966.

66. PERZYNA, P. and WOJNO, W., On the constitutive equations of elastic/viscoplastic materials at finite strain, Arch. Mech. Stos., 18, 85-100, 1968.

67. PERZYNA, P. and WOJNO, W., Thermodynamics of a rate sensitive plastic material, Arch. Mech. Stos., 20, 499-511, 1968.

68. PERZYNA, P., Thermodynamic theory of viscoplasticity, Advances Appl. Mech., 11, 313-354, 1971 (Academic Press, New York).

69. PERZYNA, P., Thermodynamics of rheological materials with internal changes, J. Mecanique, 10, 391-408,1971.

70. PERZYNA, P., A gradient theory of rheological materials with internal structural changes, Arch. Mech., 23, 845-850, 1971.

71. PERZYNA, P., Thermodynamic theory of rheological materials with internal structural changes, Symposium franco-polonais, Problemes de la Rheologie, Varsovie 1971, PWN, Varsovie 1973, pp. 277-306.

72. PERZYNA, P., Physical theory of viscoplasticity, Bull. Acad. Polon. Sci., Ser. Sci. Techn., 21, 121-139, 1973.

73. PERZYNA, P. and SAWCZUK, A., Problems of thermoplasticity, Nuclear Eng. Design, 24, 1-55, 1973.

74. PERZYNA, P., Plasticity of irradiated materials, 2nd Inter. Conf. SMiRT, Berlin, September 1973, vol.L, 3/2, 1-17.

75. PERZYNA, P. and KOSIŃSKI, W., A mathematical theory of materials, Bull. Acad. Polon. Sci., Ser. Sci. Techn., 21, 647-654, 1973.

76. PERZYNA, P., Internal variable description of plasticity, in "Problems of Plasticity", Warsaw, August 30 - September 2, 1972, ed. A. SAWCZUK, Noordhoff, pp.145-170, Leyden 1974.

77. PERZYNA, P., The constitutive equations describing thermo-mechanical behaviour of materials at high rates of strain, Inter. Symp.,Oxford, April 2-4,1974, Institute of Physics, Conf.Ser.No.21, pp.138-153.

78. PERZYNA, P., Theorie Physique de la Visco-plasticite, Acad. Pol. Sci., Centre Scien. à Paris, 104, PWN, 1974, pp. 1-25.

79. PERZYNA, P., Thermodynamics of a unique material structure, Arch. Mech., 27, 791-806. 1975.

80. PERZYNA, P., On material isomorphism in description of dynamic plasticity, Arch. Mech., 27, 1975, 473-484.

81. PERZYNA, P., Description of thermo-mechanical behaviour of irradiated materials, Symposium franco-polonais, Problemes de Rheologie et des sols, Nice 1974; PWN, 1977 pp. 375-389.

82. PERZYNA, P., Effects of irradiation on thermo-mechanical behaviour of inelastic materials, 4th Inter.Conf. SMiRT, San Francisco, August 1977, vol.L, 1/11, pp.1-15.

83. PERZYNA, P., Coupling of dissipative mechanisms of viscoplastic flow, Arch. Mechanics, 29, 607-624, 1977.

84. PERZYNA, P. and WOJNO, W., Unified constitutive equations for elastic-viscoplastic material, Bull. Acad. Polon. Scien., Ser. scien. tech., 24, 85-94, 1976.

85. PLANCK, M., Uber das Prinzip der Vermehrung der Entropie, Wied. Ann., 30, 562-582; 31, 189-203; 32, 462-503, 1887.

86. PLANCK, M., Vorlesungen über Thermodynamik, Elfte Auflage, Berlin 1964 (Erste Auflage 1897).

87. RASTALL, P., Classical thermodynamics simplified, J. Math. Phys., 11, 2955-2965, 1970.
88. ROSENFIELD, A.R. and HAHN, G.T., Numerical description of the ambient low-temperature and high strain rate flow and fracture behavior of plain carbon steel, Tran. ASM 59, 962-980, 1966.
89. SEEGER, A., The generation of lattice defects by moving dislocations and its application to the temperature dependence of the flow-stress of f.c.c. crystals, Phil. Mag., 46, 1194-1217, 1955.
90. SEEGER, A., Theory of crystal defects, Academic Press, New York 1966, pp.37-56.
91. TEODOSIUM, C. and SIDOROFF, F., A theory of finite elasto-visco-plasticity of single crystals, Int. J. Engng. Scien., 14, 165-176, 1976.
92. TRUESDELL, C. and NOLL, W., The non-linear field theories of mechanics, Handbuch derPhysik, vol.III/3, 1965, pp.1-602, Springer, Berlin.
93. TRUESDELL, C., Rational Thermodynamics, McGRAW-HILL, New York 1969.
94. VALANIS, K.C., Unified theory of thermomechanical behaviour of viscoelastic materials, Symp. Mech. Behav. Mater. Dynam. Loads, San Antonio 1967, Spinger, New York 1968, pp.343-364.

95. WANG, C.C. and BOWEN, R.M., On the thermodynamics of nonlinear materials with quasi-elastic response, Arch. Rat. Mech. Anal., 22, 79-99, 1966.

96. WILLIAMS, W.O., On internal interactions and the concept of thermal isolation, Arch. Rat. Mech. Anal., 34, 245-258, 1969.

97. WILLIAMS, W.O., Thermodynamics of rigid continua, Arch. Rat. Mech. Anal., 36, 270-284, 1970.

98. ZEMANSKY, M.W., Heat and thermodynamics, Fifth Ed: McGRAW-HILL, Tokyo 1968.

99. ZWANZIG, R., Nonlinear generalized Langevin equations, J. Stat. Phys., 9, 215-220, 1973.

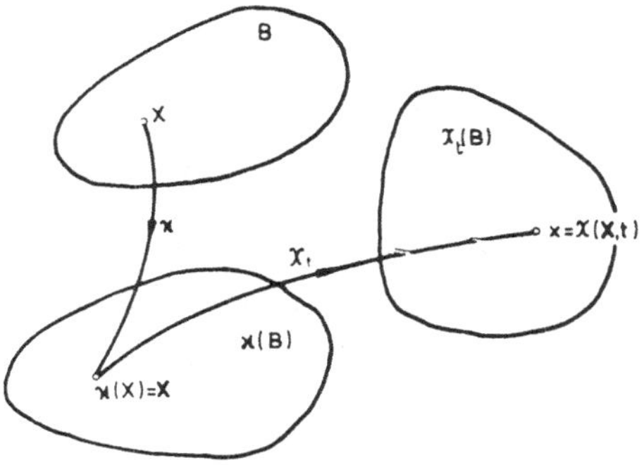

Fig. 1

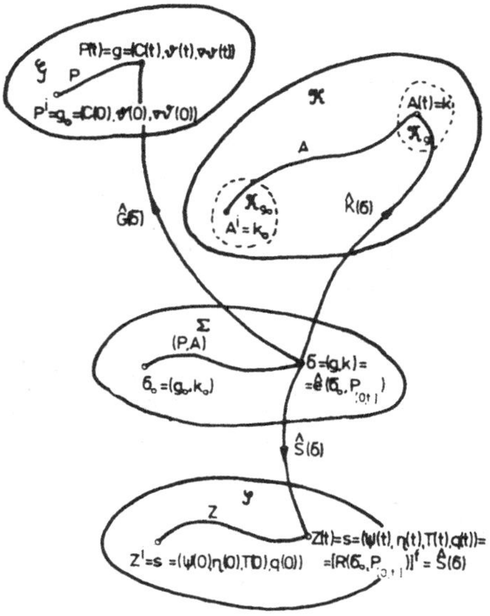

Fig. 2

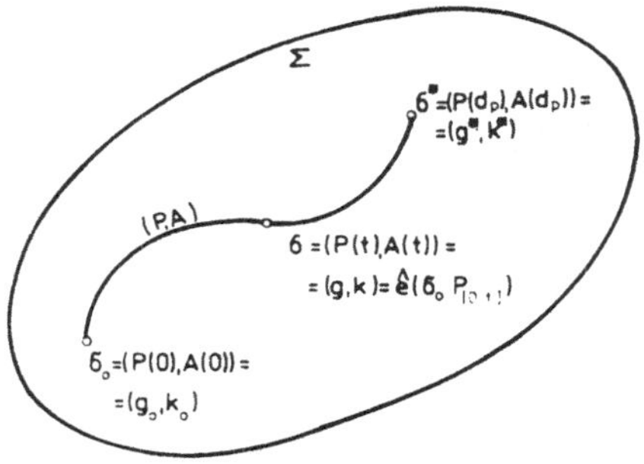

Fig. 3

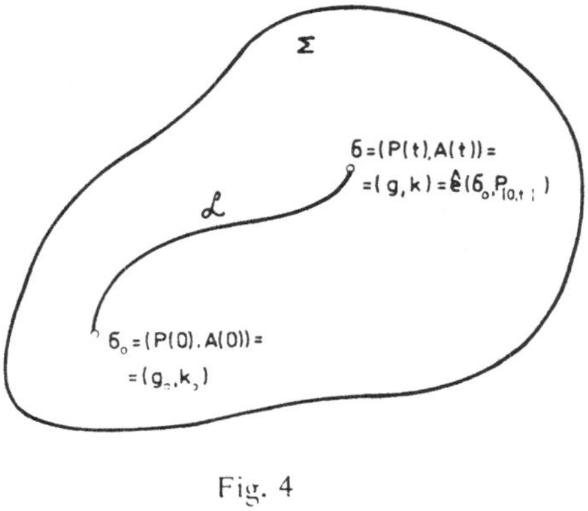

Fig. 4

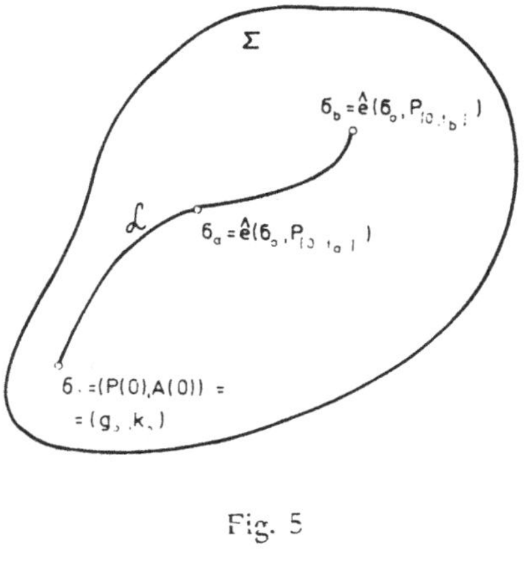

Fig. 5

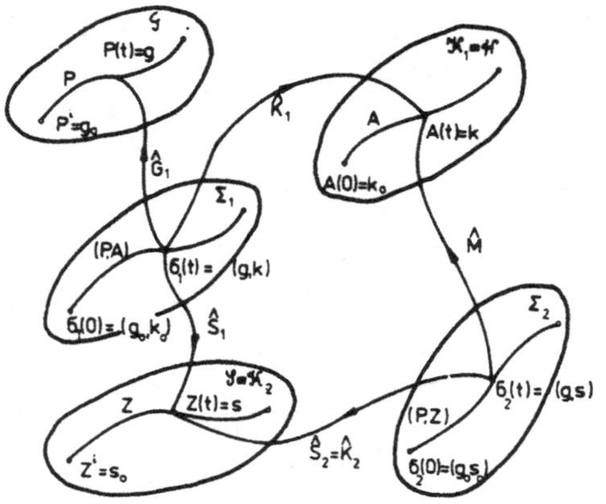

Fig. 6

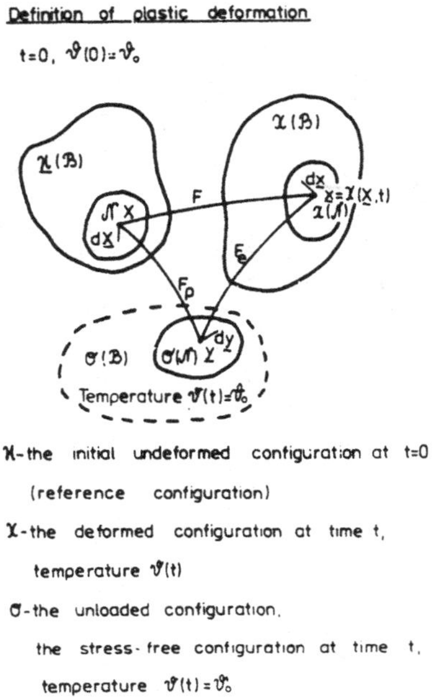
Definition of plastic deformation
t=0, ϑ(0)=ϑ₀
Temperature ϑ(t)=ϑ₀
ϰ - the initial undeformed configuration at t=0
(reference configuration)
χ - the deformed configuration at time t,
temperature ϑ(t)
σ - the unloaded configuration,
the stress-free configuration at time t,
temperature ϑ(t) = ϑ₀

Fig. 7

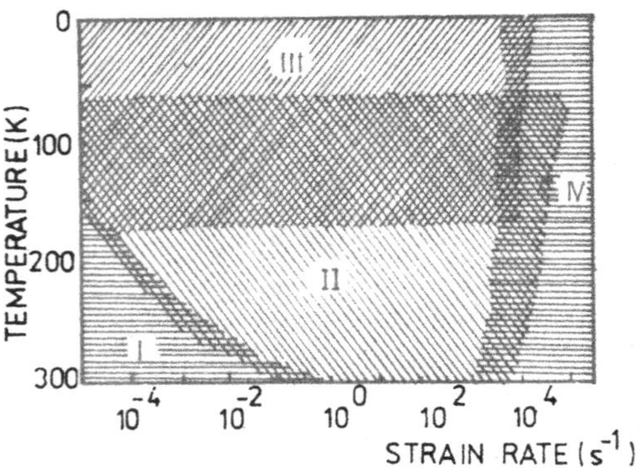
TEMPERATURE (K)
STRAIN RATE (s⁻¹)
III
IV
II
I

Fig. 8

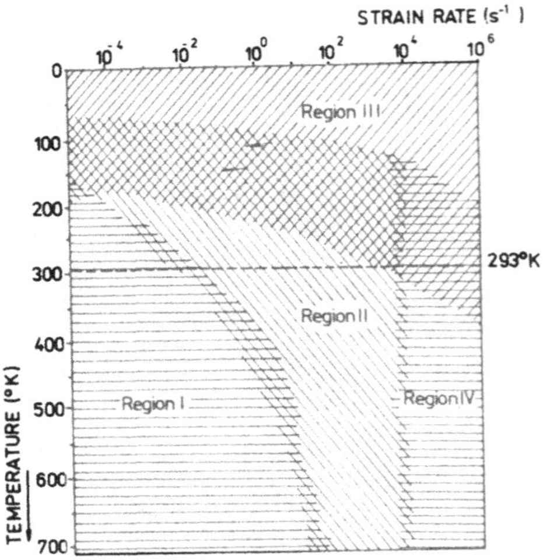
STRAIN RATE (s⁻¹)
TEMPERATURE (°K)
293°K
Region III
Region II
Region I
Region IV

Fig. 8a

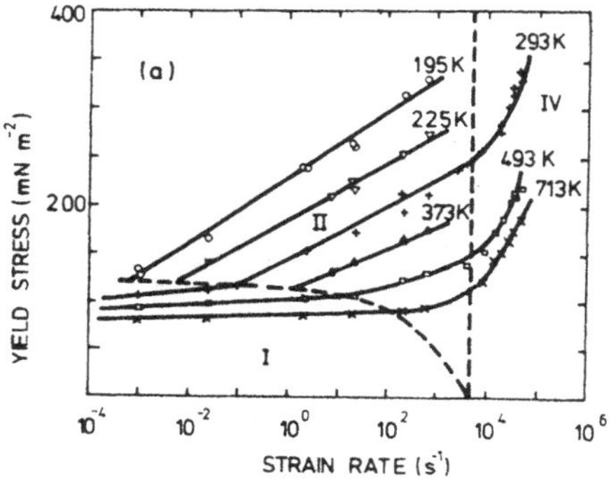

Fig. 9a

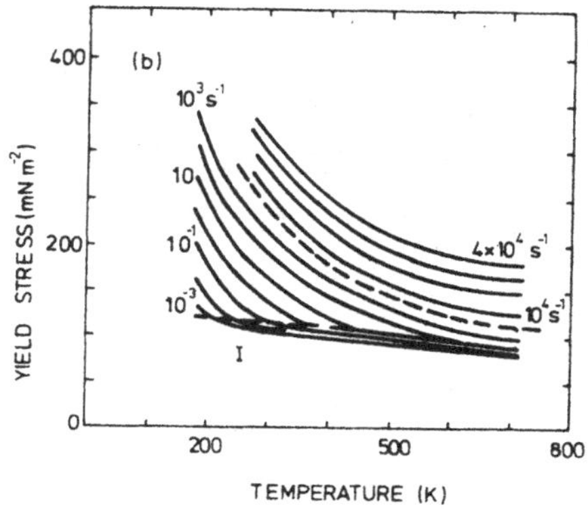

Fig. 9b

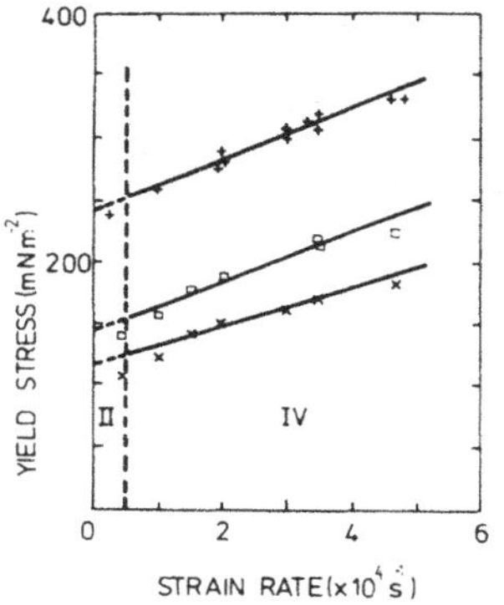

Fig. 9c

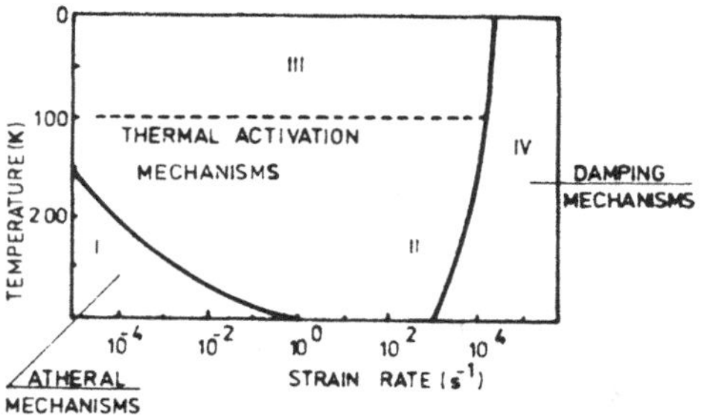

Fig. 10

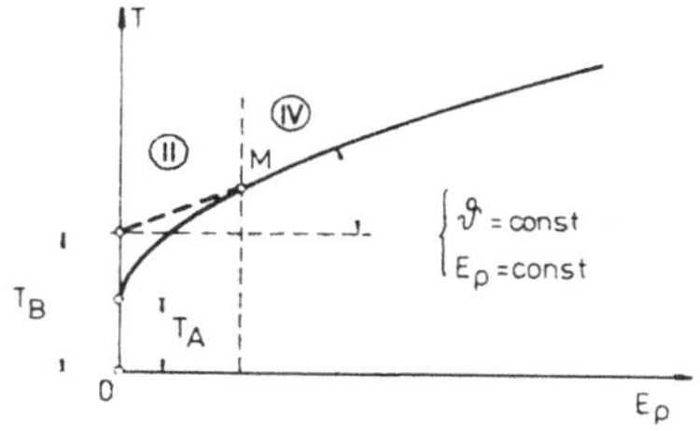

Fig. 11

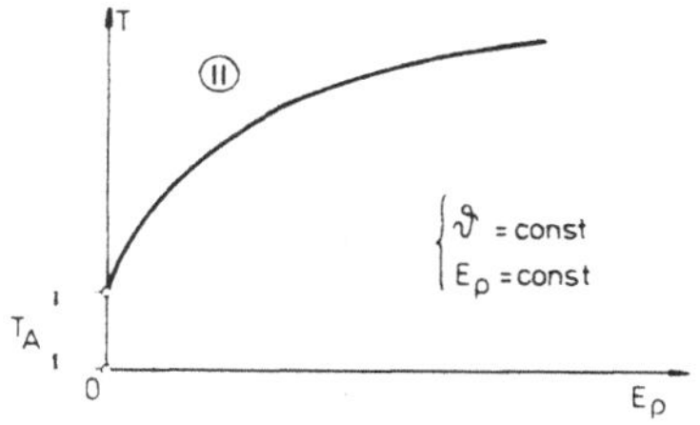

Fig. 12

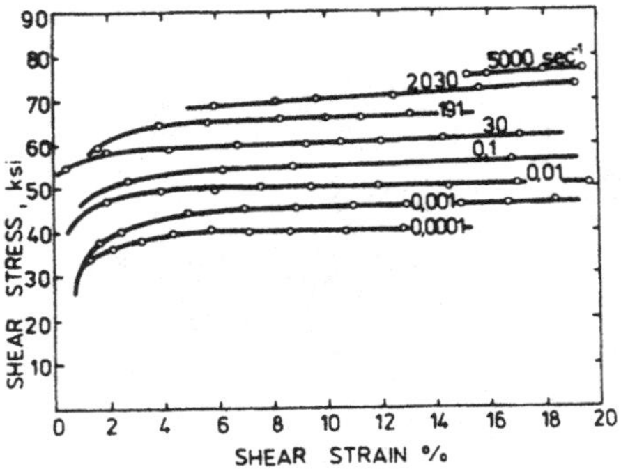

Fig. 13

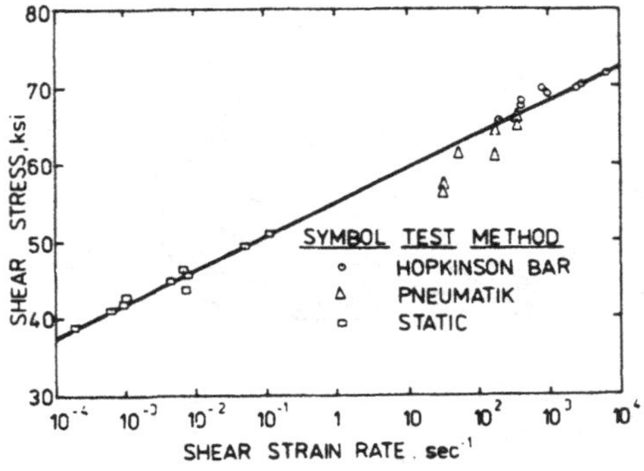

Fig. 14

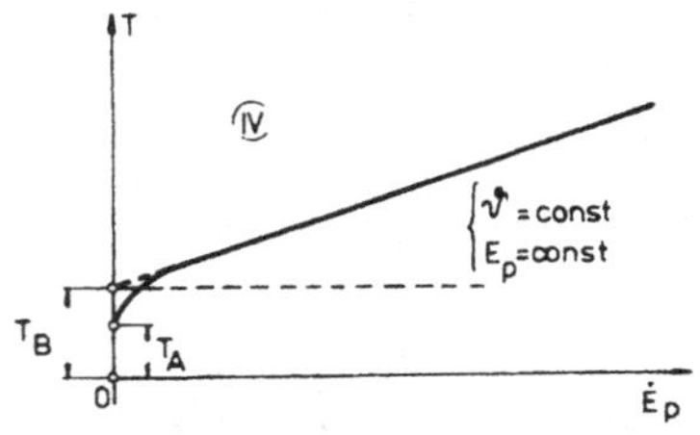

Fig. 15

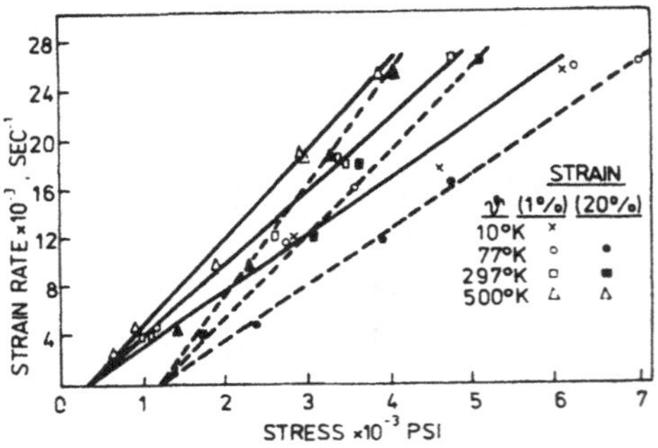

Fig. 16

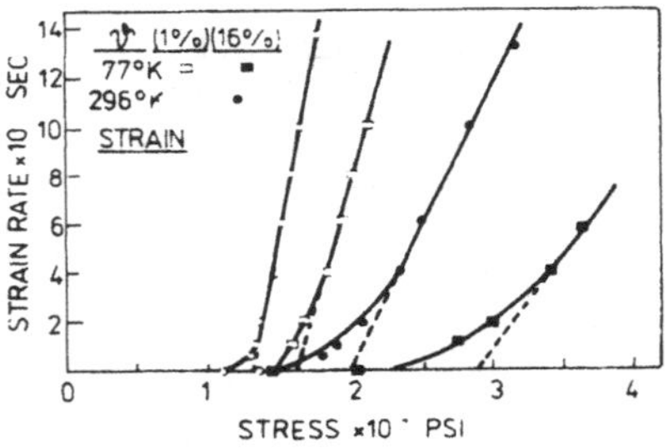

Fig. 17

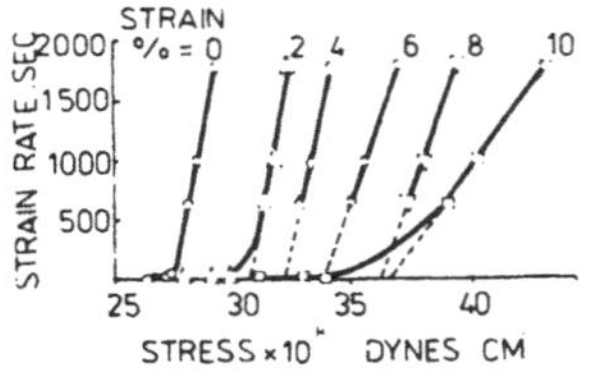

Fig. 18

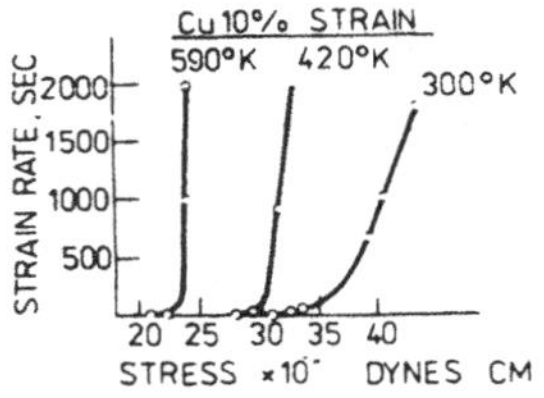

Fig. 19

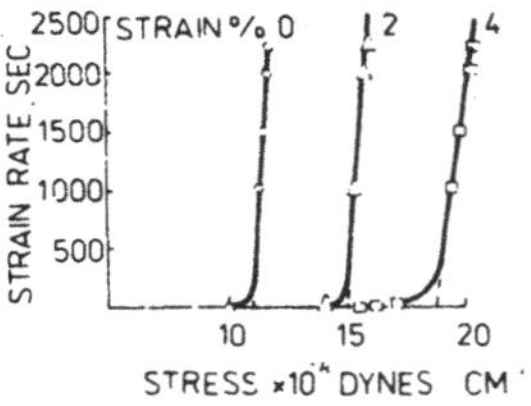

Fig. 20

VARIATIONAL PRINCIPLES IN THERMOMECHANICS

G. LEBON

LIEGE UNIVERSITY
DEPARTMENT OF MECHANICS
B-4000 SART TILMAN, LIEGE

INTRODUCTION

Variational principles have always played a central role in continuum mechanics. This interest can be justified by the following reasons:

Quoting Euler, "since the construction of the universe is the most perfect possible, being the handy-work of an all-wise Maker, nothing can be met in the world in which some maximum or minimum property is not displayed. As a consequence, no doubt that all the effects in the world can be derived by methods of minima or maxima."

It is indeed very attractive to believe that a whole class of phenomena is governed by one single law of minimum or maximum. It is very tempting to identify this single mathematical equation with the mathematical projection of the all-wise Maker of Euler. However, Euler's enthusiasm has to be moderated because as will be seen later, most of the physical phenomena cannot be interpretated in terms of minima or maxima. Nevertheless, when it exists, a variational formulation presents the advantage of concision : a single variational equation stands for a whole set of evolution equations,initial and boundary conditions.

From a more practical point of view, variational formulations provide specific methods for solving the differential equations governing the physical processes. The task of a physicist or an engineer is not only to derive equations but also to solve them. In most practical problems, exact solutions are not available so that it is of crucial importance to derive approximate solutions. Such solutions may be readily obtained when the equations can be replaced by a variational principle.

Moreover, besides their power of synthesis and their practical interest, the variational principles are also helpful for the physical interpretation of the processes. In many problems, the quantity submitted to variation possesses a physical meaning indicating which quantity is the fundamental element in the evolution of a process.

The purpose of these lectures is to provide a rather large survey of the most significant variational principles in the field of classical mechanics, elasticity and thermoelasticity. Special attention has been drawn to points which are generally omitted or erroneously presented in the literature. Due to lack of time, some basic problems like those related to stability and plasticity are not treated.

The lectures have been divided in four chapters and a general appendix.

The first chapter is devoted to a brief analysis of the calculus of variation. The classical variational methods are examined and their analogy with Galerkin's method is emphasized. The conditions under which a differential equation can be replaced by a variational principle are discussed. The concepts of quasi-variational and restricted principles are also introduced. The link between the variational methods and the finite element technique is outlined in an appendix.

The variational principles of rational mechanics and elasticity are presented in chapter two. In particular, it is seen that the law of varying action of classical mechanics can be used to provide approximate solutions to Newton's law. It is shown how to relax the constraint that the admissible function must be identical to the exact one at a given time t_1, different from the initial time t_o. The principles of classical mechanics are then extended to elastic bodies. The static and the dynamic problems, as well infinitesimally small as finite deformations, are considered.

Chapter three gathers the most significant principles describing steady and unsteady heat conduction in a rigid body. The heat conduction problems can be divided into two classes, accordingly that there exists a variational principle in the usual sense or a variational principle in the extended sense. Among this last category pertain the non-linear unsteady problems. The latter requires the formulation of a new class of variational principles, the so-called quasi-variational and restricted principles. Biot's, Vujanovic's, Glansdorff-Prigogine's and Lebon-Lambermont's formulations are presented, compared and discussed.

The last chapter concerns coupled thermoelasticity. Uncoupled thermoelasticity does not require a special treatment because it can directly be derived from the results of chapters II and III. When the coupling between thermal and mechanical effects is appreciable, some variational principles have been proposed. They cover mainly linear thermoelasticity and refer either to the class of exact variational principles like Iesan and Nickell-Sackman's formulations or to the class of quasi-variational principles like the Biot or the Lebon-Lambermont criteria.

In a general appendix, the problems of convergence and estimation of the error are briefly discussed.

To make the reader more familiar with the use of variational principles, each section contains illustrative and numerical examples. Most of them have been chosen for their pedagogic interest.

Although our objective was to be as complete as possible, we are fully aware that the list of principles presented in these notes is far from being exhaustive.

Acknowledgements. It is a pleasure to thank Prof. Olszak, Rector of the C.I.S.M., and Prof. W. Nowacki, Chairman of the Academy of Sciences of Poland, for their kind invitation to present these lectures at the C.I.S.M. Thanks are also due to Dr. J. Lambermont for fruitful discussions and to F. Hanesse, secretary, for her ability in decoding the manuscript and her careful typing of the final version.

I. BASIC CONCEPTS OF THE CALCULUS OF VARIATICN

1. Basic definitions

1.1. The variation operator δ. Let $u(x)$ be a continuously differentiable function of the independent real variable x defined on the interval $x_2 \leq x \leq x_1$. Denote by u_x its first order derivative. Let

$$F(u, u_x ; x) \tag{1.1}$$

be a real-valued function of x, u and u_x and $I(u)$ a functional on the set of all functions $u(x)$ defined by

$$I(u) = \int_{x_1}^{x_2} F(u, u_x ; x)\,dx \tag{1.2}$$

The class of functions which satisfy the above continuity restrictions are called the admissible functions. Assume that they meet the following boundary conditions:

$$u(x_1) = u_1 \quad \text{and} \quad u(x_2) = u_2 \tag{1.3}$$

The basic problem of the calculus of variation is to find the function $u(x)$ which, among all the admissible functions satisfying (1.3), makes $I(u)$ extremum, for instance a minimum.

Suppose that $u(x)$ is the function which corresponds to the smallest value for I and that $u^*(x)$ is a function which is infinitesimally different from $u(x)$. The variation δu of the function u is defined by

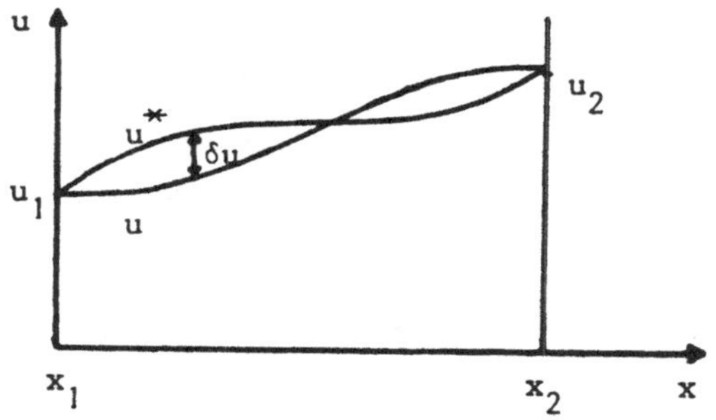

Fig. 1.1

$$\delta u(x) = u^*(x) - u(x) = \varepsilon\eta(x) \qquad (1.4)$$

ε is an arbitrary small quantity and $\eta(x)$ and arbitrary function which vanishes at x_1 and x_2 .

By definition one has

$$\delta x = 0$$

The δ operator commutes with the derivation and the integration operators and follows the same rules than the derivation operator.

The first variation of the functional I(u) is defined by

$$\delta I = \varepsilon \lim_{\varepsilon \to 0} \frac{I(u+\varepsilon\eta) - I(u)}{\varepsilon} \equiv \varepsilon \left(\frac{dI}{d\varepsilon}\right)_{\varepsilon=0} \qquad (1.6)$$

while the second variation is given by

$$\delta^2 I = \delta(\delta I) \equiv \frac{1}{2}\varepsilon^2 \left(\frac{d^2 I}{d\varepsilon^2}\right)_{\varepsilon=0} \qquad (1.7)$$

If

$$\delta I = 0 \qquad (1.8)$$

the functional is said to be *stationary*. The vanishing of the first variation of I is the *necessary* condition for I to attain an extremum, either a minimum or a maximum. An expression like (1.8) is named a *variational principle.*

1.2. The Euler-Lagrange equations. Let us determine the equation which must be satisfied by the function u(x) to make I extremum. Therefore, we calculate δI and set the result equal to zero. We have

$$\delta I = \int_{x_1}^{x_2} \delta F(u,u_x,x)dx = \int_{x_1}^{x_2} \left(\frac{\partial F}{\partial u}\,\delta u + \frac{\partial F}{\partial u_x}\,\delta u_x\right)\ dx=0 \tag{1.9}$$

and after integration of the last term by parts

$$\delta I = \int_{x_1}^{x_2} \left(\frac{\partial F}{\partial u} - \frac{d}{dx}\frac{\partial F}{\partial u_x}\right)\delta u dx + \frac{\partial F}{\partial u_x}\,\delta u\Big|_{x_1}^{x_2} = 0 \tag{1.10}$$

The last term is clearly equal to zero since u is fixed at the boundaries ($\delta u = 0$ at x_1 and x_2). Further results are derived by making use of the so-called *fundamental lemma*.

Let M(x) be a continuous function in the interval $x_1 \leq x \leq x_2$. The lemma states that if the relation

$$\int_{x_1}^{x_2} M(x)\ \delta u(x)dx = 0$$

holds for all possible variations $\delta u(x)$, such that $\delta u(x_1) = \delta(x_2) = 0$, then

$$M(x) = 0$$

Accordingly, the necessary and sufficient condition for the vanishing of I is

$$\frac{\partial F}{\partial u} - \frac{d}{dx}\frac{\partial F}{\partial u_x} = 0 \tag{1.11}$$

This equation is called the *Euler-Lagrange* equation.

When F contains higher derivatives u_{xx}, u_{xxx},...

$$F = F(u, u_x, u_{xx}, \dots ; x)$$

the Euler-Lagrange equation takes the form

$$\frac{\partial F}{\partial u} - \frac{d}{dx}\frac{\partial F}{\partial u_x} + \frac{d^2}{dx^2}\frac{\partial F}{\partial u_{xx}} + \dots = 0 \tag{1.12}$$

If F is function of the n functions u^1, u^2,... u^n

$$F = F(u_x^1, u_x^1, u^2, u_x^2, \dots u^n, u_x^n ; x)$$

the Euler-Lagrange equations corresponding to arbitrary variations δu_1,... δu_n are given by

$$\frac{\partial F}{\partial u^i} - \frac{d}{dx}\frac{\partial F}{\partial u_x^i} = 0 \qquad i = 1, \dots n \tag{1.13}$$

In the general case that F depends not only on n functions u^1,...u^n but also on m independent variables x_1, x_2,... x_m, the Euler-Lagrange equations are :

$$\frac{\partial F}{\partial u^i} - \sum_{k=1}^{m} \frac{d}{dx_k}\frac{\partial F}{\partial u_{x_k}^i} = 0 \qquad i = 1, \dots n \tag{1.14}$$

The Euler-Lagrange equations represent necessary conditions for I to be extremum. If the second variation of I is positive (respectively negative), I is a minimum (respectively a maximum) :

$$\delta^2 I \geq 0 : \text{necessary condition for a minimum} \tag{1.15}$$

$$\delta^2 I \leq 0 : \text{necessary condition for a maximum} \tag{1.16}$$

If F is positive (respectively negative) definite, then the stationary value of I is a minimum (respectively a maximum). Observe that the above inequalities are necessary but not generally sufficient conditions. As shown by Gelfand and Fomin[1], a sufficient condition for I to be a minimum (respectively maximum) is that there exists a positive number k such that for all δu, in a normed space, one has

$$\delta^2 I \geq k\|\delta u\|^2 \quad : \text{sufficient condition for a minimum}$$

$$\delta^2 I \leq k\|\delta u\|^2 \quad : \text{sufficient condition for a maximum}$$

1.3. Constrained extremals. Suppose that the function u(x) is subject to some constraints. In particular, search for the function u(x) which makes stationary.

$$I = \int_{x_1}^{x_2} F(u,u_x;x)dx \tag{1.17}$$

with the auxiliary condition that u(x) verify the relation

$$J \equiv \int_{x_1}^{x_2} G(u,u_x;x)dx = J_1 \tag{1.18}$$

where J_1 is a prescribed value of J.

This problem is easily handled with the Lagrange multipliers. Construct the new functional

$$I^{*} = I + \lambda J = \int_{x_1}^{x_2} (F + \lambda G)dx \tag{1.19}$$

where λ is a constant Lagrange multiplier. The solution of the problem is obtained by deriving the Euler-Lagrange equation corresponding to the variational equation

$$\delta I^{*} = 0 \qquad (1.20)$$

It is given by

$$\frac{\partial}{\partial u}(F + \lambda G) - \frac{d}{dx}\frac{\partial}{\partial u_x}(F + \lambda G) = 0 \qquad (1.21)$$

The equations (1.18) and (1.21) provide a set of two equations for the two unknowns $u(x)$ and λ .

If the auxiliary constraint takes the form of a differential equation

$$G(u,u_x;x) = 0 \qquad (1.22)$$

the above treatment remains valid at the condition to construct the functional

$$I^{*} = \int_{x_1}^{x_2} \{F + \lambda(x)G\}dx \qquad (1.23)$$

wherein now, contrary to (1.19), $\lambda(x)$ is no longer constant but a function of x.

1.4. Natural boundary conditions. Up to now, it was assumed that the function u was prescribed at the boundaries, so that its variation vanishes at x_1 and x_2 . In many problems, it may happen that the function is only specified on a portion of the surface or that other kinds of boundary conditions are imposed. Suppose that the admissible class of functions does not satisfy (1.3). Then in eq. (1.10), the boundary terms do not vanish since $\delta u \neq 0$ at x_1 and x_2 :

$$\delta I = \int_{x_1}^{x_2} \left(\frac{\partial F}{\partial u} - \frac{d}{dx}\frac{\partial F}{\partial u_x}\right) \delta u dx + \frac{\partial F}{\partial u_x}\delta u \Big|_{x_1}^{x_2} = 0$$

As this relation must be satisfied for arbitrary variations δu, it is necessary that both terms in the r.h.s. vanish. This implies that in addition to the Euler-Lagrange equation, the following conditions must be fulfilled at the boundaries :

$$\frac{\partial F}{\partial u_x} = 0 \quad \text{at} \quad x = x_1 \quad \text{and} \quad x = x_2 \tag{1.24}$$

These equations are called *natural boundary conditions*. It is important to distinguish them with the so-called essential boundary conditions like

$$u = u_1 \quad \text{at} \quad x = x_1 \quad , \quad u = u_2 \quad \text{at} \quad x = x_2 \tag{1.25}$$

The natural boundary conditions are not imposed a priori but arise automatically from the stationary condition expressing that δI is equal to zero.

It has been shown[1] that if the Euler-Lagrange equation is of order 2k, the essential boundary conditions are expressed by relations between the function itself and its derivatives, up to the order k-1. The natural boundary conditions involve the derivatives from the k to the 2k-1 order. If for instance, the Euler-Lagrange equation is the Laplace equation

$$d^2u/dx^2 = 0$$

then the condition

$$u = g(x) \qquad \text{at } x_1 \text{ (Dirichlet condition)}$$

is essential while the condition

$$du/dx = f(u) \quad \text{at } x_2 \text{ (Neumann condition)}$$

is natural.

1.5. Dual variational principles. Up to now, one was concerned with the problem of finding a function u(x) which renders the functional

$$I = \int_{x_1}^{x_2} F(u,v;x)dx \tag{1.26}$$

stationary, with the boundary conditions

$$u(x_1) = u_1 \qquad u(x_2) = u_2 \tag{1.27}$$

v being defined by

$$v = du/dx \tag{1.28}$$

The dual problem consists of formulating another variational equation with (1.27) and (1.28) as Euler-Lagrange equations.

The construction of this principle proceeds in three steps.

First step. Solve the above problem (1.26)-(1.28) by the introduction of Lagrange multipliers ; this procedure yields the following functional :

$$J(u,v,\lambda) = \int_{x_1}^{x_2} \left(F + \lambda(x)\left(\frac{du}{dx} - v\right)\right)dx + \eta\left(u(x_1)-u_1\right) + \zeta\left(u(x_2)-u_2\right) \tag{1.29}$$

The Euler-Lagrange equations corresponding to arbitrary variations of u, v and λ are expressed by

$$\frac{\partial F}{\partial u} - \frac{d\lambda}{dx} = 0 \tag{1.30}$$

$$\frac{\partial F}{\partial v} - \lambda = 0 \tag{1.31}$$

$$\frac{du}{dx} - v = 0 \tag{1.32}$$

The natural boundary conditions are

$$\lambda(x_2) + \zeta = 0 \tag{1.33}$$

$$-\lambda(x_1) + \eta = 0 \tag{1.34}$$

Second step. Integrate by parts the second term in (1.29), and use (1.33) and (1.34) :

$$J(u,v,\lambda) = \int_{x_1}^{x_2} (F-\lambda v - \frac{d\lambda}{dx} u)dx + \lambda(x_2)u_2 - \lambda(x_1)u_1 \tag{1.35}$$

By putting

$$p = \frac{\partial F}{\partial v} (\equiv\lambda), \qquad p' = \frac{\partial\lambda}{\partial x} = \frac{\partial F}{\partial u} \qquad (1.36),(1.37)$$

and constructing the Legendre transform of F(u,v)

$$F(p,p') = F(u,v) - pv - p'u \tag{1.38}$$

(1.35) writes as

$$J(p) = \int_{x_1}^{x_2} F(p,p')dx - p(x_1)u_1 + p(x_2)u_2 \tag{1.39}$$

Third step. Make stationary J(p), namely :

$$\delta J(p) = 0 \tag{1.40}$$

The Euler-Lagrange equation is

$$\frac{\partial}{\partial p} F\left(p, \frac{dp}{dx}\right) - \frac{d}{dx} \frac{\partial F\left(p, \frac{dp}{dx}\right)}{\partial (\frac{\partial p}{dx})} = 0 \tag{1.41}$$

while the natural boundary conditions are given by

$$\frac{\partial F(p, dp/dx)}{\partial (dp/dx)} + u_1 = 0 \qquad \text{at} \qquad x = x_1 \tag{1.42}$$

and an analogous equation at x_2.

By differentiation of the Legendre transform (1.38), one obtains

$$dF(p,p') = - vdp - udp' \tag{1.43}$$

from which follows that (1.41) can be given the form

$$- v + \frac{du}{dx} = 0 \tag{1.44}$$

In terms of u and v, the boundary conditions (1.42) read like

$$- u(x_1) + u_1 = 0 \qquad\qquad - u(x_2) + u_2 = 0 \tag{1.45}$$

The set of equations (1.44) and (1.45) are the required equations which must be satisfied in order that (1.40) holds. Dual principles are very useful for providing upper and lower bounds on the functional to be varied.

As *illustration*, consider the problem of the stationary heat conduction in an infinite wall of thickness d. The equations governing this one-dimensional problem are

$$\frac{\partial}{\partial x}\left(k(x)\,\frac{\partial T}{\partial x}\right) = 0 \qquad\qquad 0 < x < d \tag{1.46}$$

$$T = T^{*} \qquad\qquad x = 0 \tag{1.47}$$

$$- k\,\frac{\partial T}{\partial x} = q^{*} \qquad\qquad x = d \tag{1.48}$$

T is the temperature, k the heat conductivity while T^* and q^* are prescribed values of the temperature and the heat flux respectively. The variational principle associated to the above set of equations is

$$\delta I(T) = 0 \tag{1.49}$$

with

$$I(T) = \int_o^d \frac{1}{2} k(x) \left(\frac{dT}{dx}\right)^2 dx + q^* T(d) \tag{1.50}$$

Establish now the dual principle

$$\delta J(q) = 0 \tag{1.51}$$

whose Euler-Lagrange equations is

$$q = - k \frac{\partial T}{\partial x} \tag{1.52}$$

with (1.47) and (1.48) as natural boundary conditions.

Accordingly to the general procedure just developed, one has

$$\delta J(q) = \delta \int_o^d \left(- \frac{1}{2k} q\, q + T \frac{q}{x}\right) dx - \delta q(0) T^* = 0$$

from which it is directly seen that one recovers (1.52) and the boundary condition (1.47) together with

$$\delta q = 0 \quad \text{at} \quad x = d$$

2. Variational methods of approximation

So far, one has derived differential equations which have to be satisfied if some functionals are to be extremum. The objective of this section is to show how the variational formulation can be used to provide approximate solutions to the corresponding Euler-Lagrange equations. The advantages of these methods are their simplicity and their powerfulness.

One will successively examine the methods of Rayleigh-Ritz, Kantorovitch and the method of weighted residuals.[2-5] Although the latter is not subordonated to the construction of a stationary integral, it uses techniques analogous to variational methods.

2.1. The Rayleigh-Ritz method. Let

$$D\{u(x)\} = 0 \tag{2.1}$$

be the differential equation to be solved with, as boundary conditions

$$B(u) = 0 \tag{2.2}$$

Assume that (2.1) is the Euler-Lagrange equation corresponding to the variational principle

$$\delta I = 0 \qquad \text{with} \qquad I = \int_{x_1}^{x_2} F(u,u_x;x)\,dx \tag{2.3}$$

The idea of the method is to express u(x) in terms of a set of linearly independent functions $\phi_1(x) \ldots \phi_n(x)$ which are continuous, differentiable and given a priori :

$$u(x) = \sum_{i=1}^{n} a_i\phi_i(x) \tag{2.4}$$

The constant coefficients a_i are unknown and are to be determined from the procedure. The function (2.4) must satisfy the essential boundary conditions. It is not necessary that (2.4) verifies the natural boundary conditions because they are automatically fulfilled by the variational principle.

After substitution of the trial function (2.4) in the functional I and integration, one is faced with a function containing the unknowns $a_1 \dots a_n$. The latter are determined by imposing that $I(a_i)$ is extremum, namely by

$$\frac{\partial I}{\partial a_i} = 0 \qquad i=1\dots n\ . \tag{2.5}$$

The set (2.5) yields a system of n algebraic equations for the n unknowns $a_1 \dots a_n$. The approximate solution of (2.1) is obtained after substitution of the a_i's in (2.4).

The extension of the Rayleigh-Ritz method to multi-dimensional problems is straightforward : the trial function is then taken to be

$$u(x_1, x_2 \dots x_M) = \sum_{i=1}^{n} a_i \phi_i (x_1, x_2, \dots x_M) \tag{2.6}$$

In practical problems, the procedure consists in writing a sequence of approximations where

$a_1 \phi_1$ is the first approximation

$a_1 \phi_1 + a_2 \phi_2$ is the second approximation

and where (2.6) is the nth order approximation. The values of the a_n's must be recalculated at each step of the process by the method described above.

The accuracy and the convergence of such a method are of course linked to the selection of the functions ϕ_i ; this choice is dictated either by intuitive considerations or by results obtained in similar problems. It is also reasonable to believe that both convergence and precision are improved by increasing the number of terms in the approximation of u. However, in most practical applications, it is only necessary to take a reduced number of terms to obtain a satisfactory approximation of the true solution.

For many purposes, only numerical convergence of the successive approximations is investigated. Convergence is assumed to be attained when by adding more and more terms in the expansion, the answer does not vary significantly. For some particular problems, there exist mathematical criteria of convergence. In some circumstances, upper and lower bounds can be obtained. However, these subjects require rather heavy mathematical manipulations and will be considered in the general appendix.

It may happen that the trial functions ϕ_i are chosen as members of a complete set of function, that is, a set of functions such that every function can be expressed as a linear combination of the ϕ_i's. If so, the coefficients a_i obtained from the variational principle will yield the exact solution, because no other choice of the constants could satisfy the variational principle.

2.2. <u>The Kantorovitch or partial integration method.</u> Consider a partial differential equation involving two independent variables, obtained as Euler-Lagrange equation from

$$\delta I = \delta \int_x \int_y F(u, u_x, u_y; x, y)\,dx\,dy = 0 \tag{2.7}$$

The solution is approached by the following serie

$$u(x,y) = \sum_{i=1}^{n} a_i(x)\phi_i(x,y) \tag{2.8}$$

wherein, contrary to Rayleigh-Ritz'method, the a_i's are no longer constants but are functions of one of the independent variables. The ϕ's form a set of trial functions depending or not on x. Introduction of (2.8) in the expression of the stationary integral I(u) and integration with respect to y yields

$$\delta I = \delta \int_x f\left(a_i(x), \frac{da_i}{dx} ; x\right) dx = 0 \qquad (2.9)$$

The corresponding Euler-Lagrange equations are given by

$$\frac{\partial f}{\partial a_i} - \frac{d}{dx} \frac{\partial f}{\partial \left(\frac{\partial a_i}{\partial x}\right)} = 0 \qquad i = 1 \ldots n \qquad (2.10)$$

They form a set of ordinary differential equations which is generally easier to solve than the original partial differential equation. This procedure is very simple to apply and leads generally to better results than Rayleigh-Ritz's technique.

2.3. The method of weighted residuals. Like the above techniques, the method of weighted residuals (abbreviated as MWR) provides a mean of developing approximate solutions to differential or set of differential equations involving one or several dependent variables. Unlike the Rayleigh-Ritz or the Kantorovitch techniques, the Galerkin method is applicable even when no variational principle can be constructed.

The unknown function is expanded in a serie of trial functions $\phi_i(x)$ in the form

$$u(x) = \sum_{i=1}^{n} a_i \phi_i(x) \qquad (2.11)$$

where the a_i's can be chosen either as constants or as functions of one of the independent variables, when the problem is multi-dimensional ; the trial solution must in general satisfy the essential and the natural boundary conditions.

By replacing this approximate solution in the differential equation to be solved

$$D(u) = 0$$

one obtains a quantity called the residual and given by

$$R \equiv D\left(\sum_{i=1}^{n} a_i \phi_i\right) \tag{2.12}$$

Of course, the residual would be zero if the trial function were the exact solution. The unknowns a_i's are selected such that the residual is zero in an average sense, defined by

$$<f_j R> = \int_{x_1}^{x_2} f_j R \, dx = 0 \qquad j=1\ldots n \tag{2.13}$$

The f_j's are given function of x and are called weighting functions; the set (2.13) is a system of n equations, which are algebraic if the a_i's are constant, differential if the a_i's are function of one independent variable. If the f_i's are members of a complete set of functions, the orthogonality condition (2.13) implies that the residual R is equal to zero, thus providing the exact solution.(*)

The weighting function can be selected in many ways, each of them corresponding to a particular MWR.

i) If for instance, f_j is chosen as the displaced delta function

$$f_j = \delta(x - x_j) \tag{2.14}$$

(2.13) yields

(*) This result is a consequence of the general property that a continuous function is zero if it is orthogonal to every member of a complete sequence.

$$\int_{x_1}^{x_2} \delta(x-x_j)R\,dx = R\big|_{x_j} = 0 \qquad j=1\ldots n \tag{2.15}$$

This particular choice corresponds to the *collocation method* ; if n increases, the residual is zero at more and more discrete points x_j.

ii) Bv selecting as weighting function

$$f_j = \frac{\partial R}{\partial a_j} \tag{2.16}$$

the a_i's are obtained as solution of

$$\int_{x_1}^{x_2} \frac{\partial R}{\partial a_j} R\,dx = \frac{\partial}{\partial a_j}\int_{x_1}^{x_2} R^2 dx = 0 \tag{2.17}$$

i.e. by minimizing the integrated form of the square of the residual : this is nothing but the *least square method.*

iii) When a_j is identified with the trial function ϕ_j itself, the method is referred as *Galerkin's method* : the a_i's are obtained as solutions of the set

$$\int_{x_1}^{x_2} \phi_j R\,dx = 0 \qquad j=1\ldots n \tag{2.18}$$

iv) Of course, many other choices are imaginable. When f_j is taken equal to one, the method is known as the *integral method.*

v) By taking for f_j successively $1, x, x^2, \ldots$ one recovers the so-called *method of moments.*

Experience shows that the choice of a particular weighting function is not important because all the above methods leads invariably to the same results, at least at high approximations. However, it appears also that the most currently used method is Galerkin's, wherein the weighting function is the trial function. This is namely due to the connection of

Galerkin's technique with the variational one.(*)

It is easy to show this equivalence. Indeed, the variational principle whose Euler-Lagrange equation is

$$D(u) = 0$$

can be written as

$$\delta I = \int_{x_1}^{x_2} D(u)\ \delta u dx = 0 \qquad (2.19)$$

Application of the δ operator to the trial function yields

$$\delta u = \sum_{i=1}^{n} \phi_i(x)\delta a_i \qquad (2.20)$$

By substitution of this expression into (2.19), one obtains if it is assumed that the a_i's are constant,

$$\sum_{i=1}^{n} \delta a_i \int_{x_1}^{x_2} D(u)\phi_i(x)dx = 0 \qquad (2.21)$$

Since this relation must be satisfied for arbitrary and independent variations δa_i, it follows that

$$\int_{x_1}^{x_2} D(u)\phi_i(x)dx = 0 \qquad (2.22)$$

which is the central equation (2.18) of Galerkin's method.

(*) When weighted residuals are formed with the natural boundary conditions $\int f_j\ B(u)dA = 0$, it is no longer required that the trial functions verify all the boundary conditions : only the essential boundary conditions must be satisfied.

It must be realized that if a variational principle exists, it possesses some appreciable advantages on a M.W.R. method.

1. Instead of describing the behaviour of a system by one or several balance equations and boundary conditions, it can equivalently be described by *one* single variational equation.

Therefore, the first advantage of a variational formulation is its concision.

2. If the quantity submitted to variation is recognized to have a *physical* significance, the same principle can be used to describe several phenomena pertaining to a given class. For instance, all the motions in classical mechanics are governed by the principle of least action of Hamilton ; in thermostatics, the equilibrium state corresponds to a maximum entropy principle, etc..

3. The variational principles can also be used to prove *existence and uniqueness* of certain solutions. The reader interested by this aspect is recommended to consult for instance the papers of Wiener[6] and Ionescu-Casimir[7].

4. If the principle presents *minimum* or *maximum* properties, it affords upper or lower bounds for the variational functional (see the general appendix).

5. If in addition, there exists a dual principle, both *upper and lower bounds* to the approximate functionals can be specified. This provides a mean to bracket the functionals between very narrow limits and to reach a great degree of accuracy. Moreover, (see the general appendix), the use of both lower and upper bounds allows a direct measure of the error of the solution.

6. It must also be mentioned that the principles must be *invariant* under coordinate transformations. As a consequence, once the variational criterion has been written in one coordinate system, one can directly obtain the Euler-Lagrange equations in another coordinate system. This is done by writing first the invariant functional in the new coordinate system and by applying afterwards the usual procedure of derivation of the Euler-Lagrange equations.

3. The inverse problem

So far, the main objective was to derive some differential equations which had to be satisfied in order that some functionals be extremum. Now, one wishes to solve the inverse problem : given a differential or a set of differential equations, find the variational equation whose Euler-Lagrange equation is precisely the differential equation. The solution of this problem is important in that it provides an opportunity to use the variational methods to solve differential equations.

However, as will be seen, the solution of the problem is only simple for linear self-adjoint differential operators.

3.1. Adjoint and self-adjoint operators. Consider the linear differential equation

$$\left\{c_o(x) + c_1(x)\frac{d}{dx} + c_2\,\frac{d^2}{dx^2} + \ldots\right\}\,\phi(x) = L\{\phi(x)\} = 0 \tag{3.1}$$

associated with the boundary conditions

$$B\{\phi(x)\} = 0 \tag{3.2}$$

The linear operator L is said to be self-adjoint, if for two any functions u(x) and v(x) satisfying the above boundary conditions, one has

$$\int_{x_1}^{x_2} u(x)\; L\{v(x)\}dx = \int_{x_1}^{x_2} v(x)L\{u(x)\}dx \tag{3.3}$$

The linear operator L^* is adjoint to L if the relation

$$\int_{x_1}^{x_2} u(x)\; L\{v(x)\}dx = \int_{x_1}^{x_2} v(x)L^*\{u(x)\}dx \tag{3.4}$$

is satisfied. The above definitions remain of course valid in a multi-dimensional space.

3.2. The self-adjoint problem. Consider the linear and homogeneous equation (3.1) subject to the homogenous boundary conditions (3.2). The variational principle which yields this set as Euler-Lagrange equation is given by

$$\delta I = \delta \int_{x_1}^{x_2} uL(u)dx = 0 \tag{3.5}$$

By application of the δ operator to the integrant, it is seen that

$$\delta I = \int_{x_1}^{x_2} \left\{ \delta u \; L(u) + uL(\delta u) \right\} dx = 0$$

Since L is self-adjoint, this can be written in the form

$$\delta I = 2\int_{x_1}^{x_2} \delta u \; L(u) \; dx = 0$$

from which follows that for arbitrary variations δu,

$$L(u) = 0$$

If L is positive definite, the solution u(x) of the above equation gives the minimum value to I (Mikhlin[8]).

It can be easily verified that the variational principle corresponding to the non-homogeneous equation

$$L(u(x)) = g(x)$$

is

$$\delta I = \delta \int_{x_1}^{x_2} u\left\{\frac{1}{2} L(u) - g\right\} dx = 0 \tag{3.6}$$

If the operator L is of the form

$$L = W - \lambda K$$

where W and K are linear operators and λ the eigenvalue of the problem, it is clear that

$$I = \int_{x_1}^{x_2} u\, W(u)\,dx - \lambda\int_{x_1}^{x_2} u\, K(u)\,dx = 0$$

As a consequence,

$$\lambda = I_a/I_b \tag{3.7}$$

with

$$I_a = \int_{x_1}^{x_2} u\, W(u)\,dx \quad , \qquad I_b = \int_{x_1}^{x_2} u\, K(u)\,dx \ ,$$

is called the *Rayleigh quotient*[1].

Let us establish that the principle (3.5) is equivalent to finding the one which makes the Rayleigh quotient stationary. The proof is obvious : one has

$$\delta\lambda = \frac{\delta I_a}{I_b} - \frac{I_a \delta I_b}{I_b^2} = \frac{1}{I_b}\,(\delta I_A - \lambda\ \delta I_B) = \frac{1}{I_b}\,\delta I$$

from which it is clear that

$$\delta I = 0 \quad \text{implies} \quad \delta\lambda = 0 \tag{3.8}$$

This stationary property permits to obtain an estimate of the eigenvalue λ through the introduction of approximate functions for the eigenfunction u. Experience indicates that rough approximations for u gives nevertheless very accurate results for λ. It can also be shown (Courant and Hilbert[9]), that by increasing the number of terms in the approximation, the eigenvalue approaches closer the correct value from above. The convergence of eigenvalues from above is typical for self-adjoint problems.

3.3. The non self-adjoint problem. Let us show that the functional

$$I = \int_{x_1}^{x_2} u^* L(u)\,dx \tag{3.9}$$

wherein u^* is the solution of the adjoint problem, is stationary if u and u^* are solutions of their respective differential equations

$$L(u) = 0 \qquad\qquad B(u) = 0 \tag{3.10a}$$

and

$$L^*(u^*) = 0 \qquad\qquad B^*(u^*) = 0 \tag{3.10b}$$

From

$$\delta I = \delta\int_{x_1}^{x_2} u^* L(u)\,dx = 0$$

it follows that

$$\int_{x_1}^{x_2}\left\{\delta u^* L(u) + u^* L(\delta u)\right\} dx = \int_{x_1}^{x_2} \delta u^* L(u)\,dx + \int_{x_1}^{x_2} \delta u L^*(u^*)\,dx = 0$$

Since this relation must be verlfied for arbitrary and independent variations δu and δu^*, it is clear that one recovers (3.10) as Euler-Lagrange equations.

For non-homogeneous differential equations of the form

$$L\{u(x)\} = g(x), \tag{3.11}$$

the integral (3.9) must be replaced by

$$I = \int_{x_1}^{x_2} \left\{ u^* L\{u\} - u^* g - ug \right\} dx \tag{3.12}$$

3.4. General condition of existence of a variational principle.
Before deriving the general condition of existence of a variational criterion, let us introduce the notions of Gâteaux and Fréchet differentials.

Consider the operator $N(u)$ which can be non-linear.

The Gâteaux differential in the direction η is defined by

$$d_G N(u;\eta) = \lim_{\varepsilon \to 0} \frac{N(u+\varepsilon\eta) - N(u)}{\varepsilon} = \left[\frac{\partial}{\partial \varepsilon} N(u + \varepsilon\eta) \right]_{\varepsilon=0} \tag{3.13}$$

This differential presents some odd properties : it may not exist for some choices of η, it is not necessarily continuous and additive, it may be linear or non-linear in η. The condition for $d_G N(u;\eta)$ to be linear in η is to be continuous. When $d_G N(u;\eta)$ is linear in η, one can define a Gâteaux derivative $\dot{N}(u)$ as

$$\dot{N}(u)\eta = d_G N(u;\eta) \tag{3.14}$$

The Fréchet differential $d_F N$ is defined by

$$N(u + \eta) - N(u) = d_F N(u;\eta) + \omega(u;\eta) \tag{3.15}$$

where

$$\lim_{\eta \to 0} \frac{\|\omega(u;\eta)\|}{\|\eta\|} = 0 \tag{3.16}$$

and where $d_F N$ is both continuous and linear in η ,

$$d_F N(u;\eta) = N'(u)\eta \tag{3.17}$$

$N'(u)$ is the Fréchet derivative.

It has been shown that every Fréchet differentiable operator is Gâteaux differentiable.

$$d_F N(u;\eta) = \lim_{\varepsilon \to 0} \frac{N(u+\varepsilon\eta)-N(u)}{\varepsilon} \quad (\equiv d_G N) \tag{3.18}$$

but the converse is generally not true. In order that the Gâteaux be equal to the Fréchet differential, at a point u_o, the former must be continuous at u_o .

Given the continuous operator $N(u)$, the Fréchet derivative $N'(u)$ is said to be *symmetric* in η and ϕ if

$$\int \phi N'(u)\eta \, d\Omega = \int \eta N'(u)\phi \, d\Omega \tag{3.19}$$

where $d\Omega$ is the elementary volume of integration.

The necessary and sufficient condition that there exists a variational principle corresponding to the differential equation

$$N(u) = 0 \tag{3.20}$$

is that $N'(u)$ *be symmetric.* The corresponding variational functional is given by

$$I(u) = \int_\Omega \int_o^1 uN(su)\,ds\,d\Omega \tag{3.21}$$

where s is a real parameter. A demonstration of this important result can be found in Vainberg[10].

The existence of a variational principle is thus subordinated to the symmetry property of the Fréchet differential. No "exact" variational principle can be formulated if the Fréchet derivative is not symmetric.

If N is linear (N=L), one has by definition

$$L(\alpha u + \beta v) = \alpha L(u) + \beta L(v) \tag{3.22}$$

and the Fréchet differential (3.18) reads as

$$d_F L(u;\eta) = \lim_{\varepsilon \to 0} \frac{L(u+\varepsilon\eta) - L(u)}{\varepsilon} = \lim_{\varepsilon \to 0} \frac{L(\varepsilon\eta)}{\varepsilon} = L\eta \tag{3.23}$$

or according to (3.17)

$$L'(u)\eta = L\eta \tag{3.24}$$

The symmetry condition (3.19) reduces then to

$$\int_\Omega \phi L \eta d\Omega = \int_\Omega \eta L\phi \, d\Omega \tag{3.25}$$

which is automatically fulfilled if L is self-adjoint.

In virtue of (3.21), the corresponding principle is

$$\delta \int_\Omega \int_0^1 usL(u)ds \, d\Omega = \delta \frac{1}{2} \int_\Omega uL(u)d\Omega = 0 \tag{3.26}$$

which is nothing but the variational principle (3.5) derived earlier.

As next example, consider the non-linear equation[11]

$$N(u) = 2u(u_{xx} + u_{yy}) + u_x^2 + u_y^2 - f(x) = 0 \tag{3.27}$$

with, on the boundary

$$u = 0 \tag{3.28}$$

One has first to check wether N'(u) is symmetric.

$$N'(u)\eta = \left(\frac{\partial}{\partial \varepsilon} N(u + \varepsilon\eta)\right)_{\varepsilon=0} = 2\eta(u_{xx}+u_{yy}) + 2u(\eta_{xx} + \eta_{yy})$$

$$+ 2u_x\eta_x + 2u_y\eta_y \tag{3.29}$$

Taking the boundary condition (3.28) into account, one gets

$$\int_\Omega (\phi N'(u)\eta - \eta N'(u)\phi)dxdy = -2\int_\Omega \Big(\phi u(\eta_{xx} + \eta_{yy}) - \eta u(\phi_{xx} + \phi_{yy})$$

$$+ \phi(u_x\eta_x + u_y\eta_y) - \eta(u_x\phi_x + u_y\phi_y)\Big)dxdy$$

Since the line integral is zero,

$$\int_\Omega \phi u(\eta_{xx} + \eta_{yy})dxdy = -\int_\Omega \left(\eta_x \frac{\partial}{\partial x}(\phi u) + \eta_y \frac{\partial}{\partial y}(\phi u)\right)dxdy + \int_{line}$$

and a similar result when ϕ and η are permuted, one obtains

$$\int_\Omega \phi N'(u)\ \eta dxdy = \int_\Omega \eta N'(u)\phi\ dxdy$$

This indicates that the Fréchet derivative is symmetric.

The variational principle corresponding to (3.27) and (3.28) is according to (3.21) given by

$$\delta I(u) = 0$$

with

$$I(u) = \int_{\Omega}\int_{o}^{1}\left\{s^2u\ \left(2u(u_{xx} + u_{yy}) + u_x^2 + u_y^2\right) - fu\right\}dsdxdy$$

and after integration by parts

$$I(u) = \int_{\Omega}\left(u(u_x^2 + u_y^2) + uf\right)\ dxdy$$

It is easily checked that (3.27) is the corresponding Euler-Lagrange equation.

In most practical situations, it is not possible to associate a variational equation to a differential equation. This is particularly true for the non-linear transient heat conduction and the Navier-Stokes equations. To circumvent this difficulty, some authors have proposed to generalize the notion of a classical variational principle. Such a principle is characterized by the property that all the quantities appearing in the integrant of the functional are submitted to variation. It is suggested to relax somewhat this restriction and to introduce so-called quasi-variational and restricted variational principles.

4. Non-classical variational principles

4.1. Quasi-variational principles. The mathematical formulation of many linear transitory problems may be expressed by

$$L\left(u(x,t)\right) = \frac{\partial u}{\partial t} \qquad x_1 \leq x \leq x_2 \qquad \forall t > 0 \tag{4.1}$$

A typical quasi-variational principle, denoted $\widehat{\delta I}$, is

$$\widehat{\delta I} = \int_{t_1}^{t_2}\int_{x_1}^{x_2}\left\{L\left(u\right) - \frac{\partial u}{\partial t}\right\}\delta u\ dtdx = 0 \tag{4.2}$$

The arc of circle over δI indicates that there exists no functional I such that its variation vanishes ; in particular the δ symbol cannot be put in front of the integral in the r.h.s. The variational equation (4.2) contains however the original equation (4.1) as Euler-Lagrange equation.

Many examples of quasi-variational principles are found in the literature. As a first example, let us mention the Hamilton principle of classical mechanics in presence of friction, which reads as

$$\delta\int_{t_1}^{t_2} (K-V)\, dt + \sum_k \int_{t_1}^{t_2} \underline{F}^k . \delta\underline{x}^k\, dt = 0 \tag{4.3}$$

K is the kinetic energy, V the potential energy, $\underline{F}^k$ the kth frictional forces and $\delta\underline{x}^k$ the virtual displacement vector. Clearly the second integral enters in the class of the quasi-variational criteria.

Another example is Serrin's principle[12] governing the transitory fluid flow at uniform temperature. Serrin's principle can be formulated as

$$\int_\Omega \rho(\underline{F} - \frac{d^2\underline{x}}{dt^2}).\delta\underline{x}\, d\Omega - \int_\Omega \underline{\underline{\sigma}}:\nabla(\delta\underline{x})\, d\Omega + \int_A \underline{T}.\delta\underline{x}\, dA = 0 \tag{4.4}$$

$\underline{F}$ represents the body forces, $\underline{\underline{\sigma}}$ the stress tensor, $\underline{T}$ the surface tensions and ρ the density ; the integration is performed on a volume Ω of the fluid bounded by a surface A.

4.2. Restricted variational principles. Another way to construct non-classical variational principles is to keep some quantities frozen during the variational procedure. This amounts to introduce a kind of partial variation by analogy with the notion of partial derivative in mathematics.

If L in (4.1) is a self-adjoint linear operator, an example of restricted variational principle is

$$\delta_t I = \delta_t \int_{x_1}^{x_2} \left\{\frac{1}{2} L(u) - \frac{\partial u}{\partial t}\right\} u dx = 0 \qquad (4.5)$$

Subscript t indicates that only u must be varied, while its time derivative $\frac{\partial u}{\partial t}$ is kept unvaried. By application of δ_t to the integrant of (4.5), one obtains

$$\delta_t I = \int_{x_1}^{x_2} \left\{ \left(\frac{1}{2} L(u) - \frac{\partial u}{\partial t}\right) \delta_t u + \frac{1}{2} uL\left(\delta_t u\right)\right\} dx = 0$$

$$= \int_{x_1}^{x_2} \left(L(u) - \frac{\partial u}{\partial t}\right) \delta_t u dx = 0$$

For arbitrary variations $\delta_t u$, the vanishing of $\delta_t I$ implies the original equation (4.1). Like for the quasi-variational principle, there is no variational integral which is stationary. Stationarity should imply that not only

$$\delta_t I = 0 \qquad \text{but also} \qquad \delta_u I = 0$$

index u meaning that u is kept fixed.

As a corollary, one cannot infer minimum or maximum properties from the sign of the second variation $\delta_t^2 I$. Indeed, from the expression of the second variation of I, namely

$$\delta^2 I = \delta_t^2 I + \delta_u^2 I + \delta_t \delta_u I + \delta_u \delta_t I$$

it appears clearly that the sign of $\delta_t^2 I$ does not determine the sign of $\delta^2 I$; if $\delta_t^2 I > 0$, it is false to pretend that I has reached a minimum, except in the subspace wherein $\partial u/\partial t$ is a given function of x and t, and hence unvaried.

Most of the principles built in irreversible thermodynamics like the Glansdorff-Prigogine, Biot, Rosen[2], Lebon-Lambermont[13], Ziegler[14] principles are classified as restricted or quasi-variational principles. Despite their disadvantages, they have proven to be useful for solving many problems of continuum mechanics.

The main properties of classical and non-classical variational principles are summarized in the next table.

Property	Classical principle	Restricted principle	Quasi-variational principle
A stationary functional exists	YES	NO	NO
The functional may be a minimum (or a maximum)	YES	NO	NO
Applicable to linear non self-adjoint and non-linear problems	EXCEPTIONALLY	YES	YES

Appendix : The finite element technique

Introduction. The finite element method can be described by the three following steps[15].

1. The domain Ω under consideration is divided into a limited number of sub-domains called finite elements. These elements are generally contiguous and non overlapping. The corners of the sub-domains are referred as the nodal points. The shape of the elements is determined by the nature of the problem. Two dimensional problems are usually treated by means of triangles while for three dimensional problems, tetrahedrons are used.

2. The field parameters are approximated, within each finite element, by a Rayleigh-Ritz expansion involving continuous given functions and a number of unknown parameters. These parameters represent, in general, the value of the field variables at the nodal points.

Assume for instance that an element e_j contains N_e nodal points and let $\bar{u}_1, \bar{u}_2 \ldots \bar{u}_{N_e}$ be the value of the field variable u at the nodal points, i.e.

$$u(x_k) = \bar{u}_k \qquad k = 1,2\ldots N_e \tag{A.1}$$

The trial function in the sub-domain e_j is then taken as

$$u^{ej}(x) = \bar{u}_1 \psi_1^{e_j}(x) + \bar{u}_2 \psi_2^{e_j}(x) + \ldots \bar{u}_{N_e} \psi_{N_e}^{e_j}(x) \tag{A.2}$$

The function $\psi_k^{e_j}(x)$ are called weighting or element shape functions. They have to verify the conditions

$$\psi_k^{e_j}(x_j) = \delta_{kj} \tag{A.3}$$

and to be equal to zero when x falls outside the element :

$$\psi_k^{e_j}(x) = 0 \qquad \text{for} \qquad x \not\subset e_j \tag{A.4}$$

If N denotes the total number of nodes in Ω, the total trial function is taken as

$$u = \sum_{e_j} u^{e_j} = \sum_{k=1}^{N} \psi_k \bar{u}_k \left(\psi_k = \sum_{e_j} \psi_k^{e_j}\right) \tag{A.5}$$

3. The unknown coefficients $\bar{u}_k$ are derived either from a variational principle or, if the problem is not governed by a variational criterion, by the method of weighted residuals (M.W.R.). Assuming that there exists a variational principle

$$\delta I = 0 \tag{A.6}$$

the $\bar{u}_k$'s are given by

$$\frac{\partial I}{\partial \bar{u}_k} = 0 \qquad k = 1,2,\ldots N \tag{A.7}$$

In absence of a variational principle, one introduces the trial function (A.5) into the differential equation, written symbolically as

$$N\{u(x)\} = f(x)$$

This gives rise to a residual R defined by

$$N(u) - f = R \neq 0$$

According to the M.W.R., the best trial function will be the one which forces the residual to be zero by making it orthogonal to each member of a set of functions w_i :

$$\int_{\Omega} w_i R d\Omega = \int_{\Omega} \left\{ w_i N\left(\sum_k \psi_k \bar{u}_k \right) - w_i f \right\} d\Omega = 0 \qquad (i=1,2\ldots N)$$

This set permits the determination of the N unknowns $\bar{u}_1$, $\bar{u}_2 \ldots \bar{u}_N$. If w_i is taken identical to the weighting function ψ_i, one recovers Galerkin's method.

Both methods lead usually to a set of algebraic "nodal" equations of the form

$$\underline{\underline{A}}.\underline{U} = \underline{g}$$

where $\underline{U}$ is a(Nx1) column vector of the unknowns $\bar{u}_k$, while $\underline{\underline{A}}$ and $\underline{g}$ are respectively a (NxN) matrix and a (Nx1) column of known quantities.

Application. To illustrate the use of the finite elements method and its relation with Ritz's, consider the one-dimensional problem

$$L(u) \equiv \left(\frac{d^2}{dx^2} - 1\right)u = 0 \qquad 0 \leq x \leq 2 \qquad \text{(A.8)}$$

with the boundary conditions

$$u(0) = 1 \qquad u(2) = e^2 = 7.389 \qquad \text{(A.9)}$$

If L is self-adjoint, equations (2.1) and (2.2) are governed by the variational principle

$$\delta I(u) = \delta \int_0^2 uL(u)dx = 0$$

wherein I(u) is explicitely given by

$$I(u) = \int_0^2 \left(u + \left(\frac{du}{dx}\right)^2\right)dx \qquad \text{(A.10)}$$

Let us split the domain $(0,2)$ into $N = 4$ finite subdomains (finite elements) (see fig. A.1). Each element contains two nodal points x_{k-1} and x_k. In each sub-element $e_j = (x_{k-1}, x_k)$, one selects as trial function a linear function of x :

$$u^{e_j} = \alpha_1 + \alpha_2 x$$

In this form, the trial function is not convenient for the finite element technique ; the unknown constants α_1 and α_2 must be replaced by the values $\bar{u}^{e_j}_{k-1}$ and $\bar{u}^{e_j}_k$ of u^e_j at the nodal points. This is easily done, since at

$$x = x_{k-1} \qquad u^{e_j} = \bar{u}^{e_j}_{k-1} = \alpha_1 + \alpha_2\, x_{k-1}$$

while at

$$x = x_k \qquad u^{e_j} = \bar{u}^{e_j}_k = \alpha_1 + \alpha_2\, x_k$$

Elimination of α_1 and α_2 provides the form of the trial function u in terms of the desired quantities, namely

$$u^{e_j} = \psi^{e_j}_{k-1}\, \bar{u}_{k-1} + \psi^{e_j}_k\, \bar{u}_k \tag{A.11}$$

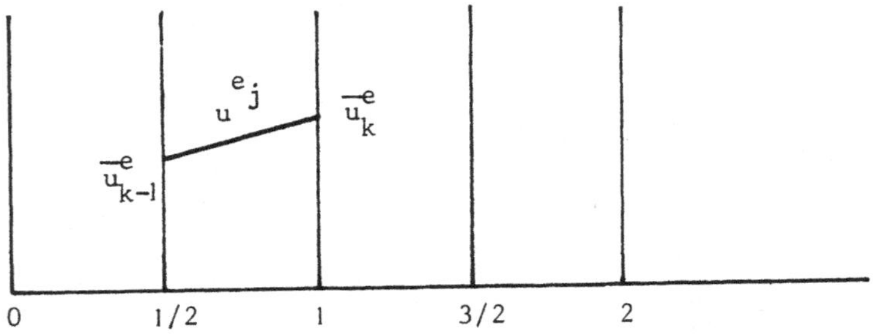

Fig. A.1.

wherein the weighting functions are respectively expressed by

$$\psi_{k-1}^{e_j} = 1 - \frac{x-x_{k-1}}{x_k-x_{k-1}} \quad , \quad \psi_k^{e_j} = \frac{x-x_{k-1}}{x_k-x_{k-1}} \tag{A.12}$$

In most cases, the form of the weighting functions is guessed. It is directly checked that

$$\psi_k^{e_j}(x_i) = \delta_{ik}$$

$$\psi_k^{e_j}(x) = 0 \qquad\qquad x \not\subset e_j$$

The trial function related to the whole domain is given by

$$u = \sum_{e_j=1}^{4} u^{e_j} = u^1 + u^2 + u^3 + u^4 \tag{A.13}$$

Similarly the functional I(u) can be partitioned into four parts

$$I = \sum_{e_j=1}^{4} I^{e_j} \tag{A.14}$$

where

$$I^{e_j} = \int_{e_j} \left\{ (u^{e_j})^2 + \left(\frac{du^{e_j}}{dx}\right)^2 \right\} dx \tag{A.15}$$

Since $x_k - x_{k-1} = 1/2$, (A.11) takes the form

$$u^{e_j} = (1-2x+2x_{k-1})\, \bar{u}_k^{e_j} + (2x-2x_{k-1})\bar{u}^{e_j} \tag{A.16}$$

Substituting (A.16) in (A.15) results in

$$I^{e_j} = (\overline{u}_k^{e_j})^2 \int_{e_j} \left\{(2x-2x_{k-1})^2 + 4\right\} dx$$

$$+ \overline{u}_k^{e_j} \, \overline{u}_{k-1}^{e_j} \int_{e_j} \left\{2(2x-2x_{k-1})(1-2x+2x_{k-1}) - 8\right\} dx$$

$$+ (\overline{u}_{k-1}^{e_j})^2 \int_{e_j} \left\{(1-2x-2x_{k-1})^2 + 4\right\} dx$$

or in a more compact form

$$I^{e_j} = a_1^{e_j} (\overline{u}_k^{e_j})^2 + a_2^{e_j} \, \overline{u}_k^{e_j} \, \overline{u}_{k-1}^{e_j} + a_3^{e_j} \, (\overline{u}_{k-1}^{e_j})^2 \qquad (e_j = 1,2,3,4)$$

It follows that I depends on the parameters $\overline{u}_k (k = 1,2 \ldots 5)$:

$$I = \sum_{e_j} I^{e_j} = I(\overline{u}_2, \overline{u}_3, \overline{u}_4) \tag{A.17}$$

$\overline{u}_1$ and $\overline{u}_5$ have been eliminated because the boundary conditions impose

$$\overline{u}_1 = 1 \qquad \overline{u}_5 = e^2 \tag{A.18}$$

The stationary conditions are

$$\frac{\partial I}{\partial \overline{u}_2} = \frac{\partial I}{\partial \overline{u}_3} = \frac{\partial I}{\partial \overline{u}_4} = 0 \tag{A.19}$$

and explicitely

$$\frac{\partial I}{\partial \bar{u}_2} = a_2^1 \bar{u}_1 + 2a_1^1 \bar{u}_2 + 2a_2^2 \bar{u}_2 + a_2^2 \bar{u}_3 = 0$$

$$\frac{\partial I}{\partial \bar{u}_3} = a_2^2 \bar{u}_2 + 2\, a_1^2 \bar{u}_3 + 2a_3^3 \bar{u}_3 + a_1^3 \bar{u}_4 = 0$$

$$\frac{\partial I}{\partial \bar{u}_4} = a_2^3 \bar{u}_3 + 2a_1^3 \bar{u}_4 + 2a_3^4 \bar{u}_4 + a_2^4 \bar{u}_5 = 0$$

These algebraic equations can also be written in matrix form

$$\begin{vmatrix} 1 & 0 & 0 & 0 & 0 \\ -46 & 104 & -46 & 0 & 0 \\ 0 & -46 & 104 & -46 & 0 \\ 0 & 0 & -46 & 104 & -46 \\ 0 & 0 & 0 & 0 & 1 \end{vmatrix} \begin{vmatrix} \bar{u}_1 \\ \bar{u}_2 \\ \bar{u}_3 \\ \bar{u}_4 \\ \bar{u}_5 \end{vmatrix} = \begin{vmatrix} 1 \\ 1 \\ 6 \\ 0 \\ e^2 \end{vmatrix}$$

from which follow directly the values of the unknowns $\bar{u}_2$, $\bar{u}_3$, $\bar{u}_4$.

The advantage of the finite element method lies in the fact that the elements can be changed in size and shape to follow complicated boundaries. The method reveals particularly useful for handling problems with irregular geometrical boundaries. This method is becoming increasingly important in continuum mechanics, and particularly in solid mechanics.

The method reveals particularly powerful when the problem is governed by a variational description because convergence may be assured. If furthermore, one can construct a dual principle, the method provides a mean to evaluate the error as will be seen in the general appendix. These advantages are generally lost when a Galerkin procedure is used. In that case, an error estimate can only be obtained for some particular problems.

Moreover for hyperbolic partial differential equations, the finite difference method is to be preferred because the numerical solution is deeply related with the physical nature of the problem, through the characteristic lines which can be identified with the Mach or the streaming lines.

.

References

1. Gelfand, I. and Fomin, S., *Calculus of Variations*, Prentice Hall, London, 1963.
2. Finlayson, B., *The Method of Weighted Residuals and Variational Principles*, Acad. Press, New-York, 1972.
3. Schechter, R., *The Variational Method in Engineering*, Mc Graw Hill, New-York, 1967.
4. Napolitano, L., *The Finite element Methods in Fluid Mechanics*, C.I.S.M. Lecture course, Udine, 1972.
5. Lee, L. and Reynolds, W.C., A Variational Method for Investigating the Stability of Parallel Flows, Tech. Report, Dpt. of Mechanical Engineering, Stantford Univ., December 1964.
6. Wiener, J., A uniqueness theorem for the coupled thermoelastic problem, *Quat. Appl. Math.*, 1, 15, 1957.
7. Ionescu, V. and Casimir, Problem of linear coupled thermoelasticity, uniqueness theorems, *Bull. Acad. Pol. Sc.*, 12,12, 1964.
8. Mikhlin, R., *Variational Methods in Mathematical Physics*, Mac Millan, New-York, 1964.
9. Courant, R. and Hilbert, D., *Methods of Mathematical Physics*, vol.1, Interscience, New-York, 1953.
10. Vainberg, M., *Variational Methods for the Study of Non-linear Operators*, Holden-Day, San Francisco, 1964.
11. Oden, J. and Reddy, J.N., *Variational methods in theoretical mechanics*, Springer-Verlag, Berlin, 1976.
12. Serrin, J., Mathematical principles of classical fluid mechanics, in *Handbuch der Physik*, Bd VIII/1, Flügge, S., Ed., Springer Verlag, Berlin, 1969.

13. Lebon, G., Thermodynamique des Phénomènes Irréversibles, E.L.E., Liège, 1979.

14. Ziegler, H., Some extremum principles in irreversible thermodynamics, in *Progress in Solid Mechanics*, vol. IV, Sneddon, I. and Hill, R., Eds., North-Holland, Amsterdam, 1963.

15. Zienkiewicz, O. and Cheung, Y., *The Finite Element Method in Structural and Continuum Mechanics*, Mc Graw Hill, New-York, 1967.

II. VARIATIONAL PRINCIPLES IN CLASSICAL MECHANICS AND IN ELASTICITY

1. The law of varying action and the Hamilton principle in classical mechanics

1.1. Formulation of the principles. Consider a material system (either a system of discrete points or a rigid body) with N degrees of freedom and submitted to holonomic constraints. *The law of varying action* states that during the time interval $t_o - t_1$, the motion of the system is governed by the variational equation

$$\delta\int_{t_o}^{t_1} K dt + \int_{t_o}^{t_1} \delta W dt - \sum_\alpha \frac{\partial K}{\partial \dot{q}^\alpha}\, \delta q^\alpha \Big|_{t_o}^{t_1} = 0 \quad (\dot{q}^\alpha = \frac{dq^\alpha}{dt}) \tag{1.1}$$

$q^1, q^2, \ldots q^N$ are the N generalized coordinates, K is the kinetic energy, W the work performed by the internal and the external forces. The quantity δW represents the virtual work

$$\delta W = \sum_k \underline{F}^k . \delta \underline{x}^k = \sum_\alpha Q^\alpha \delta q^\alpha \tag{1.2}$$

wherein the Q^α's denote the generalized forces. It is worthy to point out that δW denotes merely a differential form but is not the variation of some potential function. If some forces $\underline{F}^k$ are conservative and derive from a potential ϕ, δW may be written as

$$\delta W = - \delta\phi + \sum_\alpha Q^\alpha \delta q^\alpha \tag{1.3}$$

Q^α being now associated with the non-conservative forces ; δW represents the variation of a potential function only if all the Q^α's vanish. Using (1.3), (1.1) takes the form

$$\delta\int_{t_o}^{t_1}(K - \phi)dt + \int_{t_o}^{t_1} \sum_\alpha Q^\alpha \delta q^\alpha \, dt - \sum_\alpha \frac{\partial K}{\partial \dot{q}^\alpha} \delta q^\alpha \Big|_{t_o}^{t_1} = 0 \tag{1.4}$$

If it is imposed that the δq^α's vanish at t_o and t_1 , the law of varying action reduces to the so-called *Hamilton general principle* :

$$\delta\int_{t_o}^{t_1}(K - \phi)dt + \int_{t_o}^{t_1} \sum_\alpha Q^\alpha \delta q^\alpha dt = 0 \tag{1.5}$$

$$\delta q^\alpha(t_o) = \delta q^\alpha(t_1) = 0$$

The latter condition expresses that the real and the virtual paths are coterminus at t_o and t_1. Hamilton's general principle may be enunciated as follows : "A system moves from one configuration to another in such a way that the variation of the so-called action integral (1.5) between the actual and neighbouring virtual paths, coterminus in time, is zero."

At *equilibrium*, the kinetic energy and the time integrals vanish so that Hamilton's principle (1.5) reduces to the *principle of virtual work*

$$\delta\phi - \sum_\alpha Q^\alpha \delta q^\alpha = 0 \tag{1.6}$$

Expression (1.6) holds only for bilateral constraints, i.e. for virtual displacements which are two-directional. If the virtual displacements are only possible in one direction (unilateral constraints), the equality (1.6) must be replaced by

$$\delta\phi - \sum_\alpha Q^\alpha \delta q^\alpha < 0$$

When the system is *conservative,* the following restrictions are to be respected : the work is path independent and the total energy E is conserved :

$$W = -\phi \qquad \text{and} \qquad K + \phi = E \quad \text{(a constant)} \tag{1.7}$$

The *law of varying action* reduces then to

$$\delta\int_{t_2}^{t_1}(K - \phi)dt - \sum_{\alpha} \frac{\partial K}{\partial \dot{q}^{\alpha}} \delta q^{\alpha}\Big|_{t_2}^{t_1} = 0 \tag{1.8}$$

while *Hamilton's general principle* takes the well known form

$$\delta\int_{t_2}^{t_1} (K - \phi)dt = \delta\int_{t_2}^{t_1} L\ dt = 0 \quad , \quad \delta q^{\alpha}(t_o)=\delta q^{\alpha}(t_1) = 0 \tag{1.9}$$

L is the Lagrangian defined by

$$L = K - \phi \tag{1.10}$$

The latter principle, usually known as Hamilton's principle, was not proposed by Hamilton himself but by Lagrange, who called it *principle of least action.*

It states that the system moves from one configuration to another in such a way that the action $\int L\ dt$ is stationary, compared with adjacent virtual motions, coterminus in time and having the same total energy E as the actual motion.

It must be observed that the term least action is not adequate. Indeed it should imply that the principle presents a minimum property ; this is certainly not true in general because L is not positive definite. It has been proved by some authors (e.g. Smith and Smith[1]), that for discrete systems, the action is only minimized over short time intervals. Our purpose is not to give a general demonstration, but to present an elementary proof by taking a very simple example.

The system to be considered is the one degree of freedom harmonic oscillator consisting of a mass m and a spring of negligible mass and stiffness k. The position is denoted by the parameter q.

The principle of least action is expressed by

$$\delta I(q) = \delta \int_{t_2}^{t_1} (m\dot{q}^2 - kq^2)dt = 0 \tag{1.11}$$

whose Euler-Lagrange equation is

$$m\ddot{q} + kq = 0$$

Taking as initial conditions

$$t = 0 : q = 0 \quad , \quad \dot{q} = v_o \quad ,$$

the exact solution is

$$q^e = \frac{v_o}{\omega} \sin \omega t \qquad (\omega^2 = k/m) \tag{1.12}$$

Consider now a family a trial functions satisfying the initial conditions and coterminus with the exact motion at $t = t_1$, namely

$$q = \frac{v_o}{\omega} \sin \omega t + c \frac{1}{t_1^3} (t^3 - t^2 \; t_1) \tag{1.13}$$

c is an arbitrary constant.

By substitution of (1.12) and (1.13) in the expression (1.11) of the functional I(q), one obtains

$$\Delta I = I(q) - I(q^e) = \frac{1}{2} m \; c^2 \frac{1}{105 t_1} (14 - \omega^2 t_1^2) \tag{1.14}$$

Clearly, $\Delta I < 0$ for $t_1^2 > 14/\omega^2$ which means that for sufficiently large values of t_1, $I(q^e)$ is not minimum.

It must be realized that, at the exception of the principle of least action (1.9), all the above variational criteria are not exact but quasi-variational principles.

Let us now show that they are equivalent to the usual laws of motion. Consider more particularly (1.4) where K is a function of q^α, $\dot{q}^\alpha$ and the time t. The Euler-Lagrange equation corresponding to arbitrary and independent variations δq^α are given by

$$-\frac{d}{dt}\frac{\partial K}{\partial \dot{q}^\alpha} + \frac{\partial K}{\partial q^\alpha} - \frac{\partial \phi}{\partial q^\alpha} + Q^\alpha = 0 \qquad t_o < t < t_1 \qquad (\alpha=1,\ldots N) \quad (1.15)$$

$$\sum_\alpha \left(\frac{\partial K}{\partial \dot{q}^\alpha} - \frac{\partial K}{\partial \dot{q}^\alpha}\right)\delta q^\alpha = 0 \quad \text{at} \quad t = t_o \; , \; t = t_1 \qquad (1.16)$$

Expressions (1.15) are identical to the Lagrange equations of motion which are known to be equivalent to Newton's law.

The relation (1.16) follows from integration by parts of the term $\frac{\partial K}{\partial q^\alpha}\frac{d}{dt}(\delta q^\alpha)$; it is seen that, whatever the values taken by δq^α , (1.16) vanishes identically. It follows that the only condition for (1.4) to be satisfied is that the q^α's obey the Euler-Lagrange equations (1.15) No restrictions have to be imposed on the δq^α's at the initial and the final times. Should the last term in (1.4) be omitted, like in Hamilton's principle, then instead of (1.16), one should have obtained

$$\sum_\alpha \frac{\partial K}{\partial \dot{q}^\alpha}\,\delta q^\alpha = 0 \qquad \text{at} \qquad t = t_o \quad , \quad t = t_1$$

which implies that each δq^α vanishes at t_o and t_1. As a consequence, each q^α's must be a prescribed function, not only at the initial time but also at the arbitrary final time t_1. Since the objective of mechanics is to determine the q^α's at each instant of time, and in particular at t_1, knowing its value at t_o , it is clear that a principle like Hamilton's cannot be used to solve problems. As pointed out by Hamilton himself, it serves only to form, by the rules of the calculus of variations, the differential equations of motion. But one cannot obtain a solution directly from Hamilton's principle except in very special cases. In that respect, the law of varying action is of more practical interest for solving initial value problems, because any of the variational methods presented in chapter I, can be used to achieve approximate solutions.

1.2. An example. As an illustration of the practical interest of the law of varying action, consider the harmonic oscillator to which Rayleigh-Ritz's method will be applied.

The position of the mass is denoted by q, with at $t_o = 0$:

$$t_o = 0 : q = 1 \quad \dot{q} = 0$$

The kinetic and potential energies are respectively given by

$$K = \frac{1}{2} m\dot{q}^2 \qquad \phi = k \frac{q^2}{2}$$

so that the law of varying action is given by

$$\delta \frac{1}{2} \int_o^{t_1} (m\dot{q}^2 - kq^2)dt - m\dot{q}\delta q \Big|^{t_1} = 0$$

or

$$\int_o^{t_1} (\dot{q}\ \delta\dot{q} - \omega^2 q\delta q)dt - \dot{q}\delta q \Big|^{t_1} = 0 \ (\omega^2 = \frac{k}{m}) \tag{1.17}$$

The last term has to be evaluated only at t_1 because at t_o, this term vanishes identically ($\dot{q} = \delta q = 0$ at $t_o = 0$).

In order to apply Rayleigh-Ritz's method, one must construct a trial function ; with Bailey[2], let us select the polynome

$$q = 1 + \sum_{n=2}^{N} a_n t^n \quad ; \quad \dot{q} = \sum_{n=2}^{N} a_n n t^{n-1} \tag{1.18}$$

which satisfies the initial conditions. By application of the δ operator, one obtains

$$\delta q = \sum_{n=2}^{N} t^n \delta a_n \tag{1.19a}$$

$$\delta \dot{q} = \sum_{n=2}^{N} n t^{n-1} \delta a_n \tag{1.19b}$$

Substitution in (1.17) yields

$$\int_0^{t_1} \sum_n \left(\sum_m a_m m n t^{m+n-2} - \omega^2 (t^n + \sum_m a_m t^{m+n}) \right) \delta a_n \, dt - \sum_n \left(\sum_m a_m m t_1^{m+n-1} \right) \delta a_n = 0$$

i.e. an expression of the form

$$\sum_{n=2}^{N} A_n \delta a_n = 0 \tag{1.20}$$

Since the δa_n are arbitrary, it follows that each of the coefficients A_n must be identically zero :

$$A_n = \int_0^{t_1} \sum_m \left(a_n a_m m n t^{m+n-2} - \omega^2 t^n (1 + \sum a_m t^m) \right) dt$$

$$- \sum_m a_m m t_1^{m+n-1} = 0 \qquad (n=2,3,\ldots N) \tag{1.21}$$

and after integration

$$A_n = \sum_m \frac{m\,n\,t_1^m}{m+n-1}\,a_m - \omega^2 t_1^2\left(\frac{1}{n+1} + \sum_m \frac{t_1^m}{n+m+1}\,a_m\right)$$

$$- \sum_m m\,a_m t_1^m = 0$$

This expression can be written in matricial notation as

$$\sum_m B_{mn} a_m = b_n \tag{1.22}$$

with

$$B_{nm} = \frac{n\,m\,t_1^m}{n+m-1} - \omega^2\,\frac{1}{n+m+1}\,t_1^{m+2} - m\,t_1^m \quad , \tag{1.23}$$

$$b_n = \frac{\omega^2}{n+1}\,t_1^2$$

The solution of the set of N-2 non-homogeneous algebraic equations (1.22) is rather trivial. By substituting the values obtained for a_n in the trial function (1.17) and taking

$$\omega^2 = 1 \quad , \quad t_1 = \frac{\pi}{2} \quad , \quad N = 12$$

Bailey[2] derived the results shown in table 1. It is seen that the agreement with the exact solution is remarkably good : the relative error is only 0.0016 %. The application of Rayleigh-Ritz's direct method through the law of varying action has been performed on many problems, linear or non-linear, both conservative and non-conservative.[3,4]

$\tau = t/t_1$ Nondimensional time	τt_1 Real time (second)	q Calculated	q Exact
		First cycle of motion	
0.0	0.0	1.0 (assigned)	1.0
0.5	0.7853982	0.707107	0.707107
1.0	1.5707963	0.000000	0.0
1.5	2.3561945	-0.707107	-0.707107
2.0	3.1415927	-1.000000	-1.0
2.5	3.9269908	-0.707106	-0.707107
3.0	4.7123889	0.000000	0.0
3.5	5.4977871	0.707107	0.707107
4.0	6.2831853	1.000000	1.0
		Twenty-fifth cycle of motion	
98.0	150.796447	1.000000	1.0
96.5	151.581846	0.707095	0.707107
97.0	152.367244	-0.000016	0.0
97.5	153.152642	-0.707118	-0.707107
98.0	153.938040	-1.000000	-1.0
98.5	154.723438	-0.707095	-0.707107
99.0	155.508836	0.000016	0.0
99.5	156.294235	0.707113	0.707107
100.0	157.079633	1.000000	1.0

Table 1.

2. General results in elasticity and thermoelasticity

2.1. Notations and definitions. Let Ω be a regular (in the sense of Kellog) region of space occupied by a deformable, inhomogeneous and anisotropic medium. Assume that the latter changes its configuration continuously under external actions and heating from an original to a deformed state. Let Ω_o and Ω be the volumes, respectively in the undeformed and deformed states, with surfaces A_o and A and outwards pointing normals $\underline{n}_o$ and $\underline{n}$.

In the original state, the body is referred to a coordinate system $OX_1X_2X_3$ and one arbitrary point P of the body is assigned the coordinates X_1 , X_2 , X_3 . The body is then deformed to a new state and point P is moved to a point P' with coordinates x_1 , x_2 , x_3 , with respect to a new coordinate system $O'x_1x_2x_3$. The reference system may be curvilinear and may be different from the original one. Consider line elements connecting P and P' to neighbouring points Q and Q' respectively. The length of these elements are respectively given by

$$ds_o^2 = G_{ij}\, dX^i\, dX^j \quad (*) \qquad \text{(original configuration)}$$

$$ds^2 = g_{ij}\, dx^i\, dx^j \qquad \text{(deformed configuration)}$$

G_{ij} and g_{ij} are the Euclidean metric tensors.

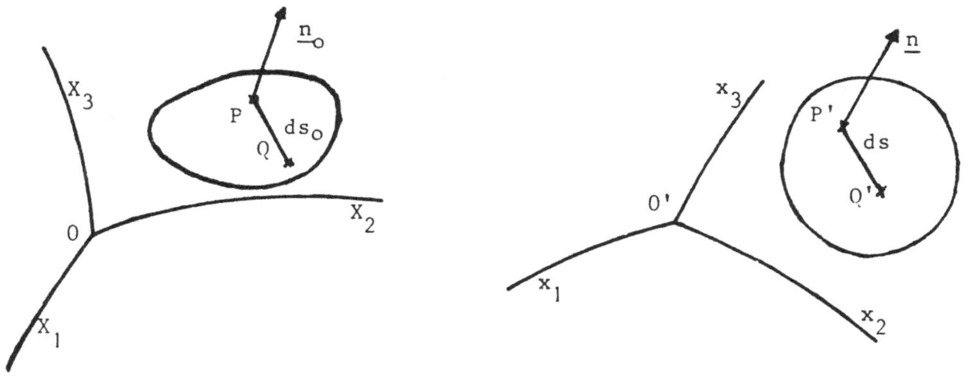

Fig. 2.1.

(*) Summation convention on repeated indices will be used throughout this work.

Strain tensors E_{ij} and ε_{ij} are defined by

$$ds^2 - ds_c^2 = 2\, E_{ij}\, dX^i\, dX^j \qquad (2.1)$$

$$ds^2 - ds_o^2 = 2\, \varepsilon_{ij}\, dx^i\, dx^j \qquad (2.2)$$

with

$$E_{ij} = \frac{1}{2}\left(g_{kl}\,\frac{\partial x_k}{\partial X_i}\frac{\partial x_l}{\partial X_j} - G_{ij}\right) \quad \text{(Lagrangian description)} \qquad (2.3)$$

$$\varepsilon_{ij} = \frac{1}{2}\left(g_{ij} - G_{kl}\,\frac{\partial X_k}{\partial x_i}\frac{\partial X_l}{\partial x_j}\right) \quad \text{(Eulerian description)} \qquad (2.4)$$

E_{ij} is the Green and ε_{ij} the Eulerian or Almansi strain tensor, both are symmetric :

$$E_{ij} = E_{ji}\,, \qquad \varepsilon_{ij} = \varepsilon_{ji}$$

Use now the same Cartesian reference system for both original and deformed configurations ($G_{ij} = g_{ij} = \delta_{ij}$). Introduce the displacement vector u_i defined by

$$u_i = X_i - x_i$$

from which follows that

$$dx_i = \frac{\partial x_i}{\partial X_j}\, dX_j = \left(\delta_{ij} + \frac{\partial u_i}{\partial X_j}\right) dX_j \qquad (2.5)$$

Under these conditions, E_{ij} and ε_{ij} take the simple form

$$E_{ij} = \frac{1}{2}\left(\frac{\partial u_i}{\partial X_j} + \frac{\partial u_j}{\partial X_i} + \frac{\partial u_k}{\partial X_i}\frac{\partial u_k}{\partial X_j}\right) \tag{2.6}$$

$$\varepsilon_{ij} = \frac{1}{2}\left(\frac{\partial u_j}{\partial x_j} + \frac{\partial u_j}{\partial x_i} - \frac{\partial u_k}{\partial x_i}\frac{\partial u_k}{\partial x_j}\right) \tag{2.7}$$

In the limit of *small deformations* (which implies that the displacement gradients are negligible compared to unity : $\partial u_i/\partial x_j << 1$) , E_{ij} and e_{ij} are identical and given by

$$E_{ij} = \varepsilon_{ij} = \frac{1}{2}\left(\frac{\partial u_i}{\partial x_j} + \frac{\partial u_j}{\partial x_i}\right) = \frac{1}{2}(u_{i,j} + u_{j,i}) \tag{2.8}$$

From now on, a comma preceding a subscript denotes derivation with respect to the Cartesian coordinates x_i .

The question then arises : how to determine the three components u_i from the six equations (2.8). In order that a solution exists, it is necessary that ε_{ij} verify the six compatibility conditions.

$$2\varepsilon_{ij,ij} = \varepsilon_{ii,jj} + \varepsilon_{jj,ii} \qquad i,j,k = 1,2,3 \tag{2.9}$$

$$\varepsilon_{ii,jk} = (-\varepsilon_{jk,i} + \varepsilon_{ki,j} + \varepsilon_{ij,k})_{ii} \qquad i,j,k = 1,2,3 \tag{2.10}$$

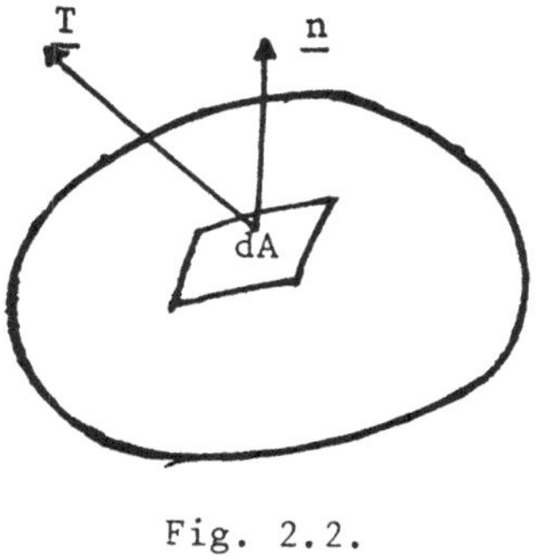

Fig. 2.2.

Denote by $\underline{T}$ the stress vector, representing the action of material exterior to a given surface element with unit normal $\underline{n}$; $\underline{T}$ is related to the stress tensor $\underline{\underline{\sigma}}$ by mean of Cauchy's formula

$$\underline{T} = \underline{n}.\underline{\underline{\sigma}}$$

2.2. Laws of balance. For systems undergoing large deformations at finite velocity, the balance equations can be expressed either in the material description (X_1 , X_2 , X_3 , t as independent variables) or in the spatial description (x_1 , x_2 , x_3 , t as independent variables). In both descriptions, the equations are given by

Material description		Spatial description	
Mass balance	: $\dfrac{\rho_o}{\rho} = J \; (J=\det \dfrac{\partial x_i}{\partial X_k})$	$\dot{\rho} = - \rho \, v_{j,j}$	(2.11)
Momentum balance	: $\rho_o \dfrac{\partial v_i}{\partial t} = \dfrac{\partial}{X_j}(T_{ji}) + F_{oi}$	$\rho \dot{v}_i = \sigma_{ij,j} + F_i$	(2.12)
Energy balance	: $\rho_o \dfrac{\partial u}{\partial t} = - \dfrac{\partial Q_i}{\partial X_i} + S_{ij} \dfrac{\partial E_{ij}}{\partial t}$	$\rho \dot{u} = - q_{i,i} + \sigma_{ij} V_{ij}$	(2.13)

Index o refers to quantities measured in the original configuration, an upper dot denotes the material time derivative

$$= \frac{\partial}{\partial t} + \underline{v}.\text{grad}$$

The following notation is used :

F_i = body force per unit volume

ρ = density

v_i = velocity

T_{ij} = first Piola-(Kirchoff) stress tensor (not symmetric) :

$$T_{ij} = \frac{\rho_o}{\rho} X_{i,m} \sigma_{mj} \qquad (T_{ij} \neq T_{ji})$$

S_{ij} = second Piola-Kirchoff stress tensor (symmetric) :

$$S_{ij} = X_{i,m} T_{jm} \qquad (S_{ij} = S_{ji})$$

σ_{ij} = Cauchy stress tensor $(\sigma_{ij} = \sigma_{ji})$

u = internal energy per unit mass

q_i = heat flux vector across the surface A

Q_i = heat flux vector across A but measured per unit area of A_o ; one has

$$q_i \, n_i \, dA = Q_i \, n_{io} dA_o$$

with

$$Q_i = J \, X_{i,k} \, q_k$$

V_{ij} = rate of deformation tensor : $V_{ij} = \frac{1}{2} (v_{i,j} + v_{j,i})$

In the limit of small deformations, one has

$$\rho = \rho_o \tag{2.14}$$

$$X_{i,j} = \delta_{ij} \tag{2.15}$$

$$S_{ij} = \sigma_{ij} \tag{2.16}$$

$$\dot{v}_i = \frac{\partial v_i}{\partial t} = \frac{\partial^2 u_i}{\partial t^2} \tag{2.17}$$

$$V_{ij} = \frac{1}{2}\frac{\partial}{\partial t}(u_{i,j} + u_{j,i}) = \frac{\partial}{\partial t}\varepsilon_{ij} \qquad (2.18)$$

and the distinction between the material and the spatial descriptions disappears.

2.3. Thermodynamic description. All the thermodynamic properties of the deformable body can be derived from Gibbs'equation, which is given by

$$du = Tds + \frac{1}{\rho_o} S_{ij} dE_{ij} \qquad \text{(large deformations)} \qquad (2.19)$$

$$du = Tds + \frac{1}{\rho}\sigma_{ij} d\varepsilon_{ij} \qquad \text{(small deformations)} \qquad (2.20)$$

T is the absolute temperature, s the entropy per unit mass.

Limiting the analysis to small deformations, it is seen that u is a state function of s and the strain ε_{ij} :

$$u = u(s,\varepsilon_{ij})$$

By derivation, one obtains the state equations

$$T = \left.\frac{\partial u}{\partial s}\right)_{\varepsilon_{ij}} \qquad \text{hence } T = T(s,\varepsilon_{ij}) \qquad (2.21)$$

$$\frac{1}{\rho}\sigma_{ij} = \left.\frac{\partial u}{\partial \varepsilon_{ij}}\right)_s \qquad \text{hence } \sigma_{ij} = \sigma_{ij}(s,\varepsilon_{ij}) \qquad (2.22)$$

It may also be convenient to work with the Legendre transform of u, namely

$$\Psi = u - Ts \qquad \text{(Helmholtz free energy)} \qquad (2.23)$$

$$g = u - Ts - \frac{1}{\rho}\sigma_{ij}\varepsilon_{ij} \qquad \text{(Gibbs free energy)} \qquad (2.24)$$

By differentiation, one gets

$$d\Psi = - sdT + \frac{1}{\rho}\sigma_{ij} d\varepsilon_{ij} \tag{2.25}$$

$$dg = - sdT - \frac{1}{\rho}\varepsilon_{ij} d\sigma_{ij} \tag{2.26}$$

from which follow the state equations

$$s = - \left.\frac{\partial\Psi}{\partial T}\right)_{\varepsilon_{ij}} \qquad \text{hence } s = s(T,\varepsilon_{ij}) \quad , \tag{2.27a}$$

$$\frac{1}{\rho}\sigma_{ij} = \left.\frac{\partial\Psi}{\partial\varepsilon_{ij}}\right)_T \qquad \text{hence } \sigma_{ij} = \sigma_{ij}(T,\varepsilon_{ij}) \tag{2.27b}$$

and

$$s = - \left.\frac{\partial g}{\partial T}\right)_{\sigma_{ij}} \qquad \text{hence } s = s(T,\sigma_{ij}) \quad , \tag{2.28a}$$

$$\frac{1}{\rho}\varepsilon_{ij} = - \left.\frac{\partial g}{\partial\sigma_{ij}}\right)_T \qquad \text{hence } \varepsilon_{ij} = \varepsilon_{ij}(T,\sigma_{ij}) \tag{2.28b}$$

In elasticity, it is common to introduce a strain energy function W and a complementary energy W_c defined respectively by

$$\frac{\partial W}{\partial\varepsilon_{ij}} = \sigma_{ij} \quad , \quad \frac{\partial W_c}{\partial\sigma_{ij}} = \varepsilon_{ij} \tag{2.29}$$

By comparison of (2.29a) with (2.22) and (2.27b), it is noted that W can be identified with u or $\rho\Psi$, depending on wether an isentropic or an isothermal deformation is considered. In the same way, by comparing (2.28b) and (2.29b), W_c is seen to coincide with minus the Gibbs free energy $-\rho g$.

Going back to (2.27b) and making the simplest assumption that σ_{ij} is a linear function of T and ε_{ij} , one recovers Neumann-Duhamel's relation

$$\sigma_{ij} = C_{ijkl}\varepsilon_{kl} - \beta_{ij}(T - T_r) \tag{2.30}$$

T_r is a reference temperature, C_{ijkl} is the tensor of the elastic moduli, β_{ij} the tensor of the thermal moduli :

$$C_{ijkl} = \rho \left.\frac{\partial^2\psi}{\partial\varepsilon_{ij}\varepsilon_{kl}}\right)_T = \frac{\partial\sigma_{ij}}{\partial\varepsilon_{kl}} , \quad \beta_{ij} = -\rho\frac{\partial^2\psi}{\partial\varepsilon_{ij}\partial T} = -\frac{\partial\sigma_{ij}}{\partial T} \tag{2.31}$$

If the temperature is uniform, (2.30) reduces to Hooke's law

$$\sigma_{ij} = C_{ijkl}\varepsilon_{kl} \tag{2.32}$$

The C_{ijkl} as well as the β_{ij} satisfy the symmetry relations

$$C_{ijkl} = C_{klij} \quad , \quad C_{ijkl} = C_{jikl} \quad , \quad C_{ijkl} = C_{ijlk}$$

and $\beta_{ij} = \beta_{ji}$.

If the body obeys Hooke's linear law, the strain energy function W takes the quadratic form

$$W = \frac{1}{2} C_{ijkl}\varepsilon_{ij}\varepsilon_{kl} \tag{2.33}$$

and is identical to the complementary strain energy.

An important consequence of the second principle of thermodynamics is that equilibrium is unconditionally stable. It follows that g and henceforth W are positive definite forms ; this result will be frequently used later on.

For an isotropic medium, (2.30) and (2.32) take the form

$$\sigma_{ij} = \lambda_L \varepsilon_{kk}\, \delta_{ij} + 2G\varepsilon_{ij} - \beta(T - T_r)\, \delta_{ij} \tag{2.34a}$$

$$\sigma_{ij} = \lambda_L \varepsilon_{kk}\, \delta_{ij} + 2G\varepsilon_{ij} \tag{2.34b}$$

wherein λ_L and G are the Lamé constants.

Another state equation is furnished by (2.27a) which reads, in differential form

$$ds = \left.\frac{\partial s}{\partial T}\right)_{\varepsilon_{ij}} dT + \left.\frac{\partial s}{\partial \varepsilon_{ij}}\right)_T d\varepsilon_{ij} \tag{2.35}$$

After introduction of the heat capacity defined as

$$\frac{1}{T}\, c_\varepsilon = \left.\frac{\partial s}{\partial T}\right)_{\varepsilon_{ij}} \tag{2.36}$$

and the use of the cross relation

$$\left.\frac{\partial s}{\partial \varepsilon_{ij}}\right)_T = -\frac{1}{\rho} \left.\frac{\partial \sigma_{ij}}{\partial T}\right)_{\varepsilon_{ij}} = \frac{1}{\rho}\, \beta_{ij} \tag{2.37}$$

(2.35) becomes

$$ds = \frac{c_\varepsilon}{T}\, dT + \frac{1}{\rho}\, \beta_{ij}\, d\varepsilon_{ij} \tag{2.38}$$

The elimination of $\dot{u}$ between the energy equation (2.13) and the Gibbs relation (2.20) yields the entropy balance

$$\rho\dot{s} = - J^s_{i,i} + \sigma \tag{2.39}$$

The entropy flux J^s_i and the entropy production per unit volume σ are respectively given by

$$J^s_i = \frac{1}{T} q_i \tag{2.40}$$

$$\sigma = - \frac{1}{T^2} q_i T_{,i} \geq 0 \tag{2.41}$$

σ is a positive quantity, according to the second principle of thermodynamics ; q_i and $T_{,j}$ are related by the Fourier law

$$q_i = -\lambda_{ij} T_{,j} \quad , \quad \lambda_{ij} = \lambda_{ji} \quad \text{(anisotropic medium)} \tag{2.42}$$

$$q_i = - \lambda T_{,i} \qquad \text{(isotropic medium)} \tag{2.43}$$

where λ is the heat conductivity coefficient, depending generally on the temperature. From (2.41), it is seen that only heat conduction contributes to heat dissipation in a thermoelastic body.

The *basic equations of thermoelasticity* are obtained by substituting the state equations (2.30), (2.38) and the Fourier law (2.42) in the momentum and entropy equations (2.12) and (2.39). For an anisotropic body, one has in the limit of small deformations,

$$\rho\, \partial^2_t u_i = C_{ijkl}\, \varepsilon_{kl,j} - \beta_{ij}\, T_{i,j} + F_i \qquad (\partial_t = \frac{\partial}{\partial t}) \tag{2.44}$$

$$\rho c_\varepsilon\, \partial_t T + T\beta_{ij}\, \partial_t \varepsilon_{ij} = (\lambda_{ij} T_{,j})_{,i} \tag{2.45}$$

For an isotropic body, one has

$$\rho\, \partial_t^2 u_i = G\, u_{j,ji} + (\lambda_L + G) u_{i,jj} - \beta\, T_{,i} + F_i \tag{2.46}$$

$$\rho c_\varepsilon\, \partial_t T + T\beta\, \partial_t\, u_{j,j} = (\lambda\, T_{,i})_{,i} \tag{2.47}$$

Of course both equations must be complemented by adequate initial and boundary conditions.

3. Basic variational principles in elastostatics

3.1. The minimum energy principle for small deformations. Consider an elastic body at uniform temperature in *equilibrium* under the action of surface tensions and body forces. The body occupies a simply connected domain Ω whose boundary A consists of two regular subsets A_u and A_σ

$$A = A_u \cup A_\sigma \qquad A_u \cap A_\sigma = 0$$

with the following mixed boundary conditions :

1° The displacement is specified on the portion A_u *(displacement boundary condition)* :

$$u_i = u_i^*(x_i) \qquad \text{on } A_u$$

2° The surface tension is specified on the portion A_σ *(traction boundary condition)* :

$$T_i = T_i^*(u_i, x_i) \qquad \text{on } A_\sigma$$

An upper asterisk denotes a prescribed function. The equilibrium equation is given by

$$\sigma_{ij,j} + F_i = 0 \qquad \text{on } \Omega \tag{3.1}$$

Our objective is to derive a solution for the mixed problem by means of a variational principle. To this end, it is necessary to specify the notion of an *admissible state*. Generally, by admissible state is meant an ordered array of functions (u_i, ε_{ij}, σ_{ij}) with the properties

$$u_i \in C^{1,2} \quad , \quad \varepsilon_{ij} \in C^{o,o} \quad , \quad \sigma_{ij} \in C^{1,o}$$

By an *admissible displacement field* is simply meant an uniform function $u_i\ (x_i,t)$ with

$$u_i \in C^{1,2}$$

A *kinematically admissible displacement field* is an admissible displacement field satisfying the displacement boundary condition.

An *admissible stress field* is defined as a symmetric second order tensor field with

$$\sigma_{ij} \in C^{1,o}$$

A *dynamically admissible stress field* is an admissible stress field obeying the traction boundary condition.

For a continuum medium, the *principle of virtual work* (1.6) takes the form

$$\int_{\Omega} \sigma_{ij} \delta\varepsilon_{ij} d\Omega - \int_{\Omega} F_i\ \delta u_i d\Omega - \int_{A_\sigma} T_i^* \ \delta u_i dA = 0 \qquad (3.2)$$

for all kinematically admissible displacement fields.

It is implicitely admitted that the strain-displacement relation (2.8) is met.

When the existence of a strain energy function is assured, the principle of virtual work (3.2) reads as

$$\delta\int_{\Omega} W(\varepsilon_{ij})\, d\Omega - \int_{\Omega} F_i\, \delta u_i d\Omega - \int_{A_\sigma} T_i^{*} \delta u_i dA = 0\,. \quad (*) \tag{3.3}$$

Application of the δ operator to the first term yields

$$\int \delta W(\varepsilon_{ij}) d\Omega = \int \frac{\partial W}{\partial \varepsilon_{ij}} \delta\varepsilon_{ij} d\Omega = \int \sigma_{ij} \delta\, \frac{1}{2}(u_{i,j} + u_{j,i}) d\Omega = \int \sigma_{ij} \delta u_{i,j} d\Omega$$

Integration by parts and substitution in (3.3) leads to

$$\int_{\Omega} (-\sigma_{ij,j} - F_i)\delta u_i d\Omega + \int_{A} \sigma_{ij} n_j \delta u_i dA - \int_{A_\sigma} T_i^{*}\, \delta u_i dA = 0$$

Since u_i is prescribed on the portion A_u ($\delta u_i = 0$), the second surface integral refers only to the part A_σ of the boundary. Because of the arbitrariness of δu_i on Ω and A_σ, one recovers the basic equations of elastostatics, namely :

$$\sigma_{ij,j} + F_i = 0 \qquad \text{on } \Omega$$

$$\sigma_{ij} n_j = T_i^{*} \qquad \text{(natural boundary condition)} \qquad \text{on } A_\sigma$$

If F_i and T_i^{*} are conservative, which means that they derive from potential functions

$$F_i = -\frac{\partial \phi_F}{\partial u_i} \quad , \quad T_i^{*} = -\frac{\partial \phi_\sigma}{\partial u_i} \tag{3.4}$$

expression (3.3) takes the form

(*) If u_i is prescribed on the whole surface, the surface integral in (3.3) disappears. If T_i is prescribed on the whole surface, the surface integral in (3.3) extends on the entire surface A.

$$\delta V(u_i) = 0 \tag{3.5}$$

where $V(u_i)$ is the potential energy of the system :

$$V(u_i) = \int_\Omega (W + \phi_F)d\Omega + \int_{A_\sigma} \phi_\sigma dA \tag{3.6}$$

Equation (3.5) indicates that among all the kinematically admissible displacements, those satisfying the equilibrium equations render V *stationary*.

It must be observed that nowhere has been invoked the restriction of a linear stress-strain relation like Hooke's law. The principles (3.3) and (3.5) are valid for any elastic body for which a strain energy function exists.

A further simplification arises if F_i and T_i^* are given functions of x_i independent of the deformation field u_i ; this situation is frequently encountered in elasticity. The potential energy V writes then as

$$V(u_i) = \int_\Omega (W - F_i u_i)d\Omega - \int_{A_\sigma} T_i^* u_i dA \tag{3.7}$$

In the general case that F_i and T_i do not derive from a potential and depend on u_i , (3.3) has to be classified as quasi-variational principle for which it is senseless to speak about minimum or maximum properties. On the contrary, if F_i and T_i^* are prescribed functions of x_i, the functional V reaches a *minimum* at equilibrium. To show that, let us calculate the second variation

$$\delta^2 V = \int \delta(\delta W)d\Omega = \int \delta(\frac{\partial W}{\partial \varepsilon_{ij}} \delta\varepsilon_{ij})d\Omega = \frac{1}{2}\int \frac{\partial^2 W}{\partial \varepsilon_{ij}\partial \varepsilon_{kl}} \delta\varepsilon_{ij}\delta\varepsilon_{kl} d\Omega$$

The integrals containing F_i and T_i^* are linear in u_i and therefore do not appear here. Since stability of equilibrium implies that W is positive definite, one must have

$$\delta^2 V > 0 \tag{3.8}$$

It follows that

$$V_{admissible} > V_{exact}$$

On the other hand, the minimum property of the potential energy V forbids that static instability takes place and as a consequence (3.8) represents a stability criterion. Therefore, if

$$\delta^2 V < 0$$

for at least one admissible set of virtual displacements, the configuration is instable. More about stability can, for instance, be found in the works of Trefftz,[5] Koiter,[6,7] Timoshenko.[8]

Uniqueness of the solution. The formulation of variational principles reveals also useful for the proof of the uniqueness of the solution. This is easily done by showing that an assumption of non-uniqueness leads to an absurdity.

Suppose that two solutions $u^{(1)}$ and $u^{(2)}$ are possible and define

$$\delta u_i = u_i^{(1)} - u_i^{(2)}$$

$$\delta \varepsilon_{ij} = \varepsilon_{ij}^{(1)} - \varepsilon_{ij}^{(2)}$$

$$\delta \sigma_{ij} = \sigma_{ij}^{(1)} - \sigma_{ij}^{(2)}$$

$$\delta V = V^{(1)} - V^{(2)}$$

Using Hooke's stress-strain relation, it can be proved that[9]

$$\int \sigma_{ij}^{(1)} \varepsilon_{ij}^{(2)} d\Omega = \int \sigma_{ij}^{(2)} \varepsilon_{ij}^{(1)} d\Omega$$

From the definition (3.7) of V and the above identity, it follows that

$$\delta V = \int_\Omega \delta\sigma_{ij} \, \delta\varepsilon_{ij} d\Omega - \int_\Omega F_i \, \delta u_i d\Omega - \int_{A_\sigma} T_i^* \, \delta u_i d\Omega$$

This quantity vanishes accordingly to the minimum energy principle. The first term in the r.h.s. can still be expressed as

$$\int_\Omega \delta\sigma_{ij} \delta\varepsilon_{ij} d\Omega = \int_\Omega \delta\sigma_{ij} \frac{1}{2} (\delta u_{i,j} + \delta u_{j,i}) d\Omega = - \int_\Omega \delta u_i \delta\sigma_{ij,j} d\Omega$$

$$+ \int_A \delta u_i \delta\sigma_{ij} n_j dA$$

Note also that by virtue of the law of equilibrium

$$\delta\sigma_{ij,j} = \sigma_{ij,j}^{(1)} - \sigma_{ij,j}^{(2)} = \rho F_i - \rho F_i = 0$$

so that expression of δV reduces to

$$\delta V = - \int_\Omega F_i \left(u_i^{(1)} - u_i^{(2)}\right) d\Omega + \int_A \left(u_i^{(1)} - u_i^{(2)}\right) \left(T_i^{(1)} - T_i^{(2)}\right) dA$$

$$- \int_{A_\sigma} T_i \left(u_i^{(1)} - u_i^{(2)}\right) dA$$

Consider now the following situations

(1) $u_i^{(1)} = u_i^{(2)}$ on A_u , $T_i^{(1)} = T_i^{(2)}$ on A_σ ,

(2) $u_i^{(1)} = u_i^{(2)}$ on A (u_i prescribed on the entire surface)

(3) $T_i^{(1)} = T_i^{(2)}$ on A (T_i prescribed on the entire surface)

Under the first condition, the vanishing of δV implies that

$$u_i^{(1)} = u_i^{(2)} \quad \text{on } \Omega \quad \text{and} \quad u_i^{(1)} = u_i^{(2)} \quad \text{on } A_\sigma$$

while for the second and third conditions, the variational principle imposes that

$$u_i^{(1)} = u_i^{(2)} \quad \text{on } \Omega$$

and

$$u_i^{(1)} = u_i^{(2)} \quad \text{on } \Omega \quad \text{and} \quad u_i^{(1)} = u_i^{(2)} \quad \text{on } A$$

These results prove the uniqueness of the thermoelastic problem, at least for linear stress-strain relations.

3.2. The Reissner and the complementary energy principles for small deformations. Instead of varying with respect to the displacement field, take the variation with respect to the stress and determine the functional which is extremum for the actual stress field. This amounts to construct the dual principle of the minimum energy principle. The procedure is the following.

a. The principle of minimum energy can be reformulated by introducing the relations

$$\varepsilon_{ij} = \frac{1}{2}(u_{j,i} + u_{i,j}) \quad \text{on } \Omega$$

$$u_j = u_j^* \quad \text{on } A_u$$

as side conditions. This yields

$$\delta\int_{\Omega}\left\{W(\varepsilon_{ij}) - \gamma_{ij}\left(\varepsilon_{ij} - \frac{1}{2}(u_{i,j} + u_{j,i})\right)\right\}d\Omega - \int_{\Omega} F_j \delta u_j d\Omega$$

$$- \int_{A_\sigma} T_j^* \delta u_j \, dA - \delta \int_{A_u} \alpha_j (u_j - u_j^*) dA = 0 \qquad (3.9)$$

A first simplification arises by restituting to the Lagrange multipliers γ_{ij} and α_j, their physical meaning. By varying with respect of ε_{ij} and u_j independently, it is seen that

$$\frac{\partial W}{\partial \varepsilon_{ij}} - \gamma_{ij} = 0 \qquad \text{on } \Omega$$

$$\gamma_{ji} n_i - \alpha_j = 0 \qquad \text{on } A_u$$

which means that γ_{ij} can be identified with the stress tensor σ_{ij} and α_j with the stress vector T_j on A_u. Moreover the quantity

$$\sigma_{ij}\varepsilon_{ij} - W(\varepsilon_{ij}) = W_c(\sigma_{ij}) \qquad (3.10)$$

which appears in the integrant of (3.9), is the complementary energy W_c, with the property

$$\frac{\partial W_c}{\partial \sigma_{ij}} = \varepsilon_{ij}$$

b. With the above results in mind, (3.9) can be written as

$$\delta\int_\Omega \left\{W_c\,(\sigma_{ij}) - \sigma_{ij}\,\frac{1}{2}\,(u_{j,i} + u_{i,j})\right\} d\Omega + \int_\Omega F_j \delta u_j d\Omega$$

$$+ \int_A T_j^* \,\delta u_j dA + \delta\int_{A_u} n_i\sigma_{ij}\,(u_j - u_j^*)dA = 0 \qquad (3.11)$$

The above principle is the so-called *canonical variational principle*. It was first proposed by *Reissner*[10] : it allows to formulate independent approximations for u_i and σ_{ij}. By variation of these quantities, one obtains the necessary conditions for (3.11) to be stationary, namely

$$\sigma_{ij,j} + F_i = 0 \qquad \text{on } \Omega$$

$$\frac{\partial W_c}{\partial \sigma_{ij}} - \frac{1}{2}\,(u_{j,i} + u_{i,j}) = 0 \qquad \text{on } \Omega$$

while at the boundaries,

$$-\,\sigma_{ij}\,n_j + T_j^* = 0 \qquad \text{on } A_\sigma$$

$$u_j = u_j^* \qquad \text{on } A_u$$

Reissner's principle is valid for non-linear stress-strain relations. However, it is important to note that minimum properties are no longer guaranteed.

c. We are now in position to formulate the complementary energy principle. From now on, let F_i and T_i^* be prescribed functions of the coordinates. Integration by parts of the second term in (3.11), namely

$$\int_\Omega \sigma_{ij}\,\frac{1}{2}\,(u_{i,j} + u_{j,i})d\Omega = \int_\Omega \sigma_{ij}u_{j,i}d\Omega = -\int\sigma_{ij,i}u_j d\Omega + \int_A n_i\,\sigma_{ij}u_j dA$$

yields the following result

$$\delta\left\{\int_{\Omega} W_c(\sigma_{ij}) + (\sigma_{ij,i} + F_j)u_j \, d\Omega + \int_{A_\sigma} (T_j^* - n_i\sigma_{ij})u_j dA - \int_{A_u} u_j^* \, n_i\sigma_{ij} dA\right\} = 0 \tag{3.12}$$

Assuming that the equilibrium and the traction relations are satisfied a priori, the above equation can be expressed as

$$\delta V_c(\sigma_{ij}) = 0 \tag{3.13}$$

where V_c is the total complementary energy depending only on the stress tensor :

$$V_c(\sigma_{ij}) = \int_{\Omega} W_c(\sigma_{ij}) d\Omega - \int_{A_u} u_j^* n_i \sigma_{ij} dA \tag{3.14}$$

If no complementary energy W_c can be defined, (3.13) must be replaced by the complementary virtual work principle :

$$\int_{\Omega} \varepsilon_{ij} \delta\sigma_{ij} d\Omega - \delta \int_{A_u} u_j^* n_i \sigma_{ij} dA = 0 \tag{3.15}$$

The Euler-Lagrange equations are obtained by setting the first variation of (3.14) equal to zero, i.e.

$$\delta V_c = \int_{\Omega} \varepsilon_{ij} \delta\sigma_{ij} d\Omega - \int_{A_u} u_j^* \, \delta(\sigma_{ij} n_i) dA = 0 \tag{3.16}$$

together with the conditions

$(\delta\sigma_{ij})_{,j} = 0$ (since $\delta F_i = 0$) on Ω

$(\delta\sigma_{ij}) \, n_i = 0$ (since $\delta T_j^* = 0$) on A_σ

These restrictions express that the equilibrium equation in Ω and the Cauchy relation on A_σ must be automatically verified.

It can be checked that (3.16) is formally satisfied by taking as solution,

$$\delta\sigma_{11} = \phi_{22,23} + \phi_{33,22} - 2\phi_{23,23}$$

$$\delta\sigma_{22} = \phi_{33,11} + \phi_{11,33} - 2\phi_{33,31}$$

$$\delta\sigma_{33} = \phi_{11,22} + \phi_{22,11} - 2\phi_{12,12}$$

$$\delta\sigma_{23} = \phi_{31,12} + \phi_{12,13} - \phi_{11,23} - \phi_{23,11}$$

$$\delta\sigma_{31} = \phi_{12,23} + \phi_{23,21} - \phi_{22,31} - \phi_{31,22}$$

$$\delta\sigma_{12} = \phi_{23,31} + \phi_{31,22} - \phi_{33,12} - \phi_{12,33}$$

where $\phi_{ij} = \phi_{ji}$ are arbitrary stress functions.

By putting

$$\phi_{12} = \phi_{23} = \phi_{31} = 0$$

one obtains Maxwell's solution while by setting

$$\phi_{11} = \phi_{22} = \phi_{33} = 0$$

one has Morera's solution. Substituting Maxwell's solution in (3.16) and integrating twice by parts leads to

$$\delta V_c = \int_\Omega \Big\{(\varepsilon_{22,33} + \varepsilon_{33,12} - 2\,\varepsilon_{23,23})\phi_{11} + (\varepsilon_{33,11} + \varepsilon_{11,33} - 2\,\varepsilon_{31,31})\phi_{22} + (\varepsilon_{11,22} + \varepsilon_{22,11} - 2\,\varepsilon_{12,12})\phi_{33}\Big\}d\Omega + \text{a surface integral}^{(*)} = 0 \qquad (3.17)$$

Since the functions ϕ_{11} , ϕ_{22} , ϕ_{33} are arbitrary, the corresponding Euler-Lagrange equations are

$$\varepsilon_{ii,jj} + \varepsilon_{jj,ii} - 2\,\varepsilon_{ij,ij} = 0 \quad i,j = 1,2,3 \qquad (3.18)$$

which is the first set of *compatibility equations* (2.9). The second set of compatibility equations (2.10) is obtained with the help of Morera's solution.

The complementary variational principle states that among all the dynamically admissible stress fields meeting the field eq., the total complementary energy function is stationary if and only if σ_{ij} is solution of the compatibility conditions.

To prove that V_c is a minimum, one calculates the second variation of V_c . From its expression (3.14), it is directly seen that

$$\delta^2 V_c = \int_\Omega \delta\left(\frac{\partial W_c}{\partial\sigma_{ij}}\,\delta\sigma_{ij}\right)d\Omega = \int_\Omega \frac{\partial^2 W_c}{\partial\sigma_{ij}\partial\sigma_{kl}}\,\delta\sigma_{ij}\delta\sigma_{kl}\,d\Omega \qquad (3.19)$$

As W_c is positive definite,

$$\delta^2 V_c > 0$$

(*) The analysis of the surface integral is rather intricate and yields some relations to be satisfied between the ϕ_{ij}'s and their derivatives.

and the stationary value of V_c is a minimum.

It follows that the principle of minimum complementary energy can be stated as follow : "Among all the stress fields satisfying the equilibrium equation and the Cauchy relation on the boundary where the stresses are prescribed, the one satisfying the compatibility equations makes V_c minimum".

To recapitulate, we have seen that a complementary variational principle can be constructed whenever a potential energy V_c exists, while the minimum property implies in addition the positive definiteness of the complementary strain energy W_c .

Castigliano theorem. Suppose that the body is subjected to a number of concentrated loads $P_1 \dots P_n$ and torques $M_1 \dots M_\ell$. After expressing W_c in terms of these loads and torques, consider a variation $\delta\sigma_{ij}$ which corresponds to a change in P_1 only. Then by virtue of (3.13) and (3.14), one has

$$\left(\frac{\partial}{\partial P_1} \int_\Omega W_c d\Omega\right) \delta P_1 = u_1 \delta P_1$$

where u_1 is the displacement of the point of application of P_1. It follows from the above relation that u_i is obtained by simple derivation of the total complementary energy :

$$u_i = \frac{\partial}{\partial P_i} \int_\Omega W_c d\Omega \qquad i = 1,\dots n$$

Similarly, it can be checked that the angles of rotation Θ_i are obtained from

$$\Theta_i = \frac{\partial}{\partial M_i} \int_\Omega W_c d\Omega \qquad i = 1,\dots \ell$$

These results were first established by Castigliano and are referred in the literature as Castigliano's theorem.

3.3. An extension of Reissner's principle : Hu-Washizu's principle. This principle generalizes Reissner's criterion and is governed by a functional wherein all the quantities, namely the displacement field u_i , the stress tensor σ_{ij} and the strain tensor ε_{ij} are varied. Hu-Washizu's principle states that

$$\delta J(u_i,\sigma_{ij},\varepsilon_{ij}) = \delta\int_\Omega \Big\{W(\varepsilon_{ij}) - \sigma_{ij}\big(\varepsilon_{ij} - \tfrac{1}{2}(u_{i,j} + u_{j,i})\big)$$

$$- F_i u_i\Big\}d\Omega - \delta\int_{A_u} \sigma_{ij} n_j (u_i - u_i^*)\,dA - \delta\int_{A_\sigma} T_i^* u_i\,dA = 0 \tag{3.21}$$

The Euler-Lagrange equations are obtained by application of the δ operator to the integrant of J :

$$\delta J = \int_\Omega \Big\{\frac{\partial W}{\partial \varepsilon_{ij}}\,\delta\varepsilon_{ij} - \delta\sigma_{ij}\big(\varepsilon_{ij} - \tfrac{1}{2}(u_{i,j} - u_{j,i})\big)$$

$$-\sigma_{ij}\big(\delta\varepsilon_{ij} - \tfrac{1}{2}(\delta u_{i,j} + \delta u_{j,i})\big) - F_i\,\delta u_i\Big\}d\Omega - \int_{A_u} (\delta\sigma_{ij}) n_j (u_i - u_i^*)\,dA$$

$$- \int_{A_u} \sigma_{ij} n_j \delta u_i\,dA - \int_{A_\sigma} T_i^*\,\delta u_i\,dA = 0 \tag{3.22}$$

Integration by parts of the terms involving $u_{i,j}$ results in

$$\frac{\partial W}{\partial \varepsilon_{ij}} = \sigma_{ij} \qquad \text{on } \Omega \tag{3.23a}$$

$$\sigma_{ij,j} + F_i = 0 \qquad \text{on } \Omega \tag{3.23b}$$

$$\varepsilon_{ij} = \tfrac{1}{2}(u_{i,j}+u_{j,i}) \quad \text{on } \Omega \tag{3.24a}$$

$$\sigma_{ij} n_j = T_i^* \qquad \text{on} \quad A_\sigma \tag{3.24b}$$

$$u_j - u_j^* = 0 \qquad \text{on} \quad A_u \tag{3.24c}$$

These are the basic equations of elastostatics.

Hu-Washizu's principle can be obtained from the principle of minimum energy

$$\delta \int_\Omega \left\{ W(\varepsilon_{ij}) - F_i u_i \right\} d\Omega - \delta \int_{A_\sigma} T_i^* u_i \, dA = 0$$

by adding the strain-displacement relation and the displacement boundary condition as subsidiary conditions by means of Lagrange multipliers. Clearly σ_{ij} and $\sigma_{ij} n_j$ play the role of Lagrange multipliers.

The above principles of minimum potential energy as well as Reissner or Hu-Washizu's principles have been extended to plasticity. A description of these criteria can be found in ref.[11,12,13].

3.4. Illustrative examples.

i) Bending of a beam under static loading. Consider a beam perfectly straight, lying along the x axis (see fig.3.1) before application of an external loading, consisting of

a distributed lateral load p(x) per unit length,

a bending moment M_0 and a shearing force Q_0 at the end $x = 0$,

a bending moment M_1 and a shearing force Q_1 at the end $x = \ell$,

axial forces of intensity P.

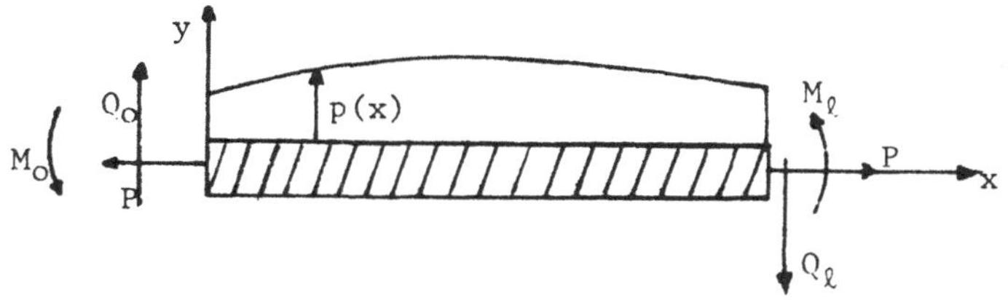

Fig. 3.1.

It is assumed that every cross-section remains plane and that the deflection is infinitesimally small. Under these hypotheses, the theory of elasticity tells us that the strain energy is given by

$$W = \frac{1}{2} EI(y'')^2 + \frac{1}{2} EA(u')^2 \quad (' = \frac{d}{dx})$$

the quantity y denotes the lateral deflection, u(x) is the displacement in the direction x, A the area of the cross section, E is Young's modulus $\left(E = \frac{G(3\lambda_\ell + 2G)}{\lambda_\ell + G}\right)$, I is the moment of inertia of a cross section with respect to the neutral axis :

$$I = \iint y^2 dydz$$

The potential energy due to the external forces is

$$-\int_0^\ell p(x)\ y(x)dx + M_o(y')_o - M_\ell(y')_\ell - Q_o y_o + Q_\ell y_\ell$$

$$- P_\ell u_\ell + P_o u_o \tag{3.26}$$

The total potential energy is given by

$$V = \frac{1}{2}\int_0^\ell \left(I(y'')^2 + EA(u')^2\right)dx - \int_0^\ell p(x)y(x)dx + M_o(y')_o - Q_o y_o$$

$$M_1(y')_\ell + Q_\ell y_\ell - P_\ell u_\ell + P_o u_o \tag{3.27}$$

At equilibrium, one has

$$\delta V = \int_0^\ell \left(EI(y'')^2 \delta y'' + EAu'\delta u - p\delta y\right)dx$$

$$+ M_o\delta(y')_o - M_\ell\delta(y')_\ell - Q_o\delta y_o - Q_\ell\delta y_\ell - F_\ell\delta u_\ell + P_o\delta u_o = 0 \tag{3.28}$$

By integrating the first term twice by parts, one gets

$$\delta V = \int_0^\ell \left((EIy'')''-p\right)\delta y dx - \int_0^\ell EAu''\delta u dx + \left(EI(y'')_\ell - M_\ell\right)\delta(y')_\ell$$

$$- \left(EI(y'')_o - M_o\right)\delta(y')_o - \left\{\left(EI(y'')_\ell\right)' - Q_1\right\}\delta y_\ell$$

$$+ \left\{\left(EI(y'')_o\right)' - Q_o\right\}\delta y_o + \left(EAu'_\ell - F_\ell\right)\delta u_\ell \tag{3.29}$$

$$- \left(EAu'_o - P_o\right)\delta u_o = 0$$

from which results

$$(EIy'')'' - p = 0 \qquad 0 < x < \ell$$

$$u'' = 0 \qquad 0 < x < \ell \tag{3.30}$$

while at the boundaries, $x = 0$ and $x = \ell$, the following conditions must be satisfied

either $EIy'' - M = 0$ (3.31a) or $\delta y' = 0$ (3.31b)

either $(EIy'')' - Q = 0$ (3.32a) or $\delta y = 0$ (3.32b)

either $EAu' - P = 0$ or $\delta u = 0$

If y and y' are specified at both ends $x = 0$ and $x = \ell$, the boundary conditions are essential and no supplementary restrictions must be imposed. In contrast, if y and y' are perfectly free and unspecified at the ends, (3.31a) and (3.32a) must be verified in order that δV vanishes : such conditions are natural boundary conditions.

ii) A beam with a clamped and a supported end, loaded at its middle. To illustrate the search for approximate solutions, consider a horizontal straight beam clamped at one extremity and simply supported at the other end : the lateral load is a constant force $\underline{F}$ acting at $x = \ell/2$ (see fig. 3.2). Since no external force is acting in the x direction, one may take

$$u_x = - yy' \qquad u_y = y \qquad u_z = 0$$

$$\varepsilon_{xx} = - yy''$$

$$\sigma_{xx} = E\ \varepsilon_{xx} = \frac{M}{I}\ y$$

where M is the bending moment. By combining the two last relations, one has

$$M = - EIy'' \tag{3.33}$$

The boundary conditions are

$$y = 0 \qquad y' = 0 \qquad \text{at } x = 0 \text{ (clamped end)} \tag{3.34}$$

$$y = 0 \qquad y'' = 0 \qquad \text{at } x = \ell \text{ (supported end)} \tag{3.35}$$

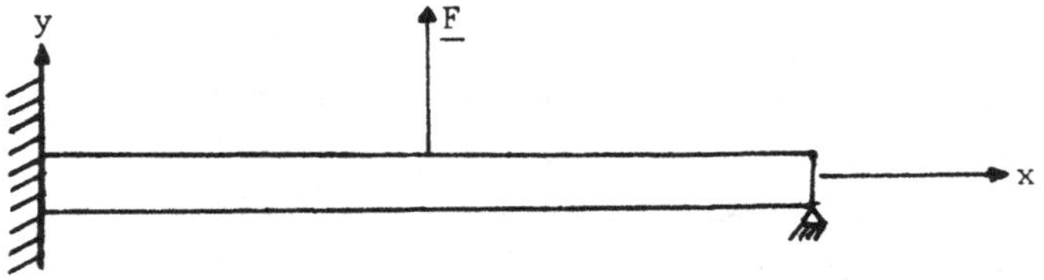

Fig. 3.2.

The principle of minimum potential energy is given by

$$\delta V = 0$$

with

$$V = \frac{1}{2}\int_0^{\ell} EI(y'')^2 dx - (Fy)_{x=\frac{\ell}{2}} \qquad (3.36)$$

As trial function for the deflection, one takes[14]

$$y = a_1 \; \ell \left(\frac{x}{\ell}\right)^2 \left(1 - \frac{x}{\ell}\right) \qquad (3.37)$$

where a_1 is the constant parameter to be determined by the Ritz method. Note that (3.37) verifies only the essential boundary conditions. After substitution of (3.37) in the expression (3.33) of V, one obtains

$$V = 2\, a_1^2 \frac{FI}{\ell} - \frac{a_1}{8} \ell F$$

Since V is a minimum

$$\frac{\partial V}{\partial a_1} = 0$$

from which results

$$a_1 = \frac{\ell}{32} \frac{\ell^2 F}{EI}$$

so that a first approximation for the deflection is

$$y = \frac{\ell^3}{32EI} F \left(\frac{x}{\ell}\right)^2 \left(1 - \frac{x}{\ell}\right) \qquad (3.38)$$

To test the quality of this solution, take a more complicated trial function involving two parameters a_1 and a_2 :

$$y = \ell\, (\frac{x}{\ell})^2 \left(a_1 + a_2 \frac{x}{\ell} - (a_1 + a_2)\, (\frac{x}{\ell})^2\right) \tag{3.39}$$

Following the same procedure as above, one obtains

$$a_1 = \frac{7}{64}\frac{\ell^2 F}{EI} \quad , \qquad a_2 = -\frac{12}{64}\frac{\ell^2}{EI} F$$

from which

$$y = \frac{\ell^3}{32EI} F \,(\frac{x}{\ell})^2 \left(\frac{7}{2} - 6\frac{x}{\ell} - \frac{5}{2}(\frac{x}{\ell})^2\right) \tag{3.40}$$

which is markedly different from (3.38) but in better agreement with exact calculations. One verifies also that

$$V(a_1) > V(a_1, a_2)$$

which shows that the second approximation is closer to the minimum value of V , attained for the true solution.

iii) The complementary energy principle. According to the general procedure of § 3.2 , introduce as side conditions

$$\varepsilon_{xx} = - yy'' \qquad\qquad 0 < x < \ell$$

$$y = y' = 0 \quad \text{at} \quad x = 0 \qquad ; \qquad y = 0 \quad \text{at} \quad x = \ell$$

and construct the variational equation

$$\delta\left\{\int\left(\frac{1}{2} E\ \varepsilon_{xx}^2 - \lambda(\varepsilon_{xx} + yy'')\right)dxdydz - Fy)_{x=\ell/2}\right.$$
$$\left. - \int dydz\left(\alpha_1(y)_{x=0} + \alpha_2(y')_{x=0} + \alpha_3(y)_{x=\ell}\right)\right\} = 0 \tag{3.41}$$

where λ , α_1, α_2, α_3 are Lagrange multipliers. By varying with respect to ε_{xx} and y, it is seen that

$$\lambda = \sigma_{xx}$$

$$\alpha_1 = - M'(o) = 0$$

$$\alpha_2 = M(o) = 0$$

$$\alpha_3 = M'(\ell)$$

Substituting these results in (3.41) and recalling that the complementary strain energy is defined by

$$\int W_c\ dxdydz = \int(\sigma_{xx}\varepsilon_{xx} - \frac{1}{2} E\ \varepsilon_{xx}^2)dxdydz = \int_o^\ell \frac{1}{2}\frac{M^2}{EI}\ dx\ , \tag{3.42}$$

one obtains Reissner's canonical principle in the form

$$\delta\left\{\int_o^\ell (\frac{1}{2}\frac{M^2}{EI} + My'')dx + F(y)_{\ell/2} + (y\ M')_\ell\right\} = 0 \tag{3.43}$$

The Euler-Lagrange equations for arbitrary variations of M and y are

$$M/EI + y'' = 0 \tag{3.44}$$

$$M'' + (F)_{y=\ell/2} = 0 \tag{3.45}$$

with as boundary condition,

$$(M)_\ell = 0 \tag{3.46}$$

With (3.45) and (3.46) as side conditions, (3.43) reduces to the complementary energy principle

$$\delta V_c \equiv \delta \int_0^\ell \frac{1}{2} \frac{M^2}{EI}\, dx = 0 \tag{3.47}$$

This principle reveals convenient to determine the unknown support reaction R acting at x = ℓ. Therefore, take for M the solution of equation (3.45), namely

$$M = (R - \frac{F}{2})\ell(1 - 2\frac{x}{\ell}) + R\ell\frac{X}{\ell} \qquad 0 < x \leq \frac{\ell}{2} \tag{3.48a}$$

$$M = R\ell(1 - \frac{x}{\ell}) \qquad \frac{\ell}{2} \leq x \leq \ell \tag{3.48b}$$

With (3.48a) and (3.48b) in (3.47), it follows after derivation with respect to R :

$$\frac{\partial V_c}{\partial R} = \frac{\ell^3}{3EI}\,(R - \frac{5}{16}F)$$

Therefore, for stationary V_c ,

$$R = \frac{5}{16}F$$

It is also verified that

$$\frac{\partial^2 V_c}{\partial R^2} > 0$$

which indicates that the total complementary energy is a minimum.

4. Variational principles for large deformations

4.1. Principle of stationary potential energy. The principle of minimum energy is readily extended to large deformations provided that the stress-strain relation can be expressed by means of a symmetric potential function W : explicitely, one must have

$$S_{ij} = \frac{\partial W}{\partial E_{ij}} \tag{4.1}$$

with

$$\frac{\partial W}{\partial E_{ij}} = \frac{\partial W}{\partial E_{ji}} \tag{4.2}$$

The body occupies a domain Ω bounded by a surface A in the deformed state to which corresponds a volume Ω_o and a surface A_o (= $A_u^o \cup A_\sigma^o$) in the original undeformed state. Consider the functional

$$V(u_j) = \int_{\Omega_o} W d\Omega_o - \int_{\Omega_o} F_{oi} u_i d\Omega_o - \int_{A^o} T^*_{oi} u_i dA_o \tag{4.3}$$

On the portion A_u^o of the surface A_o , one has

$$u_i = u_i^* \qquad \text{on } A_u^o \tag{4.4}$$

while on the complementary part A_σ^o ,

$$T_{oi} = T^*_{oi} \ (\text{with } T_{oi} = T_{ji} n_j^o) \ \text{ on } A_\sigma^o \tag{4.5}$$

Assume that the displacement boundary condition is identically satisfied. Since V is a functional of the displacements field, its variation with respect to u_i is given by

$$\delta V = \int_{\Omega_o} \frac{\partial W}{\partial E_{k\ell}} \frac{\partial E_k}{\partial(\partial u_i/\partial X_j)} \delta(\partial u_i/\partial X_j) d\Omega_o - \int_{\Omega_o} F_{oi} \delta u_i d\Omega_o$$

$$- \int_{A_\sigma^o} T^{*}_{oi} \delta u_i dA_o \tag{4.6}$$

if it is admitted that F_{oi} and T^{*}_{oi} are only function of X_i . In virtue of the relations (4.1) and (2.6), the first term in the r.h.s. writes as

$$\int_{\Omega_o} S_{k\ell}\left(\delta_{ki}\delta_{\ell j} + \delta_{kj}\delta_{\ell i} + \frac{\partial u_i}{\partial X_\ell}\delta_{kj} + \frac{\partial u_i}{\partial X_k}\delta_{\ell j}\right)\delta(\partial u_i/\partial X_j) d\Omega_o$$

$$= \int_{\Omega_o} \left(S_{j\ell}\,\delta_{i\ell} + S_{j\ell}\frac{\partial u_i}{\partial X_\ell}\right)\delta(\partial u_i/\partial X_j) d\Omega_o = \int_{\Omega_o} S_{j\ell}\frac{\partial x_i}{\partial X_\ell}\frac{\partial}{\partial X_j}(\delta u_i) d\Omega_o$$

To establish the last result, equation (2.5) has been used.

After integration by parts and substitution in (4.6), one obtains

$$\delta V = - \int_{\Omega_o} \left(\frac{\partial}{\partial X_j}\left(S_{j\ell}\frac{\partial x_i}{\partial X_\ell}\right) + F_{oi}\right) \delta u_i d\Omega_o + \int_{A^o}\left(S_{j\ell}\frac{\partial x_i}{\partial X_\ell} n_j^o\right.$$

$$\left. - T^{*}_{oi}\right)\delta u_i dA_o$$

The Euler-Lagrange equations to be satisfied in order that

$$\delta V = 0 \tag{4.7}$$

for arbitrary values of δu_i on Ω_o and A_σ^o are :

$$\frac{\partial}{\partial X_j}(T_{ji}) + F_{oi} = 0 \qquad \text{on } \Omega_o \tag{4.8}$$

$$T_{ji}n_j^o = T^{*}_{oi} \qquad \text{on } A_\sigma^o \tag{4.9}$$

i.e. the equilibrium equations. To derive the above results use has been made of the relation,

$$T_{ij} = S_{i\ell}\frac{\partial x_j}{\partial X_\ell} \tag{4.10}$$

The principle states that among all kinematically admissible displacements, the actual displacement is the one which makes the potential energy stationary. Of course, it is no question of a minimum principle because a Taylor expansion of W around its equilibrium value cannot be limited to the second order terms.

When no strain energy W exists, the variational equation (4.7) must be replaced by a generalized principle of virtual work, taking the form

$$\int_{\Omega_o} S_{ij}\delta E_{ij}d\Omega_o - \int_{\Omega_o} F_{oi}\delta u_i d\,\Omega_o - \int_{A_\sigma^o} T^{*}_{oi}\delta u_i dA_o = 0$$

4.2. A canonical principle. Starting from expression (4.3), a canonical principle extending Reissner's criterion can be formulated. By using

$$E_{ij} = \frac{1}{2}\left(\frac{\partial u_i}{\partial X_j} + \frac{\partial u_j}{\partial X_i} + \frac{\partial u_k}{\partial X_i}\frac{\partial u_k}{\partial X_j}\right) \tag{4.11}$$

as side condition, the energy principle reads as

$$\delta\int_{\Omega_o}\left\{W(E_{ij}) + \frac{1}{2} S_{ij}\left(\frac{\partial u_i}{\partial X_j} + \frac{\partial u_j}{\partial X_j} + \frac{\partial u_k}{\partial X_i}\frac{\partial u_k}{\partial X_j} - 2 E_{ij}\right)\right\}d\Omega_o$$

$$- \delta\int_{\Omega_o} F_{io}u_i d\Omega_o - \delta\int_{A_\sigma^o} T^*_{oi}u_i dA_o = 0 \tag{4.12}$$

wherein the Lagrange multiplier has been identified with the second Piola-Kirchoff tensor. Equation (4.12) is a two-field principle since the field of the Green tensor E_{ij} is made independent of the displacement field.

The introduction of the following Legendre (or contact) transform

$$S_{ij}E_{ij} - W(E_{ij}) = W_c(S_{k\ell}) \qquad (E_{ij} = \frac{\partial W_c}{\partial S_{ij}}) \tag{4.13}$$

rules out E_{ij} as independent variable and furnishes the canonical form :

$$\delta\int_{\Omega_o}\left\{W_c(S_{ij}) - S_{ij}\frac{1}{2}\left(\frac{\partial u_i}{\partial X_j} + \frac{\partial u_j}{\partial X_i} + \frac{\partial u_k}{\partial X_i}\frac{\partial u_k}{\partial X_j}\right)\right\}d\Omega_o$$

$$+ \delta\int_{\Omega_o} F_{oi}u_i d\Omega_o + \delta\int_{A_\sigma^o} T^*_{oi}u_i dA_o = 0 \tag{4.14}$$

Consider now arbitrary variations of both S_{ij} and u_j. It is easy to check that the Euler-Lagrange equations corresponding to (4.14) are

$$\frac{\partial W_c}{\partial S_{ij}} = E_{ij} \qquad \text{on } \Omega_o \tag{4.15}$$

$$\frac{\partial}{\partial X_j}\left(S_{j\ell}\frac{\partial x_i}{\partial X_\ell}\right) + F_{oi} = 0 \qquad \text{on } \Omega_o \tag{4.16}$$

while on the boundaries, one has

$$\delta u_i = 0 \qquad \text{on } A_u^o \tag{4.17}$$

$$- S_{j\ell} \frac{\partial x_i}{\partial X_\ell} n_j^o + T_{oi}^* = 0 \qquad \text{on } A_\sigma^o \tag{4.18}$$

Equations (4.16) to (4.18) are the basic equations of finite elasticity while (4.15) is the state equation corresponding to (4.1). Clearly, the canonical principle (4.14) constitutes a generalization to finite displacements of the Reissner principle of linear elasticity.

4.3. <u>A complementary energy principle.</u> By discarding in (4.14) the quadratic term $\frac{\partial u_k}{\partial X_i} \frac{\partial u_k}{\partial X_j}$ and identifying $\frac{\partial x_i}{\partial X_j}$ with the Kronecker symbol, one would recover, after integrating by parts and prescribing S_{ij} to obey the equilibrium equations of the linear theory, the complementary principle (3.13). Unfortunately, this procedure cannot be extended to large deformations.

However, when instead of S_{ij}, one uses T_{ji} as variable, a complementary principle for large deformations can be generated.

If W is regarded as a function of the deformation gradient, it is seen from (4.6) that by derivating it with respect to $\partial u_j/\partial X_i$, one obtains the first Piola-Kirchoff tensor, namely :

$$T_{ij} = \frac{\partial W}{(\partial u_j/\partial X_i)} \tag{4.19}$$

Instead of (4.13) define now the Legendre transform

$$\overline{W}_c(T_{k\ell}) = T_{ij} \frac{\partial u_j}{\partial X_i} - W \tag{4.20}$$

where

$$\frac{\partial \overline{W}_c}{\partial T_{ij}} = \frac{\partial u_j}{\partial X_i}$$

We are then in position to write a new canonical principle, analogous to (4.14), and given by

$$\delta\left\{\int_{\Omega_o} \left(- T_{ij} \frac{\partial u_j}{\partial X_i} + \overline{W}_c\right) d\Omega_o + \int_{\Omega_o} F_{oj} u_j d\Omega_o + \int_{A^o_\sigma} T^{*}_{oj} u_j dA_o\right\} = 0 \quad (4.21)$$

Integrating by parts and imposing the equilibrium equations (4.8) and (4.9) to be fulfilled a priori, one derives the complementary energy principle in the form

$$\delta\overline{V}_c(T_{ij}) = \delta\left\{\int_{\Omega_o} - \overline{W}_c d\Omega_o + \int_{A^o_u} T_{ij} n^o_i u^{*}_j dA\right\} = 0 \quad (4.22)$$

More about the construction of complementary principles in the non-linear theory of elasticity can be found in the references[15-18].

5. The elastodynamic problem.

5.1. The law of varying action and Hamilton's principle. For the sake of simplicity, we shall limit our considerations to *small deformations*.

The governing equation of elastodynamics is

$$\rho \partial^2_t u_i = \sigma_{ij,j} + F_i \quad (5.1)$$

with

$$\sigma_{ij} = C_{ijk\ell}\, \varepsilon_{k\ell} \quad (5.2)$$

and

$$\varepsilon_{ij} = \frac{1}{2} (u_{i,j} + u_{j,i}) \tag{5.3}$$

ρ is the density of the material assumed to be constant.

To this equation, one adds the following boundary conditions :

$$u_i(x_i,t) = u_i^*(x_i,t) \qquad \text{on } A_u x (t_o,t_1) \tag{5.4}$$

$$T_i(x_i,t) = T_i^*(x_i,t) \qquad \text{on } A_\sigma x (t_o,t_1) \tag{5.5}$$

and the initial conditions

$$u_i(x_i,t) = u_i^o(x_i) \qquad \text{at } t = t_o \tag{5.6}$$

$$\partial_t u_i(x_i,t) = v_i^o(x_i) \qquad \text{at } t = t_o \tag{5.7}$$

The variational principle which produces the above set of equations is the law of varying action (1.4) :

$$\delta \int_{t_o}^{t_1} (K-\phi)dt + \int_{t_o}^{t_1} \sum_\alpha Q^\alpha \delta q^\alpha dt - \sum_\alpha \frac{\partial K}{\partial \dot{q}^\alpha} \delta q^\alpha \Big|^{t_1} = 0 \tag{5.8}$$

wherein K is the kinetic energy of deformation

$$K = \frac{1}{2}\int \rho\, \partial_t u_i \partial_t u_i d\Omega \ , \tag{5.9}$$

ϕ is the total strain energy function

$$\phi = \int W d\ \Omega \tag{5.10}$$

while the virtual work $\sum_\alpha Q^\alpha \delta q^\alpha$ is given by

$$\sum_\alpha Q^\alpha \delta q^\alpha = \int_\Omega F_i \delta u_i d\Omega + \int_{A_\sigma} T_i^* \delta u_i dA \tag{5.11}$$

Substituting (5.9) to (5.11) in (5.8) yields the explicit form of the principle for elasticity, namely :

$$\int_{t_o}^{t_1} dt \left\{ \delta \int_\Omega \left(\frac{1}{2} \rho \partial_t u_i \partial_t u_i - W(\varepsilon_{ij}) \right) d\Omega + \int_\Omega F_i \delta u_i d\Omega + \int_{A_\sigma} T_i^* \delta u_i dA \right\}$$

$$- \int_\Omega \rho \partial_t u_i \, \delta u_i \Big|^{t_1} d\Omega = 0 \tag{5.12}$$

If there exists no strain energy function, the term W must be replaced by $\sigma_{ij} \, \delta\varepsilon_{ij}$. If the existence of W is assured, (5.12) becomes :

$$\int_{t_o}^{t_1} dt \left\{ \int_\Omega \left(\rho \partial_t u_i \delta \partial_t u_i - \sigma_{ij} \, \delta u_{i,j} \right) d\Omega + \int_\Omega F_i \delta u_i d\Omega + \int_{A_\sigma} T_i^* \delta u_i dA \right\}$$

$$- \int_\Omega \rho \partial_t u_i \delta u_i \Big|^{t_1} d\Omega = 0$$

Integration of the two first terms by parts gives

$$\int_{t_o}^{t_1} \int_\Omega (- \rho \partial_t^2 u_i + \sigma_{ij,j} + F_i) \, \delta u_i d\Omega dt + \int_\Omega \left(\rho \partial_t u_i \, \delta u_i \Big|_{t_o}^{t_1} - \rho \partial_t u_i \delta u_i \Big|^{t_1} \right) d\Omega$$

$$- \int_{t_o}^{t_1} \int_{A_u} \sigma_{ij} n_j \, \delta u_i dA dt - \int_{t_o}^{t_1} \int_{A_\sigma} (\sigma_{ij} n_j - T_i^*) \delta u_i dA dt = 0$$

which implies that the equation of motion (5.1) is satisfied together with the boundary condition on A_σ . In addition, the admissible fields u_i must satisfy

$$\delta u_i(x_i,t) = 0 \qquad \text{on } A_u x(t_o,t_1)$$

and

$$\delta u_i(x_i,t_o) = 0 \qquad \text{on } \Omega$$

Observe that nowhere it has been imposed that u_i must be prescribed at t_1, as is generally required in most formulations (Washizu[11], Fung[19]). In the special case that

$$\delta u_i(x_i,t_1) = 0 \qquad \text{on } \Omega \tag{5.13}$$

the last term in (5.12) disappears and one recovers Hamilton's principle

$$\int_{t_o}^{t_1} dt\left\{\delta\int_\Omega\left(\frac{1}{2}\rho(\partial_t u_i)^2 - W\right)d\Omega + \int_\Omega F_i\delta u_i d\Omega + \int_{A_\sigma} T_i^*\delta u_i dA\right\} = 0 \tag{5.14}$$

In general, there is no reason for δu_i to vanish at t_1 since the solution is not known at that instant of time. Therefore (5.14) is not suitable for obtaining approximate solutions via direct methods, like Rayleigh-Ritz's.

In the particular situation that F_i and T_i^* derive from potentials, i.e.,

$$\int_\Omega F_i\delta u_i d\Omega = -\delta\Phi_F \quad , \quad \int_{A_\sigma} T_i^*\delta u_i dA = -\delta\Phi_\sigma \tag{5.15}$$

(5.14) reduces to

$$\delta\int_{t_o}^{t_1} L dt = 0 \tag{5.16}$$

with

$$L = K - \phi - \phi_F - \phi_\sigma$$

Equation (5.16) is the classical form of Hamilton's principle in elastodynamics : it states that among all kinematically admissible displacements which meet prescribed limit conditions at t_o and t_1 , the actual solution makes the time integral of the Lagrangien L stationary.

Like (5.14), the form (5.16) is not accessible to direct methods of calculation.

By analogy with elastostatics, a canonical and a complementary variational principle can be constructed. Likewise, the extension of the above results to large deformations does not raise fundamental difficulties.

5.2. Gurtin's principle of linear elasticity[20]. Insofar, the initial condition on the velocity did not play any role in the formulation of the variational principles. It is the purpose of the next formulation to incorporate both initial conditions on the displacement and the velocity into the variational equation. The formulation to be presented is rather general in that it can be extended to any linear initial value problem in thermoelasticity and thermo-viscoelasticity.

Let us introduce some preliminary definitions. Denote by $u(x_i,t)$ and $v(x_i,t)$ two functions defined on $\bar{\Omega} \times (0,\infty)$, continuous on $(0,\infty)$ with respect to the time for each $x_i \in \bar{\Omega}$. Let Ω be the interior of $\bar{\Omega}$, with a boundary A such that

$$A = A_1 \cup A_2 \quad , \quad A_1 \cap A_2 = 0 \tag{5.17}$$

The convolution product of u and v is defined by

$$u * v = \int_o^t u(x_i,t-\tau)\, v(x_i,\tau)d\tau \tag{5.18}$$

with the properties

i) $u * v = v * u$ (5.19)

ii) $u * v = 0$ implies either $u = 0$ or $v = 0$ (5.20)

iii) $u * (v * \omega) = (u * v) * \omega = u * v * \omega$ (5.21)

iv) $u * (v + \omega) = u * v + u * \omega$ (5.22)

For later purpose, let us recall the following important lemmas established by Gurtin[16] :

- Lemma 1 : If u is continuous on $\bar{\Omega} \times (0,\infty)$ then

$$\int_{\Omega} (u * v)\, d\Omega = 0$$

implies that

$$u = 0 \qquad \text{on } \bar{\Omega} \times (0,\infty)$$

for every $v \in C^{\infty,\infty}$, which together with all its space derivatives vanishes on $A \times (0,\infty)$.

- Lemma 2 : If u is a piecewise regular function on $A_2 \times (0,\infty)$ then

$$\int_{A_2} (u * v)\, dA = 0$$

implies that

$$u = 0 \qquad \text{on } A_2 \times (0,\infty)$$

for every $v \in C^{\infty,\infty}$ that vanishes on $A_1 \times (0,\infty)$.

- Lemma 3 : If the components u_i are continuous on $A_1 \times (0,\infty)$, then

$$\int_{A_1} u_i * (\gamma_{ij} n_j)\, dA = 0$$

implies that

$$u_i = 0 \qquad \text{on } A_1 \times (0,\infty)$$

for every symmetric $\gamma_{ij} \in C^{\infty,\infty}$, which together with all its space derivatives vanishes on $A_2 \times (0,\infty)$.

In order to express the field equations in terms of the convolution product, apply an operator $g(t)*$ to equation (5.1) :

$$g*\rho\partial_t^2 u_i = g*\sigma_{ij,j} + g*F_i \qquad \text{on } \Omega \times (0,\infty) \tag{5.23}$$

and identify g(t) with the time :

$$g(t) = t \tag{5.24}$$

Accordingly, the l.h.s. of (5.23) writes, after integrations by parts

$$g*\rho\partial_t^2 u_i = \rho\int_0^t (t-\tau)\partial_t^2 u_i(x_i,\tau)d\tau = \rho\, u_i(x_i,t) - \rho\, t\, v_i^o(x_i) \tag{5.25}$$

$$- \rho u_i^o(x_i)$$

Substituting in (5.23) yields

$$\rho u_i = g*\sigma_{ij,j} + f_i \tag{5.26}$$

where f_i is defined by

$$f_i = g * F_i + \rho\left(tv_i^o(x_i) + u_i^o(x_i)\right) \tag{5.27}$$

By reversing the above argumentation, it is directly seen that any solution of (5.26) satisfies the equation of motion (5.1) and the initial conditions (5.6) and (5.7).

With (5.2) and (5.3), expression (5.26) can still be written as

$$\left(\rho\delta_{ki} - g * C_{ijk\ell} \frac{\partial^2}{\partial x_\ell \partial x_j}\right) u_k = f_i \tag{5.28}$$

This has the general form

$$Lu = f \tag{5.29}$$

where L is a linear operator, symmetric in convolution which means that

$$\int_\Omega Lu * v \, d\Omega = \int u * Lv d\Omega$$

By extending the result (3.6) of chapter I, it has been shown[21] that for homogeneous boundary conditions, the variational principle associated to (5.29) is given by

$$\delta I = 0$$

with

$$I = \int_\Omega \left(\frac{1}{2} u * L(u) - u * f\right) d\Omega \tag{5.30}$$

It must be observed that the above expression is only valid if the displacements are specified over the entire boundary ; otherwise, surface integrals should be added to (5.30).

Replacing in (5.30), L by the quantities between brackets in (5.28), yields

$$I = \int_\Omega \left\{\frac{1}{2} u_i * \left(\rho\delta_{ki} - g * C_{ijk\ell} \frac{\partial^2}{\partial x_\ell \partial x_j}\right) u_k - u_i * f_i\right\} d\Omega$$

and after integration by parts of the second term

$$I = \int_\Omega \left(\frac{1}{2} \rho u_i * u_i + \frac{1}{2} g * C_{ijk\ell} u_{k,\ell} u_{\ell,j} - u_i * f_i\right) d\Omega \tag{5.31}$$

In order to incorporate the mixed boundary conditions (5.4) and (5.5), an additional boundary term must be introduced and the final expression of the variational principle is given by :

$$\delta I(u_i) = 0$$

where

$$I(u_i) = \int_\Omega \left(\frac{1}{2}\rho u_i * u_i + \frac{1}{2} g * C_{ijk\ell} u_{k,\ell} * u_{\ell,j} - u_i * f_i\right) d\Omega$$

$$- \int_{A_\sigma} g * T_i^* * u_i dA \tag{5.32}$$

To derive the Euler-Lagrange equations, apply the δ operator to (5.32) :

$$\int_\Omega \Big(\rho u_i * \delta u_i + \frac{1}{2} g * \delta u_{i,j} * C_{ijk\ell} + \frac{1}{2} g * C_{ijk\ell} u_{k,\ell} * \delta u_{i,j}$$

$$- f_i * \delta u_i\Big) d\Omega - \int_{A_\sigma} g * T_i^* * \delta u_i dA = 0 \tag{5.33}$$

Using the divergence theorem and the stress-strain relation, the above expression reduces to

$$\int_\Omega (\rho u_i - g * \sigma_{ij,j} - f_i) * \delta u_i d\Omega + \int_{A_\sigma} g * (\sigma_{ij} n_j - T_i^*) * \delta u_i dA = 0$$

This equation must hold for every δu_i that vanishes on $A_u \times [0,\infty)$. Of course, every function that vanishes on $A \times [0,\infty)$ will automatically meet this requirement ; as a consequence, equation (5.34) must also hold for arbitrary variations δu_i vanishing on $A \times [0,\infty)$. By applying lemma 1, it is concluded that the momentum equation

$$\rho \ddot{u}_i = g * (C_{ijk\ell} u_{k,\ell})_{,j} + f_i \qquad \text{on } \Omega$$

is identically satisfied. In the next step, consider equation (5.34) but containing now the surface integral over A_σ. Recalling that δu_i vanishes on $A_u \times (0,\infty)$ and using lemma 2, one verifies that

$$\sigma_{ij} n_j = T_i^* \qquad \text{on } A_\sigma$$

It is easy to construct principles which are the counterpart of the canonical and the complementary energy principles of elastostatics. For further details, the reader is referred to Gurtin's original work[15].

5.3. Numerical applications of the law of varying action to the vibrations of a beam.

a) The Rayleigh-Ritz method. Consider the lateral free vibrations of a beam of length ℓ and section A : let y be the small deflection measured from the initial straight configuration. Let ρ be the mass per unit length and ignore the rotary contribution to the kinetic energy, given by

$$K = \frac{1}{2} \int_0^\ell \rho \, (\partial_t y)^2 dx \tag{5.35}$$

For small deflections the strain energy of the beam is

$$W = \frac{1}{2} EI (y'')^2 \qquad (' = \frac{d}{dx}) \tag{5.36}$$

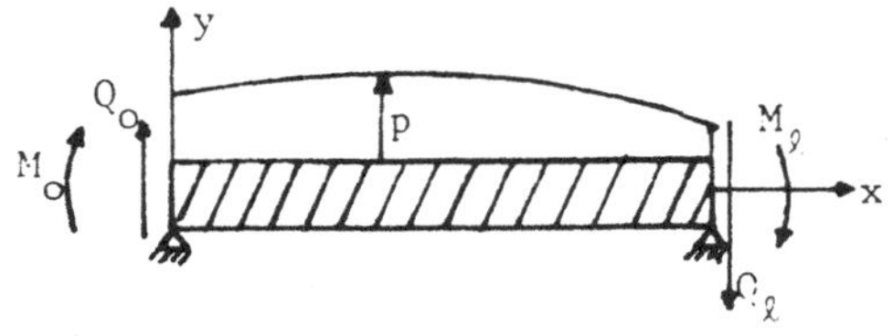

Fig. 5.1.

Suppose that the beam is loaded with a distributed lateral load of intensity p(x,t) per unit length and submitted to a moment M and a shear Q, respectively at the two ends of the bar ; the corresponding virtual work is

$$\int_0^\ell p(x,t)\delta y dx - M_o\ \delta(y')_o + M_\ell \delta(y')_\ell + Q_o\ y_o - Q_\ell \delta y_\ell$$

Initially,one has

$$y(x,o) = y_o(x) \quad , \quad \partial_t y(x,o) = \partial_t y_o(x)$$

Under these conditions, the law of varying action can be formulated as

$$\int_{t_o}^{t_1}\int_o^\ell \left(\rho\partial_t y\delta\partial_t y - EI\ y''\delta y'' + p\delta y\right)dxdt - \int_{t_o}^{t_1} Q\delta y\Big|_o^\ell\ dt$$

$$+ \int_{t_o}^{t_1} M\delta\ y'\Big|_o^\ell\ dt - \int_o^\ell \rho\partial_t y\delta y\Big|^{t_1}\ dx = 0 \qquad (5.37)$$

Approximate solutions will be derived from the Rayleigh-Ritz method. With Bailey[22] , select as trial function

$$y(x,t) = y_o(x) + \partial_t y_o(x)t + \sum_{i=0}^{N}\sum_{k=2}^{M} A_{ik}x^i t^k \qquad (5.38)$$

from which follows

$$\delta y = \sum_{i=0}\sum_{k} x^i t^k\ \delta A_{ik} \qquad (5.39)$$

Substituting (5.38) and (5.39) in (5.37) and performing all the calculations yields the following set of algebraic equations wherein the non-dimensional quantities $X = \frac{x}{\ell}$, $Y = \frac{y}{\ell}$ and $\tau = t/t_1$ have been introduced :

$$\sum_{i=0} \sum_{k=2} \left\{ - \frac{k(k-1)}{t_1^2 (k+s-1)} \int_o^1 \rho X^{i+j} dX - \frac{1}{k+s+1} \int_o^1 \frac{EI}{\ell^4} \frac{\partial^2 X^i}{\partial X^2} \right.$$

$$\left. \frac{\partial^2 X^j}{\partial X^2} dX \right\} A_{ik} = - \frac{1}{\ell} (\frac{1}{2})^j \int_o^1 p\tau^s d\tau + \frac{1}{s+1} \int_o^1 \frac{EI}{p^4} \frac{\partial^2 Y_o}{\partial X^2} \frac{\partial^2 X^j}{\partial X^2} dX$$

$$+ \frac{1}{s+2} \int_o^1 \frac{EI}{\ell^4} \frac{\partial^2 Y_o}{\partial X^2} \frac{\partial^2 X^j}{\partial X^2} dX \qquad j = 0,1...N \quad ; \quad s = 2,3...M \quad (5.40)$$

When 10 terms are retained in the expansion (5.38), one obtains results which are in excellent agreement with the exact solution[22].

b) Kantorovitch's method. Consider now a simply supported beam without lateral load and constraints at the ends.

The boundary conditions are

$$y = 0 \quad \text{and} \quad y'' = 0 \quad \text{at} \quad x = 0 \quad \text{and} \quad x = \ell$$

while the initial conditions are

$$y = 0 \quad , \quad \partial_t y = v_o \sin \frac{nx}{\ell}$$

The law of varying actions reads simply

$$\delta \frac{1}{2} \int_o^{t_1} \int_o^{\ell} \left(\rho (\partial_t y)^2 - EI \, (y'')^2 \right) dtdx - \int_o^{\ell} \rho \partial_t y \delta y \Big|^{t_1} dx = 0 \qquad (5.41)$$

The trial function for the deflection y is taken to be

$$y(x,t) = \frac{4}{\ell^4} (x\ell - x^2) q(t) \qquad (5.42)$$

where q(t) is the unknown function to be determined by Kantorovitch's method. Expression (5.42) does not satisfy the traction boundary condition y" = 0 but it meets the essential displacement boundary conditions, which is all that is required by the variational principle. Substitution of (5.42) into (5.41), and integration with respect to the space variables, gives

$$\int_{o}^{t_1} (\partial_t^2 q + \omega^2 q)\,\delta q \; dt = 0 \qquad (5.43)$$

where

$$\omega^2 = \frac{120}{\rho \ell^4} EI$$

is the square of the frequency of vibration. Since equation (5.43) holds for arbitrary δq, it follows that

$$\partial_t^2 q + \omega^2 q = 0 \qquad (5.44)$$

The initial conditions impose that

$$q(0) = 0$$

$$\partial_t q(0) = \frac{\ell^4}{4x(\ell - x^2)} \sin \frac{\pi x}{\ell}$$

so that the solution of (5.44) is

$$q = \frac{\ell^4 v_o \sin(\pi x/\ell)}{4\omega\ell(x-1)} \sin \omega t$$

The approximate solution is finally given by

$$y(x,t) = \frac{v_o}{\omega} \sin (\pi x/\ell) \sin \omega t \qquad (5.45)$$

It is interesting to compare with the exact solution

$$y_{exact} = \frac{v_o}{\omega_e} \sin (\pi x/\ell) \sin \omega_e t \qquad (5.46)$$

where

$$\omega_e^2 = \frac{\pi^4}{\rho \ell^4} EI < \omega^2 \qquad (5.47)$$

The approximate solution is very close to the exact one. Moreover, an upper bound for the frequency has also been obtained.

c) Calculation of the frequencies of a vibrating system. The determination of the natural frequencies of a vibrating solid is of great interest in engineering practice and therefore, methods which yields such frequencies are very important. Hamilton's principle is the most commonly used technique.

To illustrate the procedure, consider the free lateral vibration of a beam clamped at its origin and simply supported at its end. The boundary conditions are

$$y = 0 \qquad y' = 0 \qquad \text{at } x = 0 \qquad (5.48)$$

$$y = 0 \qquad y'' = 0 \qquad \text{at } x = 1 \qquad (5.49)$$

Approximate the deflection by

$$y(x,t) = y_1(x)\, e^{i\omega t} \qquad (5.50)$$

where

$$y_1(x) = a_1 x^2 (x-1) + a_2 x^3 (x-1) \qquad (5.51)$$

This function satisfies the essential boundary conditions but not (5.49b).

In terms of y , the equation of motion of the beam is

$$EI\, y^{IV} - \rho\, \omega^2 y = 0 \tag{5.52}$$

Substitution of (5.50) in Hamilton's principle (5.41) yields after integration with respect to t ,

$$\delta I = 0$$

with

$$I = \lambda \int_0 \rho y^2\, dx - \int_0^{\ell} EI(y'')^2 dx \tag{5.53}$$

where $\lambda = \omega^2$.

By virtue of Rayleigh-Ritz's method, the values of a_1 and a_2 which make I stationary are given by

$$\frac{\partial I}{\partial a_1} = 0 \qquad \frac{\partial I}{\partial a_2} = 0$$

This yields two homogeneous algebraic equations in a_1 and a_2 . The solution is non trivial if the determinant of the coefficients is zero which leads to

$$\omega_1 = 15.45\, (EI/\rho \ell^4)^{1/2} \qquad \omega_2 = 75.33\, (EI/\rho \ell^4)^{1/2}$$

The exact eigenvalues are

$$\omega_1 = 15.42\, (EI/\rho \ell^4)^{1/2} \qquad \omega_2 = 49.96\, (EI/\rho \ell^4)^{1/2}$$

THE MAIN PRINCIPLES IN ELASTICITY

Elastostatics

Principle	Analytic form	Euler-Lagrange eqs.
Virtual work :		
small def.	$\int_\Omega \sigma_{ij}\delta\varepsilon_{ij}d\Omega - \int_\Omega F_i\delta u_i d\Omega - \int_{A_\sigma} T_i^* \delta u_i dA = 0$	$\sigma_{ij,j} + F_i = 0$ in $\Omega, T_i = T_i^*$ on A_σ
large def.	$\int_{\Omega_o} S_{ij}\delta E_{ij}d\Omega - \int_{\Omega_o} F_{io}\delta u_i d\Omega - \int_{A_\sigma^o} T_{oi}^* \delta u_i dA_o = 0$	$T_{ji,j} + F_{oi} = 0$ in $\Omega_o, T_{oi} = T_{oi}^*$ on A_σ^o
Min. potential energy :		
small def.	$\delta\int_\Omega (W - F_i u_i)d\Omega - \delta\int_{A_\sigma} T_i^* u_i dA = 0$	equilibrium eq. + traction b.c.
large def.	$\delta\int_{\Omega_o} (W - F_{oi}u_i)d\Omega_o - \delta\int_{A_\sigma^o} T_{oi}^* u_i dA_o = 0$	equilibrium eq. + traction b.c.
Complementary virtual work :		
small def.	$\int_\Omega \varepsilon_{ij}\delta\sigma_{ij}d\Omega - \delta\int_{A_u} u_j^* n_i \sigma_{ij} dA = 0$	compatibility eqs.

Elastostatics

Principle	Analytic form	Euler-Lagrange eqs.
Complementary energy principle:		
small def.	$\delta\left\{\int_\Omega W_c\, d\Omega - \int_{A_u} u_j^* n_j \sigma_{ij}\, dA\right\} = 0$	compatibility eqs.
large def.	$\delta\left\{\int_{\Omega_o} \overline{W}_c\, d\Omega_o - \int_{A_u^o} u_j^* n_j T_{ij}\, dA_o\right\} = 0$	compatibility eqs.
Reissner principle :		
small def.	$\delta\left\{\int_\Omega \left(W_c - \sigma_{ij}\frac{1}{2}(u_{i,j} + u_{j,i})\right)d\Omega\right\} + \int_\Omega F_j \delta u_j\, d\Omega$ $+ \int_A dA T_j^* \delta u_j + \int_{A_u} dA\, \delta\left(n_i \sigma_{ij}(u_j - u_j^*)\right) = 0$	$\left.\begin{array}{l}\sigma_{ij,j} + F_i = 0 \\ \frac{\partial W_c}{\partial \sigma_{ij}} = \varepsilon_{ij}\end{array}\right\}$ on Ω $T_i = T_i^*$ on A_σ , $u_i = u_i^*$ on A_u
large def.	$\delta\left\{\int_{\Omega_o} \left(W_c - \frac{1}{2} S_{ij}(u_{i,j} + u_{j,i} + u_{k,i} u_{k,j})\right) d\Omega_o\right\}$ $+ \int_{\Omega_o} F_{jo} \delta u_j\, d\Omega + \int_{A_\sigma^o} dA T_{ij}^* \delta u_j + \int_{A_u^o} dA\, \delta\left(n_i S_{ij}(u_j - u_j^*)\right)$ $= 0$	$\left.\begin{array}{l}(S_{j,\ell} x_{i,\ell})_{,j} + F_{io} = 0 \\ \frac{\partial W_c}{\partial S_{ij}} = E_{ij}\end{array}\right\}$ on Ω $T_{oi} = T_{oi}^*$ on A_σ^o , $u_i = u_i^*$ on A_u^o

Elastodynamics (small deformations)

Principle	Analytic form	Euler-Lagrange eqs.
Law of varying action	$\delta\int_{t_o}^{t_1}(K-\phi)dt + \sum_\alpha\int_{t_o}^{t_1}\Omega^\alpha\delta q^\alpha dt - \sum_\alpha \frac{\partial K}{\partial \dot{q}^\alpha}\delta q^\alpha\Big\vert_{t_o}^{t_1} = 0$	$\rho\partial_t^2 u_i = \sigma_{ij,j} + F_i$ in Ω + traction b.c.
Least action	$\delta\int_{t_o}^{t_1}(K - \phi - \phi_F - \phi_\sigma)dt = 0$	idem previous case
Gurtin	$\delta\Big\{\int_\Omega(\frac{1}{2}\rho u_i * u_i + \frac{1}{2} t * \sigma_{ijk\ell}u_{k,\ell} * u_{i,j} - u_i * f_i)d\Omega$ $- \int_{A_\sigma} t * \overset{*}{T}_i * u_i dA\Big\} = 0$	$\rho u_i = t * u_{ij,j} + f_i$ in Ω $T_i = \overset{*}{T}_i$ on A_σ

References

1. Smith, D. and Smith, C., When is Hamilton's principle an extremum principle, *A.I.A.A. Journal*, 12, 1573, 1974.
2. Bailey, C.D., The method of Ritz applied to the equation of Hamilton, *Comp. Meth. in Appl. Mech. and Eng.*, 7, 135, 1976.
3. Bailey, C.D., Application of Hamilton's law of varying action, *A.I.A.A. Journal*, 13, 1154, 1975.
4. Bailey, C.D., A new look at Hamilton's principle, *Found. of Phys.*, 5, 433, 1975.
5. Trefftz, E., Zur Theorie der Stabilität des elastischen Gleichgewichts, *Zeit. Ang. Math. Mech.*, 13, 160, 1933.
6. Koiter, W., On the concept of stability of equilibrium for continuum bodies, *Proc. Kon. Ned. Ak. Wet.*, B66, 173, 1963.
7. Koiter, W., *Lecture notes on elastic stability*, C.I.S.M., Udine, 1971.
8. Timoshenko, S. and Gere, J., *Theory of elastic stability*, Mac Graw Hill, New-York, 1961.
9. Pearson, C.E., *Theoretical elasticity*, Harvard Univ. Press, Cambridge, 1959.
10. Reissner, E., On a variational theorem in elasticity, *J. Math. Phys.*, 29, 90, 1950.
11. Washizu, K., *Variational methods in elasticity and plasticity*, Pergamon, New-York, 1968.
12. Hill, R., *Mathematical plasticity*, Oxford Univ. Press, Oxford, 1950.
13. Prager, W. and Hodge, P., *Theory of perfectly plastic solids*, Wiley, New-York, 1951.
14. Lippmann, H., *Extremum and variational principles in mechanics*, C.I.S.M. Lecture n°54, Springer Verlag, Berlin, 1970.
15. Koiter, W., On the complementary energy theorem in nonlinear elasticity theory, in: *Trends in applications of pure mathematics to mechanics*, 6, Fichera, Ed., Pitman Pb., London, 1976.

16. Nemat-Nasser, S., General principles in nonlinear and linear elasticity with applications, in: *Mechanics Today*, vol. 1, Nemat-Nasser, Eds., Pergamon, New-York, 1972.
17. Stumpf, H., Generating functionals and extremum principles in non-linear elasticity with applications to nonlinear plate and shallow shell theory, in I.U.T.A.M. - I.M.U. *Symposium on applications of the methods of functional analysis to problems of mechanics*, Lecture Notes in Mathematics, 503, Springer-Verlag, Berlin, 1976.
18. Stumpf, H., Dual extremum principles and error bounds in nonlinear elasticity theory, *J. of Elasticity*, 8, 425, 1978.
19. Fung, Y., *Foundations of solid mechanics*, Prentice Hall, New-York, 1965.
20. Gurtin, M., Variational principles for linear elastodynamics, *Arch. Rat. Mech. Anal.*, 16, 34, 1964.
21. Hlavacek, I., Variational formulation of the Cauchy problem for equations with operator coefficients, *Aplik. Math.*, 16, 46, 1971.
22. Bailey, C., Hamilton, Ritz and elastodynamics, *J. Appl. Mech.*, 98, 684, 1976.

III. VARIATIONAL THEORY OF HEAT CONDUCTION

1. Introduction

This chapter is devoted to the problem of heat conduction in rigid bodies. A representative collection of variational principles is examined. According to the nature of the problem to be handled, these criteria will be classical or not.

Classical variational principles wherein all the quantities are varied, have been formulated not only for the steady but also for the transient problem. However in the latter case, only linear situations, characterized by constant density, heat capacity and heat conductivity are covered. The non-linear transient problem appeals to quasi-variational or restricted principles.

The classical variational formulations are presented in section 2 (steady problems) and section 3 (unsteady problems). The restricted and quasi-variational criteria are analyzed in section 4 : we shall more particularly discuss the principles of Biot,[1] Vujanovic,[2] Glansdorff-Prigogine,[3,4] and Lebon-Lambermont.[5,6]

Consider an isotropic rigid body of volume Ω bounded by a surface A,

$$A = A_T \cup A_q \cup A_r$$

with the following boundary conditions :

i) the temperature is prescribed on the portion A_T :

$$T = T^*(\underline{x},t) \qquad \text{on } A_T \qquad \text{(Dirichlet condition)} \qquad (1.1)$$

ii) the flux is prescribed on the portion A_q :

$$\underline{q}.\underline{n} \equiv -\lambda \text{ grad } T.\underline{n} = q^*(\underline{x},t) \quad \text{on} \quad A_q \quad \text{(Neumann condition)} \tag{1.2}$$

iii) radiation type boundary condition on the portion A_r :

$$\underline{q}.\underline{n} \equiv -\lambda \text{ grad } T.\underline{n} = h_s\left(T(\underline{x},t) - T_s\right) \text{ on } A_r \text{ (Newton cooling law)} \tag{1.3}$$

λ denotes the heat conductivity, $\underline{n}$ the unit normal pointing outwards, h_s the heat transfer coefficient assumed to be positive, T_s the temperature of the medium surrounding the body, and $\underline{q}$ the heat flux vector.

The initial condition is

$$T(x_i,0) = T_o(x_i)$$

Inside the body, the temperature obeys the heat conduction equation

$$c\,\frac{\partial T}{\partial t} = \text{div }(\lambda \text{ grad } T) + Q \quad \text{on } \Omega \tag{1.5}$$

Q is a source term depending generally on $\underline{x}$ and t, c is the heat capacity per unit volume related to the heat capacity per unit mass c_m by

$$c = \rho c_m$$

2. Steady heat conduction

2.1. The linear problem. The governing equation is

$$\text{div }\left(\lambda(\underline{x}) \text{ grad } T(\underline{x})\right) + Q(\underline{x}) = 0 \quad \text{on } \Omega \tag{2.1}$$

to which one adjoins the boundary conditions (1.1) to (1.3). The variational principle which restitutes the above equations is defined by the functional[6]

$$I(T) = \frac{1}{2}\int_{\Omega} \left(\lambda(\underline{x}) \text{ grad } T(\underline{x}).\text{grad } T(\underline{x}) - 2Q(\underline{x})T(\underline{x})\right) d\Omega$$

$$+ \int_{A_q} q^{*}(\underline{x})T(\underline{x})dA - \int_{A_T} \left(T(\underline{x}) - T^{*}(\underline{x})\right)\underline{n}.\lambda(x) \text{ grad } T(\underline{x})dA$$

$$+ \frac{1}{2}\int_{A_r} h_s(\underline{x}) \left(T(\underline{x}) - T_s(\underline{x})\right)^2 dA \qquad (2.2)$$

Observe that the trial function does not need to satisfy the boundary conditions since they follow directly from the variational equation. In the particular event that the trial function T is taken identical to $T^{*}(\underline{x})$ on A_T, it is clear that the corresponding surface integral in (2.2) disappears. In this case, the second variation of I(T) takes the form

$$\delta^2 I(T) = \int_{\Omega} \lambda(\text{grad } \delta T)^2 d\Omega + \int_{A_r} h_s(\delta T)^2 dA > 0 \quad , \qquad (2.3)$$

This result implies a minimum principle with as a consequence

$$I(T \text{ exact}) \leq I(T \text{ admissible}) \qquad (2.4)$$

If T is prescribed on the whole boundary, (2.2) writes simply, in absence of a source term,

$$I(T) = \frac{1}{2}\int_{\Omega} \lambda(\text{grad } T)^2 d\Omega$$

The extension of the above results to non-isotropic systems is trivial; in cartesian coordinates, (2.1) and (2.2) must be replaced by

$$(\lambda_{ij} T_{,i})_{,j} + Q = 0 \qquad (\lambda_{ij} = \lambda_{ji}) \qquad (2.1')$$

and

$$I(T) = \frac{1}{2}\int_{\Omega}\left(\lambda_{ij}\ T_{,i}\ T_{,j}\ - 2\ QT\right)d\Omega + \int_{A_q} q_i^{*} T dA + \int_{A_T} (T - T^{*})$$

$$n_i \lambda_{ij}\ T_{,j}\ dA + \frac{1}{2}\int_{A_r} h_s (T - T_s)^2 dA \qquad (2.2')$$

If $T = T^{*}$ is satisfied a priori on A_T , $I(T)$ is minimum for the exact solution.

2.2. The non-linear problem. Before writing the criterion corresponding to a temperature dependent conductivity, let us introduce the notations

$$K = \int_{T^{*}}^{T} \lambda(\xi) d\xi \qquad (2.5)$$

$$R = \int_{0}^{T} \lambda(\xi)\left(\xi - \xi_s\right) d\xi \qquad (2.6)$$

from which follow, according to Leibniz's derivation rule,

$$\delta K = \frac{\partial K}{\partial T}\ \delta T = \lambda(T)\delta T$$

$$\delta R = \frac{\partial R}{\partial T}\ \delta T = \lambda(T)\ \left(T - T_s\right)\delta T$$

The variational principle (2.2) must be replaced by

$$\delta I_1(T) = 0$$

with

$$I_1(T) = \frac{1}{2}\int_\Omega (\lambda^2 \text{ grad } T.\text{grad } T - 2QK)d\Omega$$

$$+ \int_{A_q} q^* K dA - \int_{A_T} \underline{n}.\lambda(\text{grad } T)KdA + \int_{A_r} h_s R dA \tag{2.7}$$

The Euler-Lagrange equation is

$$\lambda \frac{\partial\lambda}{\partial T} (\text{grad } T)^2 - \text{div } (\lambda^2 \text{ grad } T) - \lambda Q = 0$$

which after simplifications, reduces to (2.1).

At the surface, one obtains the boundary conditions

$$\lambda^2 \text{ grad } T.\underline{n} + q^*\lambda = 0 \qquad \text{on } A_q \tag{2.8}$$

$$\underline{n}.(\lambda^2 \text{ grad } T - \lambda^2 \text{ grad } T)\delta T - \left(\delta(\lambda \text{ grad } T.\underline{n})\right)\int_{T^*}^{T} \lambda(\xi)d\xi = 0 \qquad \text{on } A_T \tag{2.9}$$

$$\lambda^2 \text{grad } T.\underline{n} + h_s\lambda(T - T_s) = 0 \qquad \text{on } A_r \tag{2.10}$$

Clearly (2.8) and (2.10) are identical to Neumann's and Newton's laws while the vanishing of (2.9) implies the Dirichlet condition

$$T = T^* \tag{2.11}$$

If furthermore, T is prescribed everywhere on the boundary, (2.7) writes, in absence of source terms,

$$I_1(T) = \frac{1}{2}\int_\Omega \underline{q}.\underline{q} \, d\Omega \tag{2.12}$$

Contrary to the linear problem, (2.7) and (2.12) cannot be extended directly to anisotropic systems.

2.3. Canonical and dual principles. Consider firstly the linear problem. By introduction of the relation

$$q(\underline{x}) = -\lambda(x)\ \mathrm{grad}\ T(\underline{x})$$

as side condition by means of a Lagrange multiplier α , the variational equation (2.2) becomes

$$\delta\left\{\int_\Omega \left(\frac{1}{2}\frac{\underline{q}\cdot\underline{q}}{\lambda} - QT - \underline{\alpha}\cdot(\underline{q} + \lambda\ \mathrm{grad}\ T)\right) d\Omega \right.$$

$$\left. + \int_{A_q} q^* T dA + \frac{1}{2}\int_{A_r} h_s (T - T_s)^2 dA - \int_{A_T} (T - T^*)\underline{n}.\lambda\ \mathrm{grad}\ T dA\right\} = 0 \qquad (2.14)$$

for compactness of the notation, argument $\underline{x}$ has been omitted. The variation with respect to $\underline{q}$ permits to identify the coefficient $\underline{\alpha}$ as

$$\underline{\alpha} = \frac{1}{\lambda}\underline{q}$$

Substitution in (2.14) leads to

$$\delta I_c(T,\underline{q}) = 0$$

with

$$I_c(T,\underline{q}) = \int_\Omega \left(-\frac{1}{2\lambda}\underline{q}\cdot\underline{q} - QT - \underline{q}\cdot\mathrm{grad}\ T\right) d\Omega$$

$$+ \int_{A_q} q^* T dA + \frac{1}{2}\int_{A_r} h_s (T - T_s)^2 dA + \int_{A_T} \underline{q}\cdot\underline{n}(T - T^*) dA \qquad (2.15)$$

This represents the canonical form of the principle which allows the simultaneous variations of the temperature and the heat flux. The Euler-Lagrange equations corresponding to arbitrary variations of $\underline{q}$ and T are respectively

$$-\frac{1}{\lambda}\,\underline{q} - \text{grad } T = 0 \qquad \text{on } \Omega \tag{2.16}$$

$$\text{div } \underline{q} - Q = 0 \qquad \text{on } \Omega \tag{2.17}$$

while at the boundaries,

$$\underline{q}\cdot\underline{n} - q^{*} = 0 \qquad \text{on } A_q \tag{2.18}$$

$$T - T^{*} = 0 \qquad \text{on } A_T \tag{2.19}$$

$$\underline{q}\cdot\underline{n} - h_s(T - T_s) = 0 \qquad \text{on } A_r \tag{2.20}$$

Integrate now by parts the third term in (2.15). By imposing that $\underline{q}$ obeys the heat conduction equation (2.17) on Ω, and the boundary condition (2.18) on A_q, one obtains

$$\delta\left\{\int_{\Omega} -\frac{1}{2\lambda}\,\underline{q}\cdot\underline{q}\;d\Omega - \int_{A_r}\left(T\underline{q}\cdot\underline{n} - \frac{1}{2}\,h_s(T - T_s)^2\right)dA - \int_{A_T}\underline{q}\cdot\underline{n}\;T^{*}dA\right\} = 0 \tag{2.21}$$

The surface integral on A_r is easily expressed in term of $\underline{q}$ by observing that

$$T = T_s + \frac{1}{h_s}\,\underline{q}\cdot\underline{n}$$

hence

$$T\,\underline{q}\cdot\underline{n} - \frac{1}{2}\,h_s(T - T_s)^2 = T_s\,\underline{q}\cdot\underline{n} + \frac{1}{2h_s}\,(\underline{q}\cdot\underline{n})^2$$

It follows that the dual principle can be expressed as

$$\delta J(\underline{q}) = 0$$

with

$$J(\underline{q}) = \int_{\Omega} - \frac{1}{2\lambda}\, \underline{q}\cdot\underline{q}\, d\Omega - \int_{A_T+A_r} \underline{n}\cdot\underline{q}\, T^{*} dA - \int_{A_r} \frac{1}{2h_s}(\underline{q}\cdot\underline{n})^2 dA \qquad (2.22)$$

For anisotropic bodies,

$$J(\underline{q}) = \int_{\Omega} - \frac{1}{2}\, r_{ij} q_i q_j d\Omega - \int_{A_T+A_r} q_i n_i T^{*} dA - \int_{A_r} \frac{1}{2h_s}(q_i n_i)^2 dA \qquad (2.23)$$

where r_{ij} is the heat resistivity tensor defined by

$$r_{ik}\lambda_{kj} = \delta_{ij} \qquad (2.24)$$

Since the volume integral in (2.22) is negative definite, J attains its maximum value when $\underline{q}$ is solution of the field equations (2.16) to (2.20) :

$$J(\underline{q}_{exact}) \geq J(\underline{q}_{admissible}) \qquad (2.25)$$

Moreover, when $\underline{q}$ and T are solution of the field equations, it is easily seen that

$$J(\underline{q}_{exact}) = I(T_{exact}) \qquad (2.26)$$

By combining (2.11) and (2.25), upper and lower bounds for the functional I (or J) are found :

$$I(T_{admissible}) \geq I(T_{exact}) = J(\underline{q}_{exact}) \geq J(\underline{q}_{admissible}) \qquad (2.27)$$

In the non-linear case, the canonical principle takes the form

$$\delta I_1(T,\underline{q}) = 0$$

with

$$I_1(T,\underline{q}) = \int_{\Omega} \left(-\frac{1}{2}\,\underline{q}\cdot\underline{q} - \lambda\,\underline{q}\cdot\text{grad } T + \cap K(T)\right) d\Omega$$

$$+ \int_{A_q} q^{*} K(T)\, dA + \int_{A_T} \underline{q}\cdot\underline{n}\ \lambda (T - T^{*})\, dA + \int_{A_r} h_s R(T)\, dA \tag{2.28}$$

The Euler-Lagrange equations are the Fourier law and the heat conduction equation ; the correct boundary conditions are also recovered. Expression (2.28) has been constructed by following the procedure outlined for the linear problem. But contrarily to the linear case, it is not possible to derive from (2.28) a principle involving only $\underline{q}$ as variable. Indeed, by integrating by parts the second term in the volume integral of (2.28), the heat conduction equation is not reproduced, and therefore the variable T cannot be eliminated.

3. Unsteady linear heat conduction

3.1. The parabolic equation. Like in elastodynamics, the given initial-value problem will be replaced by an equivalent boundary-value problem by using the concept of the convolution product.

It is well known that the linear heat conduction equation

$$c(\underline{x})\,\frac{\partial T(\underline{x},t)}{\partial t} = \text{div}\left(\lambda(\underline{x})\,\text{grad } T(\underline{x},t)\right) + \cap(\underline{x},t) \tag{3.1}$$

can be transformed into the integro-differential equation,

$$c(\underline{x})T(\underline{x},t) - c(x)T_o(x) =$$

$$\mathrm{div}\{\lambda(\underline{x})H(t)*\mathrm{grad}\ T(\underline{x},t)\} + \hat{Q}(\underline{x},t) \tag{3.2}$$

$T_o(x)$ is the initial temperature field, $H(t)$ the unit step function

$$H(t) = 0 \quad \text{for } t \leq 0, \qquad H(t) = 1 \quad \text{for } t > 0$$

while $\hat{Q}(\underline{x},t)$ is defined by

$$\hat{Q}(\underline{x},t) = \int_o^t Q(\underline{x},\tau)d\tau \tag{3.3}$$

A variational principle associated with (3.2) can be constructed by using Gurtin's method of convolution.[7] It gives

$$\delta I_G(T) = 0$$

with

$$I_G(T) = \int_\Omega \{c(\underline{x})T(\underline{x},t)*T(\underline{x},t) + \lambda(\underline{x})H(t)*\mathrm{grad}\ T(\underline{x},t) \overset{*}{.}\ \mathrm{grad}\ T(x,t)$$

$$- 2\ \hat{Q}(\underline{x},t)*T(\underline{x},t) - 2(\underline{x})T_o(\underline{x})*T(\underline{x},t)\}d\Omega \tag{3.4}$$

where

$$\mathrm{grad}\ T \overset{*}{.} \mathrm{grad}\ T = \int_o^t \mathrm{grad}\ T(t-\tau).\mathrm{grad}\ \tau\ d\tau$$

By permutation of the δ and the integral symbol, (3.4) becomes

$$\int_\Omega \Big\{ c(\underline{x})\big(\delta T(\underline{x},t)*T(\underline{x},t) + T(\underline{x},t)*\delta T(\underline{x},t)\big) + \lambda(\underline{x})H(t)*\big($$

$$\text{grad } T(\underline{x},t)\overset{*}{\cdot}\text{grad } T(\underline{x},t) + \text{grad } T(\underline{x},t)\overset{*}{\cdot}\text{grad } \delta T(\underline{x},t)\big)$$

$$- 2\,\hat{Q}(\underline{x},t)*\delta T(\underline{x},t) - 2\,c(\underline{x})T_o(\underline{x})*\delta T(\underline{x},t)\Big\}\, d\Omega = 0$$

Using the permutativity property of the convolution product, and integrating by parts the third and fourth terms yields

$$\int_\Omega \Big\{2c(\underline{x})T(\underline{x},t)*\delta T(\underline{x},t) - 2\,\text{div}\big(\lambda(\underline{x})H(t)*\text{grad } T(\underline{x},t)\big)*\delta T(\underline{x},t)$$

$$- 2\hat{Q}(\underline{x},t)*\delta T(\underline{x},t) - 2c(\underline{x})T_o(\underline{x})*\delta T(\underline{x},t)\Big\}d\Omega + \int_A 2\lambda(\underline{x})*\text{grad } T(\underline{x},t)*\delta T(\underline{x},t)dA$$

$$= 0 \tag{3.5}$$

If T is prescribed all over the boundary (Dirichlet condition), (3.5) ends in

$$\int_\Omega \Big\{\underline{c}(x)*T(\underline{x},t) - \text{div}\big(\lambda(\underline{x})H(t)*\text{grad } T(\underline{x},t) - \hat{Q}(\underline{x},t) - c(\underline{x})T_o(x)\Big\}$$

$$*\delta T(\underline{x},t)d\Omega = 0$$

This relation is clearly verified for arbitrary variations $\delta(T(\underline{x},t)$ if and only if the integro-differential equation (3.2) is satisfied.

An alternate variational principle was derived by Tonti[8] in the form

$$\delta\int_\Omega \Big(\lambda(\underline{x})\text{grad } T(\underline{x},t)\overset{*}{\cdot}\text{grad } T(\underline{x},t) + c(\underline{x})T(\underline{x},t)\,\frac{\partial T(\underline{x},t)}{\partial t}$$

$$- c(\underline{x})T_o(\underline{x})T(\underline{x},t) - 2Q(\underline{x},t)*T(\underline{x},t)\Big)d\Omega = 0 \tag{3.6}$$

Instead of (3.2), the Euler-Lagrange equation is now the original partial differential equation (3.1).

In both Gurtin's and Tonti's formulations, the initial condition

$$T(\underline{x},0) = T_o(x) \qquad x \in \Omega \qquad t = 0 \tag{3.7}$$

follows directly from the criterion, but not Dirichlet's condition which must be considered as a constraint to be imposed on the admissible function.

Recently, Reddy[9] proposed a criterion similar to the previous ones but wherein T and $\underline{q}$ are varied independently. The heat conduction equation (3.1) is replaced by the so-called canonical equations

$$\underline{q}(\underline{x},t) = -\lambda(\underline{x})\,\text{grad}\,T(\underline{x},t) \tag{3.8}$$

$$c(\underline{x})\frac{\partial T(\underline{x},t)}{\partial t} + \text{div}\,\underline{q}(\underline{x},t) - Q(\underline{x},t) = 0 \tag{3.9}$$

to which are adjoined Dirichlet's and Neumann's boundary condition as well as the initial condition (3.7). It is a simple matter to verify that these equations derive from the principle

$$\delta I_R(T,\underline{q}) = 0$$

with

$$I_R(T,\underline{q}) = \int_\Omega \Big(\frac{1}{2}\,c(\underline{x})T(\underline{x},t)*\frac{\partial T(\underline{x},t)}{\partial t} - \underline{q}(\underline{x},t)\overset{*}{\cdot}\text{grad}\,T(\underline{x},t)$$

$$-\frac{1}{2\lambda(\underline{x})}\,\underline{q}(\underline{x},t)\overset{*}{\cdot}\underline{q}(\underline{x},t) - T(\underline{x},t)*Q(\underline{x},t) + \frac{1}{2}c(\underline{x})T(\underline{x},t) - c(\underline{x})T_o(\underline{x})T(\underline{x},t)\Big)d\Omega$$

$$+\int_{A_T}\underline{q}(\underline{x},t)\cdot\underline{n}(\underline{x},t)*\big(T(\underline{x},t) - T^*(\underline{x},t)\big)dA + \int_{A_q}T(\underline{x},t)*q^*(\underline{x},t)dA \tag{3.10}$$

3.2. The hyperbolic equation. The equation (3.1) is of parabolic type ; as a consequence, heat signal perturbations propagate through the medium with an infinite speed. To avoid this unpleasant physical behaviour, some authors, Maxwell,[10] Cattaneo,[11] Vernotte[12] suggested to replace the parabolic by a hyperbolic equation

$$c\tau\ddot{T}(\underline{x},t) + c\dot{T}(\underline{x},t) = \lambda \text{ div grad } T(\underline{x},t) \tag{3.11}$$

to which corresponds a finite speed of propagation equal to

$$(\lambda/\tau c)^{1/2}$$

A superposed dot represents the derivation with respect to time, τ is the so-called relaxation time ; all the quantities τ, c and λ are supposed to be positive constants. Vujanovic[2] presented a variational principle of the form

$$\delta\int_{t_1}^{t_2}\int_{\Omega}\left\{c\,\tau\,\dot{T}(\underline{x},t)^2 - \lambda\left(\text{grad } T(\underline{x},t)\right)^2\right\}e^{t/\tau}\, dt d\Omega = 0 \tag{3.12}$$

It is left as an exercise to show that (3.12) reproduces the hyperbolic heat equation (3.11) at the condition that the admissible function obeys Dirichlet's condition on A and that

$$\delta T = 0 \quad \text{at} \quad t = t_1 \quad \text{and} \quad t = t_2$$

Unfortunately, the criteria of Gurtin, Tonti and Reddy as well as Vujanovic's principle cannot be applied to the non-linear problem. Moreover, their range of application is limited to isotropic bodies. To our knowledge, there exists no *exact* variational principle for the non-linear case and (or) anisotropic bodies. All the principles which have been constructed are either of *restricted* or *quasi-variational* character. This means that during the variational procedure, some quantities are frozen, i.e., not subjected to variation.

In the next section, we shall successively analyze the criteria proposed by Biot,[1] Glansdorff-Prigogine,[3] Vujanovic[2] and Lebon-Lambermont.[5,6] Other principles have been formulated by Onsager,[23] Gyarmati,[14] Rosen,[15] etc..., but they will not be discussed here, because they fall in the same class and share the same characteristics as the above mentioned ones.

4. Unsteady non-linear heat conduction

4.1. Biot's principle.[1] Biot proposed a variational principle which bears some resemblance with Hamilton's principle in mechanics and which leads to a lagrangian formulation in terms of generalized coordinates. The quantity submitted to variation is no longer the temperature field or the heat flux vector, but the integrated flux

$$\underline{H} = \int_0^t \underline{q}\, dt \tag{4.1}$$

$\underline{H}$ is called the *heat displacement vector*. In term of $\underline{H}$, Fourier's law takes the form

$$\dot{\underline{H}} = -\lambda(T)\ \text{grad}\ T \qquad (\dot{\underline{H}} = \frac{d}{dt}\underline{H} = \frac{\partial}{\partial t}\underline{H}) \tag{4.2}$$

Recalling that the internal energy per unit volume is expressed by

$$u_v = \int_0^T c(\xi)d\xi \tag{4.3}$$

the heat conduction equation reads, after integration with respect to the time and omission of the source term,

$$u_v = -\ \text{div}\ \underline{H} \tag{4.4}$$

The construction of Biot's principle proceeds as follows. Multiply (4.2) by $\delta\underline{H}$ and integrate on the volume such that

$$\int_\Omega (\text{grad } T + \frac{1}{\lambda}\dot{\underline{H}}).\delta\underline{H}\, d\Omega = 0 \qquad (4.5)$$

while after integration by parts,

$$\int_\Omega (- T \text{ div } \delta H + \frac{1}{\lambda}\dot{\underline{H}}.\delta\underline{H})d\Omega = - \int_A T\, \underline{n}.\delta\underline{H}\, dA$$

Using eq.(4.4) as a constraint to be fulfilled a priori, one obtains Biot's principle

$$\int_\Omega T\,\delta u_v\, d\Omega + \int_\Omega \frac{1}{\lambda}\dot{\underline{H}}.\delta\underline{H}\, d\Omega = - \int_A T\, \underline{n}.\delta\underline{H}dA \qquad (4.6)$$

Since the δ operator cannot be put in front of the integrals, (4.6) is a quasi-variational principle.

In the particular case that c is temperature independent, one has

$$u_v = c\, T$$

and the first term in the l.h.s. of (4.6) can be expressed in term of the thermal potential

$$V = \frac{1}{2}\int_\Omega c\, T^2 d\Omega$$

Reversing the procedure, one can go from (4.6) to (4.5) and conclude that the Euler-Lagrange equation corresponding to arbitrary $\delta\underline{H}$ is Fourier's equation (4.2), under the stipulation that (4.4) is satisfied a priori.

As further step, Biot introduces the concept of *generalized coordinates* ; the displacement vector is supposed to depend on some parameters $a_1 \ldots a_n$, called generalized coordinates :

$$\underline{H} = \underline{H}\,(a_1\ ,\ a_2\ ,\ \ldots\ a_n\ \ ;\ \ \underline{x},t) \tag{4.7}$$

The a_α's are unknown functions of the time. They are strictly equivalent to the unknown parameters introduced in Kantorovitch's variational method. It is commonly assumed[1] that $\underline{H}$ depends linearly on the a_α's :

$$\underline{H} = \sum_{\alpha=1}^{n} a_\alpha(t)\ \underline{f}_\alpha(\underline{x}) \tag{4.8}$$

where $\underline{f}_\alpha(\underline{x})$ are a-priori given functions of $\underline{x}$.

In view to express the variational equation (4.6) in terms of the a_α's, observe from (4.7) that

$$\delta\underline{H} = \sum_{\alpha=1}^{n} \frac{\partial\underline{H}}{\partial a_\alpha}\,\delta a_\alpha \tag{4.9}$$

and

$$\dot{\underline{H}} = \sum_{\alpha=1}^{n} \frac{\partial\underline{H}}{\partial a_\alpha}\,\dot{a}_\alpha + \frac{\partial\underline{H}}{\partial t} \tag{4.10}$$

from which follows

$$\frac{\partial\dot{\underline{H}}}{\partial\dot{a}_\alpha} = \frac{\partial\underline{H}}{\partial a_\alpha} \tag{4.11}$$

With this result in mind, one can write

$$\dot{\underline{H}}.\delta\underline{H} = \sum_{\alpha=1}^{n} \dot{\underline{H}}.\frac{\partial\dot{\underline{H}}}{\partial\dot{a}_\alpha}\,\delta a_\alpha = \sum_{\alpha=1}^{n} \delta a_\alpha\,\frac{\partial}{\partial a_\alpha}\left(\frac{1}{2}\,\dot{H}^2\right) \tag{4.12}$$

By putting

$$D = \int_\Omega \frac{1}{\lambda} \dot{H}^2 d\Omega \tag{4.13}$$

the second volume integral in (4.6) reads as

$$\int_\Omega \frac{1}{\lambda} \dot{\underline{H}}.\delta\underline{H} d\Omega = \sum_{\alpha=1}^{n} \delta a_\alpha \frac{\partial}{\partial \dot{a}_\alpha} \int_\Omega \frac{1}{2\lambda} \dot{H}^2 d\Omega = \sum_{\alpha=1}^{n} \frac{\partial D}{\partial \dot{a}_\alpha} \delta a_\alpha \tag{4.14}$$

Moreover, the first volume integral can be given the form

$$\int_\Omega T\delta u_v d\Omega = -\int_\Omega T \operatorname{div} \delta\underline{H}\, d\Omega = -\sum_{\alpha=1}^{n} \int_\Omega T \frac{\partial}{\partial a_\alpha}(\operatorname{div} \underline{H})\delta a_\alpha d\Omega$$

With Biot, let us make the following formal identifications :

$$\sum_{\alpha=1}^{n} \delta a_\alpha \int_\Omega - T \frac{\partial}{\partial a_\alpha} \operatorname{div} \underline{H}\, d\Omega \equiv \sum_{\alpha=1}^{n} \delta a_\alpha \frac{\partial V}{\partial a_\alpha} \tag{4.15}$$

and

$$-\int_\Omega T\, \delta\underline{H}.\underline{n}\, dA = -\sum_{\alpha=1}^{n} \delta a_\alpha \int_\Omega T \frac{\partial \underline{H}}{\partial a_\alpha} .\underline{n}\, dA \equiv \sum_{\alpha=1}^{n} Q_\alpha \delta a_\alpha \tag{4.16}$$

Using the results (4.14) to (4.16), equation (4.6) becomes

$$\sum_{\alpha=1}^{n} \left(\frac{\partial V}{\partial a_\alpha} + \frac{\partial D}{\partial \dot{a}_\alpha} - Q_\alpha\right)\delta a_\alpha = 0$$

Since δa_α is arbitrary, it follows that

$$\frac{\partial V}{\partial a_\alpha} + \frac{\partial D}{\partial \dot{a}_\alpha} = Q_\alpha \qquad \alpha = 1, 2, \ldots n \tag{4.17}$$

in analogy with the Lagrange equation of a dissipative mechanical system with a potential energy V, a dissipative function D and generalized forces Q_α . The same relation (4.17) holds true for anisotropic conductivity at the condition to replace in expressions (4.6) and (4.13), λ^{-1} by the thermal resistivity tensor r_{ij} . Biot's principle is also applicable to moving boundaries. Indeed, the volume and surface integrals have been evaluated at instantaneous geometric configurations which are not necessarily the same during the time evolution of the process.

Application : the semi-infinite solid with constant temperature at the surface. Consider a semi-infinite solid initially at temperature T = 0. At time zero, the face x = 0 is suddenly changed to a temperature T_o and maintained at this value. The thermodynamic properties of the solid are constant. The temperature profiles respectively at t = 0 and at a given time are shown on fig.4.1. The temperature field comprises a region wherein the temperature drops from T_o to its initial value T = 0. The depth to which the thermal effects are felt is the penetration length a(t), it is a time dependent quantity.

Select as trial function for the temperature a parabolic distribution of the form[1]

$$T = T_o \left(1 - \frac{x}{a(t)}\right)^2 \qquad x \leq a$$

$$T = 0 \qquad x > a \qquad (4.18)$$

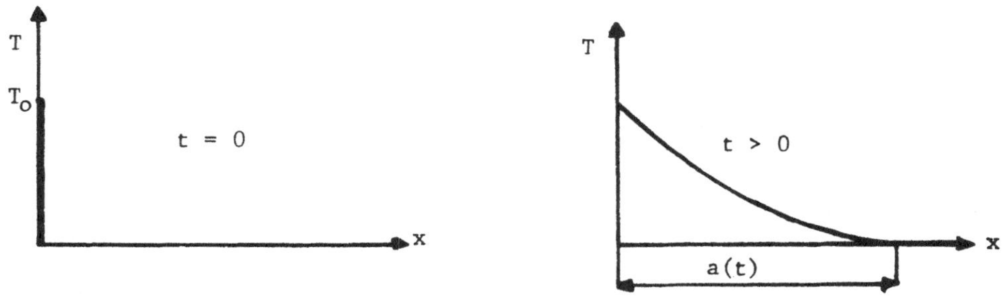

Fig. 4.1.

wherein a(t) plays the role of the generalized coordinate to be determined from the variational criterion. Since c is constant, the heat displacement is related to T by the energy law

$$cT = - dH/dx \tag{4.19}$$

After integration on x, one obtains with the help of (4.18)

$$H(x) - H(a) = - cT_o \int_a^x (1 - \frac{x}{a})^2 dx$$

and since H vanishes at x = a ,

$$H(x) = cT_o \left(\frac{a}{3} - x + \frac{x^2}{a} - \frac{x^3}{3a^2}\right)$$

Use of the Lagrange equation (4.17) requires the calculation of the thermal potential V, the dissipation function D and the generalized force Q_α; these quantities are here respectively given by

$$V = \frac{1}{2} c \int_o^a T^2 dx = \frac{1}{10} cT_o^2 a$$

$$D = \frac{1}{2\lambda} \int_o^a \dot{H}^2 dx = \frac{13}{630} \frac{c^2 T_o^2}{\lambda} a \dot{a}^2$$

$$Q\delta a = (T \delta H)_{x=o} = \frac{1}{3} c T_o^2 \delta a$$

Substituting these results in

$$\frac{\partial V}{\partial a} + \frac{\partial D}{\partial a} = 0$$

yields the differential equation

$$\frac{13}{315}\, a\, \dot{a} = \frac{7}{20}\,\frac{\lambda}{c}$$

whose solution is

$$a = 3.36\ (\chi\ t)^{1/2} \tag{4.20}$$

$\chi = \lambda/c$ is the thermal diffusivity. The approximate temperature distribution is thus

$$T = T_o \left(1 - \frac{x}{3.36(\chi\ t)^{1/2}}\right)^2 \tag{4.21}$$

This solution is compared with the exact one

$$T = T_o \ \mathrm{erfc}\ x/2(\chi\ t)^{1/2} \tag{4.22}$$

in fig.4.2., the agreement is very satisfactory :

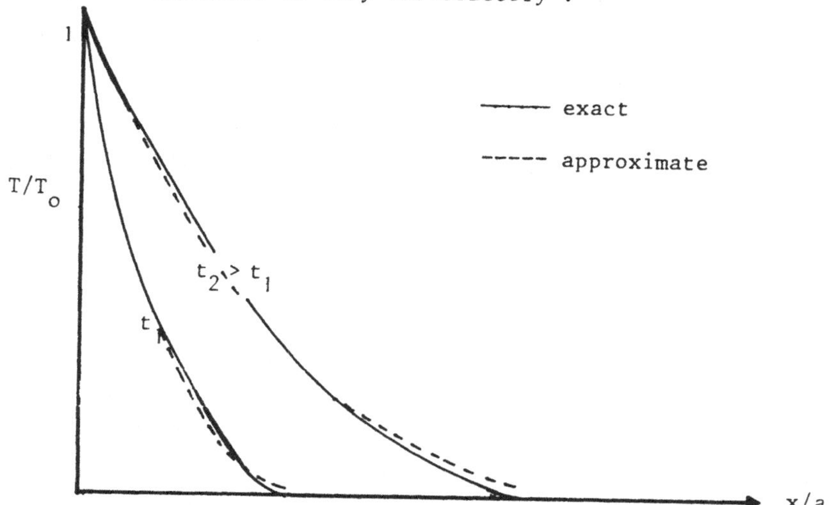

Fig.4.2.

Although Biot's principle has been used with success to obtain approximate solution for a large class of problems,[23,24] it has raised some criticisms.[25] Clearly, Biot's criterion is not a classical variational principle, but rather a quasi variational formulation. Moreover, as shown by Finlayson and Scriven,[25] the approximation method proposed by Biot is equivalent to Galerkin's technique. However, this result is true for any variational method, when trial expansions like (4.8) are used (see chapter I, § 2.3).

The procedure that consists of using the heat displacement $\underline{H}$ instead of the temperature T as central quantity, may conduce to heavy developments. As a matter of fact, the integration of

$$u_v(T) = - \operatorname{div} \underline{H}$$

may become difficult for non-linear and three dimensional problems. Nonetheless, it must be observed that, as in the above application, it is the temperature which is given an approximate form and not $\underline{H}$.

Although equation (4.7) bears some resemblance with the Lagrange equation of analytical mechanics, the analogy is not complete. In mechanics, the generalized coordinates are unknowns of the *physical* problem. In Biot's formulation, they appear as unknowns of the *numerical* problem, in the same way as the coefficients a_α in Kantorovitch's method.

4.2. Vujanovic's principle[2]. As mentioned in section 3, the hyperbolic heat conduction equation can be associated with the following "exact" variational principle

$$\delta V_u = \delta \int_{t_1}^{t_2} \int_\Omega \left(\tau \dot{T}^2 - \frac{\lambda}{c} (\operatorname{grad} T)^2\right) e^{t/\tau} \, dt d\Omega = 0 \tag{4.23}$$

But the major interest of Vujanovic's criterion lies in its possibility to be applied to classical problems (where the speed of propagation is infinite) by allowing the relaxation time τ to approach zero.

Approximate solution for the classical linear heat conduction equation

$$c\dot{T} = \lambda \text{ div grad } T$$

is obtained by applying the usual variational methods to (4.23) and letting $\tau \to 0$.

The principle remains valid in the case of temperature dependent thermal properties at the condition to take as functional

$$Vu = \frac{1}{2}\int_{t_1}^{t_2}\int_{\Omega} \left(\tau c(T)\lambda(T)\ \dot{T}^2 - \lambda^2(T)\ (\text{grad } T)^2\right)\ e^{t/\tau}\ dt d\Omega$$

The corresponding Euler-Lagrange eq. is

$$- e^{t/\tau} \text{ div } (\lambda^2 \text{grad } T) + e^{t/\tau} \lambda \frac{\partial\lambda}{\partial T} (\text{grad } T)^2 - \tau e^{t/\tau} \frac{\partial}{\partial T} (c\lambda)\dot{T}$$

$$(\tau e^{t/\tau}\ c\lambda\ \dot{T}) = 0$$

After simplifying by $\lambda e^{t/\tau}$ and requiring that $\tau \to 0$, the above expression reduces to

$$- \text{ div } (\lambda \text{ grad } T) + c\ \dot{T} = 0$$

The principle has also been extended to hydrodynamics[16,17] and magnetohydrodynamics[18] and reveals very useful for handling general problems of continuum mechanics. However, it must be realized that the extension to anisotropic bodies is not trivial. Moreover, the criterion has been formulated for one particular boundary condition (namely Dirichlet's) to be satisfied by admissible functions ; in particular, it does not yield natural boundary conditions. Finally, some variational techniques like Rayleigh-Ritz method are not applicable because they give infinite results when $\tau \to 0$.

Application : Heat conduction in an insulated cylindrical rod. Take a cylindrical body of length 2ℓ laterally insulated, with constant thermal properties. The ends of the rod are kept at a constant temperature equal to zero

$$T(-\ell,t) = T(\ell,t) = 0 \qquad \forall t$$

while the initial distribution is given by

$$T(x,o) = T_c\left(1 - \left(\frac{x}{\ell}\right)^2\right)$$

T_c is a constant.

Vujanovic's functional is

$$V_u = \int_o^{t_1}\int_{-\ell}^{+\ell}\left(\tau(\dot{T})^2 - \chi\left(\frac{\partial T}{\partial x}\right)^2\right) e^{t/\tau}\, dtdx = 0 \tag{4.24}$$

where t_1 is an arbitrary instant of time.

With

$$T(x,t) = T_c\left(1 - \left(\frac{x}{\ell}\right)^2\right)a(t) \qquad \left(a(0) = 1\right) \tag{4.25}$$

as trial function, V_u yields, after integration with respect to x :

$$V_u = \int_o^{t_1}\left(\frac{8}{15}\tau\ell\dot{a}^2 - \frac{4}{3}\chi\left(\frac{a}{\ell}\right)^2\right) e^{t/\tau}\, dt \equiv \int_o^{t_1} F(a,\dot{a},t)dt$$

In order that V_u be stationary, it is necessary that the following Euler-Lagrange equation be satisfied :

$$\frac{2}{5}\tau\ell\ddot{a} + \frac{2}{5}\ell\dot{a} + \chi\frac{a}{\ell} = 0$$

Since it is wanted to solve the classical parabolic heat equation, set $\tau = 0$ so that the above relation reduces to

$$\frac{2}{5}\ell\dot{a} + \chi\frac{a}{\ell} = 0$$

Its solution is

$$a = e^{-5/2\,\frac{\chi}{\ell^2}\,t}$$

so that

$$T = T_c\left(1 - \left(\frac{X}{\ell}\right)^2\right) e^{-5/2\,\frac{\chi}{\ell^2}\,t} \tag{4.26}$$

Comparison with the exact result

$$T = \frac{32}{\Pi^3} T_c \sum_{n=0}^{\infty} \frac{(-1)^n}{(2n+1)^3} e^{-(2n+1)^2\Pi^2\chi t/4\ell^2} \cos\left(\frac{2n+1}{2}\frac{X}{\ell}\right)$$

indicates an error lesser than 10 % .

4.3. Glansdorff-Prigogine's local potential.[3,4] This method has gained considerable interest in the last years because its connection with a general evolution criterion and the theory of dissipative structures. The *construction* of the so-called local potential is systematic and proceeds as follows. Let us firstly concentrate on the *steady heat conduction* problem. The starting equation is the energy balance

$$\rho\dot{u} = -\operatorname{div}\underline{q} \tag{4.27}$$

where u is the internal energy per unit mass. Multiply it by an arbitrary variation δT^{-1} and integrate on the volume

$$\int_\Omega \rho\dot{u}\,\delta T^{-1}\,d\Omega = -\int_\Omega (\operatorname{div}\underline{q})\,\delta T^{-1}\,d\Omega$$

From now on, concentrate on the r.h.s. Integrating it by parts and taking $\delta T = 0$ on the boundary yields

$$\int_\Omega \underline{q}.\mathrm{grad}\ (\delta T^{-1})d\Omega$$

Introduction of Fourier's law

$$\underline{q} = - \lambda(T)\ \mathrm{grad}\ T = \lambda(T)\ T^2\ \mathrm{grad}\ T^{-1}$$

results in

$$\frac{1}{2}\int_\Omega \lambda(T)\ T^2\ \delta(\mathrm{grad}\ T^{-1})^2 d\Omega \tag{4.28}$$

Let $T^o(x)$ be the exact steady solution, assumed to be given a priori. The admissible function T can be written as

$$T = T^o + \delta T \qquad\qquad (\delta T^o = 0)$$

Moreover, by expanding $\lambda(T)$ around its steady value $\lambda^o(\underline{x})$, one obtains

$$\lambda T^2 = (\lambda^o + \delta\lambda)\ \{T^{o2} + 0\ (\delta T)\}$$

$$= \lambda^o T^{o2} + 0\ (\delta T)$$

Substitution in (4.28) leads to

$$\frac{1}{2}\ \delta\int_\Omega \lambda^o T^{o2}(\mathrm{grad}\ T^{-1})^2 d\Omega \equiv \delta\emptyset\ (T,T^o)$$

wherein $\emptyset\ (T,T^o)$ is defined by

$$\emptyset\,(T,T^o) = \frac{1}{2}\int_\Omega \lambda^o T^{o2}(\text{grad } T^{-1})^2 d\Omega \tag{4.29}$$

This functional is called the *local potential* ; it depends on two kinds of variables : the true, but not determined, solution T^o and the function T, submitted to variation.

It is interesting to determine under which conditions $\emptyset(T,T^o)$ is stationary. In other words, derive the Euler-Lagrange equation corresponding to

$$\delta\emptyset(T,T^o) = 0 \tag{4.30}$$

Clearly, the solution is

$$-\,\text{div}(\lambda^o T^{o2}\text{grad } T^{-1}) = 0 \tag{4.31}$$

In order to recover the correct balance law, it is necessary to introduce the alias condition

$$T = T^o \tag{4.32}$$

expressing that the function T is identical with the exact solution.

The nature of the extremum is determined by calculating the quantity

$$\Delta\emptyset = \emptyset(T,T^o) - \emptyset(T^o,T^o) \tag{4.33}$$

Setting

$$\theta = T^{-1} - (T^o)^{-1},$$

it is seen that

$$\Delta\emptyset = \frac{1}{2}\int_\Omega \lambda^o T^{o2}\left(\text{grad }(T^{o-1} + \theta)^2 - (\text{grad } T^{o-1})^2\right) d\Omega$$

$$= \frac{1}{2}\int_\Omega \lambda^o T^{o2}\left((\text{grad }\theta)^2 + 2 \text{ grad } T^{o-1} \text{ grad }\theta\right) d\Omega$$

After integration by parts of the term involving grad T^{o-1} and recalling that T is fixed on the boundary ($\theta = 0$ on A), $\Delta\emptyset$ reduces to

$$\Delta\emptyset = \frac{1}{2}\int_\Omega \lambda^o T^{o2}(\text{grad }\theta)^2 \, d\Omega > 0 \tag{4.34}$$

This indicates that $\emptyset(T,T^o)$ attains its minimum value at the steady state, at the condition that in the expression of the local potential, T^o is replaced by its true steady value.

For a given problem, the local potential is not unique. By multiplication of the energy balance (4.26) by $\delta(\ln T)$ or by δT instead of δT^{-1}, one would have obtained the following local potentials :

$$\emptyset_1 = \frac{1}{2}\int_\Omega \lambda^o T^o (\text{grad } \ln T)^2 \, d\Omega \tag{4.35}$$

$$\emptyset_2 = \frac{1}{2}\int_\Omega \lambda^o (\text{grad } T)^2 d\Omega \tag{4.36}$$

The choice of a particular local potential is linked with the temperature dependence of λ. If λ is constant, $\emptyset_2$ is recommended ; if λ varies like T^{-1}, $\emptyset_1$ is more appropriate. The above results can be directly generalized to anisotropic system.

With the following hypotheses

1) $\lambda_{ij} = \frac{1}{T^2} L_{ij} \qquad (L_{ij} = \text{constant})$ (4.37)

2) $L_{ij} = L_{ji}$ (4.38)

the local potential behaves like an usual functional. It involves only one kind of variable and reads as

$$\emptyset = \int_\Omega L_{ij}\, T^{-1}_{,i}\, T^{-1}_{,j}\, d\Omega \tag{4.39}$$

According to irreversible thermodynamics, $\emptyset$ is equal to the total entropy production inside the body due to heat condition. Since L_{ij} is positive definite, as a consequence of the second principle of thermodynamics, $\emptyset$ is positive definite and

$$\delta\emptyset = 0 \tag{4.40}$$

represents a minimum principle. This principle is known as the *minimum entropy production principle* and was first proposed by Prigogine.[19]

The above considerations are easily extended to the *unsteady case*. It suffices to substract from (4.29) the term $\rho^o \dot{u}^o T^{-1}$ so that the corresponding local potential is

$$\Psi(T,T^o) = \int_{t_o}^{t_1}\int_\Omega \left(\frac{1}{2}\,\lambda^o T^{o2}(\mathrm{grad}\ T^{-1})^2 - \rho^o \dot{u}^o T^{-1}\right) d\Omega dt \tag{4.41}$$

superscript zero refers to the exact solution of the unsteady problem. Setting $T^o = T$ after that the variation has been performed, one verifies that the local potential Ψ is minimum when T satisfies the energy balance ; this is expressed by

$\delta\Psi = 0$ (stationary property)

and

$\Delta\Psi = \Psi(T,T^o) - \Psi(T^o,T^o) > 0$ (minimum property)

Application : Unsteady heat conduction in an insulated cylindrical rod. The problem to be handled is the same as in section 4.2. As λ is constant, it is convenient to use the following potential

$$\Psi_2(T,T^o) = \int_o^{t_1}\int_{-\ell}^{\ell} \left(\frac{1}{2}\lambda^o(\text{grad } T)^2 - \rho^o\dot{u}^o T\right)dxdt \qquad (4.42)$$

with

$$\rho\dot{u}^o = c\dot{T}^o \qquad (4.43)$$

Since Ψ_2 involves two classes of functions T and T^o, one must select two trial functions, namely

$$T = T_c\left(1 - (\frac{x}{\ell})^2\right)a(t) \qquad \left(a(o) = 1\right) \qquad (4.44a)$$

$$T^o = T_c\left(1 - (\frac{x}{\ell})^2\right)a^o(t) \qquad \left(a^o(o) = 1\right) \qquad (4.44b)$$

T^o has the same dependence with respect to x than T, superscript o affecting a indicates that the corresponding quantity must be kept constant during the variational procedure. Notice that (4.44a) and (4.44b) meet the boundary and initial conditions. After substitution of (4.44a) and (4.44b) into (4.42), one arrives at

$$\Psi_2(T,T^o) = T_c^2\int_o^{t_1}\int_{-\ell}^{+\ell}\left\{2\lambda\frac{x^2}{\ell^2}a^2 - c\left(1 - (\frac{x}{\ell})^2\right)^2 a\dot{a}^o\right\}dxdt$$

$$= T_c^2\int_o^{t_1}\left(\frac{4}{3}\lambda\frac{a^2}{\ell^2} + \frac{16}{15}ca\dot{a}^o\right)dt$$

The quantity a must be chosen so that Ψ_2 is stationary and therefore must obey the Euler-Lagrange equation

$$\frac{a}{\ell} + \frac{2}{5\chi}\,\dot{a}^{o} = 0$$

Setting $a = a^o$ and integrating, yields

$$a = e^{-\frac{5\chi t}{2\ell^2}}$$

One recovers the result of Vujanovic's approach.

4.4. Lebon-Lambermont's principle.[5] Let us go back to the most general problem of heat conduction in a rigid isotropic dy with internal sources $Q(\underline{x},t)$ and temperature dependent thermal properties. The boundary and initial conditions are expressed by the relations (1.1) to (1.4). This problem can be described by the following variational equation, proposed by Lebon and Lambermont :

$$\delta_t I_L(T) = 0$$

where

$$I_L(T) = \int_o^t\int_\Omega \left(\frac{1}{2}\,\lambda^2(T)(\text{grad } T)^2 + \Gamma(T)\dot{T} - Q\,K(T)\right)d\Omega dt$$

$$+ \int_o^t\int_{A_q} q^* K(T)\,dAdt + \int_o^t\int_{A_r} h_s R(T)\,dAdt - \int_o^t\int_{A_T} \underline{n}.\lambda(T)\text{grad } T\,K(T)\,dAdt \tag{4.45}$$

$K(T)$ and $R(T)$ have been defined earlier while $\Gamma(T)$ is given by

$$\Gamma(T) = \int_o^T c(\xi)\lambda(\xi)\,d\xi \tag{4.46}$$

subscript t on δ indicates that the time derivative of the temperature must not be submitted to variation ; by definition

$$\delta_t \dot{T} = 0 \tag{4.47}$$

In that respect, (4.45) is a restricted variational principle. Any admissible function must satisfy the initial conditions but not the boundary conditions. It is a simple matter to verify that the corresponding Euler-Lagrange equation and natural boundary conditions are the correct field equations.

The canonical form of (4.45) wherein not only T but also $\underline{q}$ is submitted to variation is given by

$$\delta_t I_L(T,\underline{q}) = 0$$

with

$$I_L(T,\underline{q}) = \int_0^t\int_\Omega \left(-\frac{1}{2}\,\underline{q}\cdot\underline{q} - \lambda \text{ grad } T.\underline{q} + \Gamma(T)\dot{T} - Q\,K(T)\right) d\Omega dt$$

$$+ \int_0^t\int_{A_q} q^* K(T)\,dAdt + \int_0^t\int_{A_r} h_s R(T)\,dAdt - \int_0^t\int_{A_T} \underline{q}.\underline{n}\ K(T)\,dAdt \tag{4.48}$$

By varying with respect to $\underline{q}$, one recovers the Fourier law as supplementary Euler-Lagrange equation.

For an *anisotropic* body, the variational equation (4.45) must be replaced by the following quasi-variational criterion.

$$\int_o^t\int_\Omega \left(\frac{1}{2}\lambda_{ij}\delta(T_{,i}\ T_{,j}) + \dot{T}\delta u_v\right) d\Omega dt$$

$$+ \delta\left\{-\int_o^t\int_\Omega QTd\Omega dt + \int_o^t\int_\Omega \dot{q}^{*}TdAdt + \int_o^t\int_{A_r} \frac{1}{2} h_s(T - T_s)^2 dAdt\right.$$

$$\left. - \int_o^t\int_{A_T} (T - T^{*})\lambda_{ij}\ T_{,j}\ n_i\ dAdt\right\} = 0 \qquad (4.49)$$

The above variational criteria (4.45) and (4.49) can be extended to thermoelasticity, as shown in the next chapter. Lebon and Lambermont proposed also several alternate variational principles covering a large class of dissipative[20] and convective phenomena.[21]

Application : heat conduction in an infinite slab. An isotropic solid bounded by two parallel planes ($-\ell \leq x \leq \ell$) maintained at zero temperature, is initially at an uniform temperature T_o. The heat diffusivity is assumed to vary linearly with the temperature

$$\chi(T) = \frac{\lambda(T)}{c} = \chi_o \left(1 + \alpha \frac{T}{T_o}\right) \qquad (\alpha = \text{constant}) \qquad (4.50)$$

With the non-dimensional notation

$$X = \frac{x}{\ell}\ ,\quad \tau = \frac{\chi_o t}{\ell^2}\ ,\quad \theta = \frac{T}{T_o} \qquad (4.51)$$

the initial and boundary conditions are

$$\theta(X,o) = \qquad (4.52)$$

$$\theta(\pm 1, \tau) = 0 \qquad (4.53)$$

while the heat equation is

$$\frac{\partial\theta}{\partial\tau} = \frac{\partial}{\partial X}\left(\frac{\chi}{\chi_o}\frac{\partial\theta}{\partial X}\right)$$

The function Γ defined by (4.46) is explicitely given by

$$\Gamma = c^2\int_o^T \chi\, d\xi = c\ \chi_o T_o \theta(1 + \frac{\alpha}{2}\theta)$$

By taking trial functions satisfying (4.53), the functional (4.45) writes simply

$$I_L(T) = \int_o^\infty\int_{-1}^{+1}\left(\frac{1}{2}(1+\alpha\theta)\frac{\partial\theta}{\partial X} + \theta(1 + \frac{\alpha}{2}\theta)\frac{\partial\theta}{\partial\tau}\right)dXdt \tag{4.54}$$

Assume that

$$\theta = \frac{4}{\pi}\sum_o^N \frac{(-1)^n}{2n+1}\cos\left(\frac{2n+1}{2}\pi X\right)e^{-a\pi^2\frac{(2n+1)^2}{4}\tau} \tag{4.55}$$

where a is a constant coefficient to be determined in such a way that I_L is stationary ; it is clear that (4.55) obeys the initial and the boundary conditions. According to Rayleigh-Ritz's method, the best value of a is obtained by derivating $I_L(T)$ with respect to a and setting the result equal to zero :

$$0 = \frac{\partial I_L}{\partial a} = \int_o^\infty\int_{-1}^{+1}\left[\alpha(1+\alpha\theta)\frac{\partial\theta}{\partial a}\left(\frac{\partial\theta}{\partial X}\right)^2\right.$$

$$\left. + (1+\alpha\theta)\frac{\partial\theta}{\partial X}\frac{\partial}{\partial a}\frac{\partial\theta}{\partial X} + \frac{\partial\theta}{\partial a}\frac{\partial\theta}{\partial\tau} + \alpha\theta\frac{\partial\theta}{\partial a}\frac{\partial\theta}{\partial\tau}\right]dXd\tau$$

α / a	0	0.2	0.4	0.5
Lebon-Lambermont	1	1.0254	1.0356	1.0356
Local potential	1	1.0403	1.0806	1.1008

The parameter a has been computed for various values of α (see above), it is compared with the corresponding values furnished by the local potential method.

It is seen that the calculated values are very close for a rather large range of α . By comparison with a finite difference technique, one obtains an agreement within a range of ± 1.5 %.[22]

By increasing α, there appears appreciable differences between the variational and the finite difference methods essentially for large values of α. This indicates that a trial function, involving only one single parameter, is too simple and must be substituted by a more complicated expression, with a greater number of parameters like for instance,

$$\theta = \frac{4}{\pi} \sum_{n=o}^{N} \frac{(-1)^n}{2n+1} \cos \frac{2n+1}{2} \pi X \, e^{-\left(a_n \frac{\pi^2(2n+1)}{4} \tau\right)}$$

References

1. Biot, M., *Variational principles in heat transfer*, Oxford Univ. Press, Oxford, 1970.
2. Vujanovic, B., An approach to linear and non-linear heat transfer problems using a Lagrangian, *A.I.A.A. Journal*, 9, 131, 1971.
3. Glansdorff, P. and Prigogine, I., On a general evolution criterion in macroscopic physics, *Physica*, 30, 351, 1964.
4. Glansdorff, P. and Prigogine, I., *Structure, Stability and fluctuations*, J. Wiley, New-York, 1971.

5. Lebon, G. and Lambermont, J., Some variational principles pertaining to non-stationary heat conduction and coupled thermoelasticity, *Acta Mech.*, 15, 121, 1972.
6. Lebon, G., A new variational principle for the non-linear unsteady heat conduction problem, *Quart. J. Mech. Appl. Math.*, 29, 499, 1976.
7. Gurtin, M., Variational principles for linear initial value problems, *Quart. Appl. Math*, 22, 252, 1964.
8. Tonti, E., On the variational formulation for linear initial value problems, *Annali di Math. pura ed appl.*, 95, 331, 1973.
9. Reddy, J., A note on mixed variational principles for initial-value problems, *Quart. J. Mech. Appl. Math.*, 28, 123, 1975.
10. Maxwell, J., On the dynamical theory of gases, *Phil. Trans. Roy. Soc. London*, 157, 49, 1867.
11. Cattaneo, C., Sur une forme de l'équation de la chaleur éliminant le paradoxe d'une propagation instantanée, *C.R.Acad. Sc. Paris*, 247, 431, 1958.
12. Vernotte, P., Les paradoxes de la théorie continue de l'équation de la chaleur, *C.R.Acad. Sc. Paris*, 247, 3154, 1958.
13. Onsager, L., Reciprocal relations in irreversible processes, *Phys. Rev.*, 37, 405, 1931.
14. Gyarmati, I., *Non-equilibrium thermodynamics*, Springer-Verlag, Berlin, 1971.
15. Rosen, P., Variational approach to magnetohydrodynamics, *Phys. Fluids*, 1, 251, 1958.
16. Djukic, D. and Vujanovic, B., On a new variational principle of Hamiltonian type for classical field theory, *Z.A.M.M.*, 21, 611, 1971.
17. Djukic, D., Vujanovic, B., Tatic, N. and Strauss, A., On two variational methods for obtaining solutions to transport problems, *Chem. Eng. J.*, 5, 145, 1973.
18. Djukic, D., Hiemenz magnetic flow of power-law fluids, *J. Appl. Mech.*, September 1974, 822, 1974.
19. Prigogine, I., *Etude thermodynamique des phénomènes irréversibles*, Desoer, Liège, 1947.

20. Lebon, G. and Lambermont, J., A rather general variational principle for purely dissipative processes, *Annalen Phys.*, 7, 15, 1972.
21. Lebon, G. and Lambermont, J., Generalization of Hamilton's principle to continuous dissipative systems, *J. Chem. Phys.*, 59, 2929, 1973.
22. Lebon, G., Distribution non-stationnaire de la température dans les milieux dont la conductibilité thermique dépend de la température, *Ann. Soc. Sc. Bruxelles*, 84, 304, 1970.
23. Lardner, T.J., Biot's variational principle in heat conduction, *J. A.I.A.A.*, 1, 1, 1963.
24. Richardson, P.D., Unsteady heat conduction with a non-linear boundary condition, *J. Heat Transfer*, 86, 298, 1964.
25. Finlayson, B.A. and Scriven, L., On the search for variational principles, *Int. J. Heat Mass Transfer*, 10, 799, 1967.

IV. COUPLED THERMOELASTICITY

1. Basic equations

In this chapter, the variational principles of elasticity and heat conduction are extended to the initial boundary value problem of fully coupled thermoelasticity. A great variety of criteria have been published. However, due to a lack of time, only the most significative formulations of Iesan,[1,2] Nickell - Sackman,[3] Biot,[4-7] Lebon-Lambermont[8,9] and Parkus[10] will be examined.

The body under consideration is an inhomogeneous anisotropic elastic solid which is heated and submitted to mechanical deformations. All the properties of the body are evaluated with respect to a virgin free-stress state at uniform reference temperature T_r. Let Ω denote the volume of the body and A its boundary.

The *material properties* of the body are determined by the state equations

$$s = s\,(\varepsilon_{ij},\theta) \text{ or equivalently } u = u(\varepsilon_{ij},\theta) \tag{1.1}$$

$$\sigma_{ij} = \sigma_{ij}\,(\varepsilon_{k\ell},\theta) \text{ or equivalently } \varepsilon_{ij} = \varepsilon_{ij}\,(\sigma_{k\ell},\theta) \tag{1.2}$$

and the Fourier law

$$q_i = -\lambda_{ij}\theta_{,j} \tag{1.3}$$

θ denotes the temperature $T - T_r$ above the reference value T_r, the other quantities have the same meaning as before.

The laws of evolution are

balance of mass : $\frac{d\rho}{dt} = -\rho \, v_{i,i}$ (1.4)

balance of momentum : $\rho \frac{dv_i}{dt} = \sigma_{ij,j} + F_i$ (1.5)

balance of entropy(*) : $\rho \frac{ds}{dt} = -\frac{1}{T} q_{i,i} + \frac{Q}{T}$ (1.6)

One must also add the *initial conditions*

$$u_i(x_i\,,0) = u_{oi}(x_i) \tag{1.7}$$

$$v_i(x_i\,,0) = v_{oi}(x_i) \tag{1.8}$$

$$\theta(x_i\,,0) = \theta_o(x_i) \tag{1.9}$$

$$\rho(x_i\,,0) = \rho_o(x_i)$$

and the mixed *boundary conditions :*

displacement b.c. : $u_i(x_i\,,t) = u_i^*(x_i\,,t)$ on A_u (1.10)

traction b.c. : $\sigma_{ij}(x_i\,,t)\eta_j \equiv T_i(x_i\,,t) = T_i^*(x_i\,,t)$ on A_σ (1.11)

temperature b.c. (Dirichlet) : $\theta(x_i\,,t) = \theta^*(x_i\,,t)$ on A_T (1.12)

heat flux b.c. (Neumann) : $-\lambda_{ij}\theta_{,j}\eta_i \equiv q_i(x_i,t)\eta_i = q^*(x_i,t)$ on A_q (1.13)

(*) Instead of (1.6), it is equivalent to work with the energy balance

$$\rho \frac{du}{dt} = -q_{i,i} + \sigma_{ij} \frac{1}{2}(v_{i,j} + v_{j,i}) + Q \tag{1.6b}$$

where A_u and A_σ (respectively A_T and A_q) are disjoint sets whose union is A :

$$A = A_u \cup A_\sigma = A_T \cup A_q$$

$$A_u \cap A_\sigma = A_T \cap A_q = 0$$

Expressions (1.1) to (1.13) are the relevant equations of the coupled dynamic problem of thermoelasticity.

The linear theory. Assume that the displacements, the strains, the rotations and the temperature variations are small of the first order. Suppose also that the thermo-elastic properties are constant, i.e. independent of u_i and θ, but function of x_i. The state equations (1.1) and (1.2) read then simply (see the appendix for their derivation and definition of the coefficients) :

$$\rho s = \beta_{ij}\varepsilon_{ij} + \rho \frac{c_\varepsilon}{T_r} \theta \qquad (1.14a)$$

$$\rho u = T_r \beta_{ij}\varepsilon_{ij} + \rho c_\varepsilon \theta \qquad (1.14b)$$

$$\sigma_{ij} = C_{ijk\ell}\varepsilon_{k\ell} - \beta_{ij}\theta \quad \text{(Neumann-Duhamel's law)} \qquad (1.15a)$$

$$\varepsilon_{ij} = B_{ijk\ell}\sigma_{k\ell} + \alpha_{ij}\theta \qquad (1.15b)$$

where

$$\beta_{ij} = C_{ijk\ell}\alpha_{k\ell} \quad , \quad C_{ijk\ell}B_{k\ell mn} = \delta_{im}\delta_{jn} \qquad (1.16)$$

Substituting these results in the balance equations (1.5) and (1.6) and assuming that T remains close to T_r so that

$$T\,\beta_{ij} = T_r\beta_{ij}$$

one obtains

$$\rho\, \partial_t^2 u_i = (C_{ijk\ell}\varepsilon_{k\ell})_{,j} - (\beta_{ij}\theta)_{,j} + F_i \quad \left(\partial_t \equiv \frac{\partial}{\partial t}\right) \tag{1.17}$$

$$\rho\, c_\varepsilon \partial_t \theta + T_r \beta_{ij} \partial_t \varepsilon_{ij} = (\lambda_{ij}\theta_{,j})_{,i} + Q \tag{1.18}$$

i.e. a system of two vectorial equations for the two unknowns u_i and θ .

As in Chap. II, ∂_t represents a partial derivation with respect to the time.

In most applications, it is legitimate to omit the mechanical term in (1.18) without appreciable error so that the problem becomes uncoupled. The formulation of the corresponding variational equation is then straightforward : the temperature distribution is derived from the principles of Chap. III., the strain (or stress) field is obtained from the principles of elastodynamics given in Chap. II. The only difference with pure elasticity is that the strain (or stress) energy functions W (or W_c) contains a contributions arising from the temperature field. In the linear theory, the expressions of W and W_c are respectively given by (see appendix)

$$W(\varepsilon_{ij}\,,\theta) \equiv \rho f = \frac{1}{2}\, C_{ijk\ell}\varepsilon_{ij}\varepsilon_{k\ell} - \beta_{ij}\varepsilon_{ij}\theta - \rho\,\frac{c_\varepsilon}{2T_r}\,\theta^2 \tag{1.19}$$

$$W_c(\sigma_{ij}\,,\theta) \equiv -\rho g = \frac{1}{2}\, B_{ijk\ell}\sigma_{ij}\sigma_{k\ell} + \alpha_{ij}\sigma_{ij}\theta + \rho\,\frac{c_\varepsilon}{2T_r}\,\theta^2 \tag{1.20}$$

where

$$B_{ijk\ell} = B_{jik\ell} = B_{k\ell ij} \tag{1.21}$$

$$\alpha_{ij} = \alpha_{ji} \tag{1.22}$$

Equations (1.15a) and (1.15b) are obtained by direct derivation of (1.19) and (1.20) with respect to ε_{ij} and σ_{ij} respectively.

2. The Iesan-Nickell-Sackman principle.

This criterion is an extension of Gurtin's principle to the linear problem of coupled dynamic thermoelasticity. It was formulated independently by Iesan[1] and Nickell and Sackman.[3]

In terms of the convolution products, the basic equations (1.17) and (1.18) are expressed by

$$\rho u_i = g * (C_{ijk\ell} u_{k,\ell} - \beta_{ij}\theta)_{,j} + f_i \tag{2.1}$$

$$\rho c_\varepsilon \theta = g' * (\lambda_{ij}\theta_{,j})_{,i} - T_r \beta_{ij} u_{i,j} + h \tag{2.2}$$

with

$$g(t) = t \tag{2.3}$$

$$g'(t) = 1 \tag{2.4}$$

$$f_i = g * F_i + \rho(t\, v_{oi} + u_{oi}) \tag{2.5}$$

$$h = g' * Q + \rho\, c_\varepsilon \theta_o + T_r \beta_{ij} u_{oi,j} \tag{2.6}$$

Construct the functional

$$I(u_i\,,\,\theta) = I_{mech} + I_{therm} + I_{inter}$$

where

$$I_{mech} = \frac{1}{2}\int_\Omega \left(C_{ijk\ell}(g * u_{i,j} * u_{k,\ell}) + \rho u_i * u_i - 2f_i * u_i\right) d\Omega - \int_{A_\sigma} g * T_i^* * u_i \, dA \tag{2.7}$$

$$I_{therm} = \int_\Omega \left(- \frac{\rho c_\varepsilon}{T_r} g*\theta*\theta - \frac{\lambda_{ij}}{T_r} g*g'*\theta_{,i}*\theta_{,j} + \frac{1}{T_r} g*h*\theta \right) d\Omega$$

$$- \int_{A_q} \frac{1}{T_r} g*g'*q^{*}*\theta \quad dA \tag{2.8}$$

$$I_{inter} = - \int_\Omega \beta_{ij} g*\theta*u_{i,j} d\Omega \tag{2.9}$$

Following the procedure outlined in Chap. II and III, it can be proved that the kinematically and thermally admissible fields which make $I(u_i, \theta)$ stationary are the solutions of the field equations (2.1) and (2.2) and the following boundary conditions

$$(C_{ijk\ell} u_{k,\ell} - \beta_{ij}\theta) n_j = T_i^{*} \text{ on } A_\sigma \quad ; \quad - \lambda_{ij}\theta_{,j} n_i = q^{*} \text{ on } A_q$$

A dual principle admitting the stress tensor and the heat flux as variables has also been enounced by Iesan and Nickell and Sackman. The admissible functions for σ_{ij} and q_i must now satisfy the traction and heat flux boundary conditions. The principle reads as

$$\delta J(\sigma_{ij}, q_i) = \delta(J_{mech} + J_{therm} + J_{inter}) = 0 \tag{2.10}$$

with

$$J_{mech} = \frac{1}{2} \int_\Omega \left(x_{ijk\ell} (\sigma_{ij} * \sigma_{k\ell}) + \frac{1}{\rho} (g * \sigma_{ij,j} * \sigma_{im,m} + 2 f_i * \sigma_{ij,j}) \right) d\Omega$$

$$- \int_{A_u} u_i^{*} * T_i dA \tag{2.11}$$

$$J_{therm} = -\frac{1}{2}\int_{\Omega} \Big(\frac{r_{ij}}{T_r}\, g' * q_i * q_i + \frac{1}{\rho\, c_\varepsilon T_r}\,(g' * g' * q_{i,i} * q_{m,m}$$

$$- 2g' * h * q_{i,i})\Big)\, d\Omega - \int_{A_T} \frac{1}{T_r}\, g' * \theta^* * q_i n_i\, dA \qquad (2.12)$$

$$J_{inter} = \int_{\Omega} \frac{\gamma_{ij}}{T_r}\, \sigma_{ij} * (h - g' * q_{m,m})\, d\Omega \qquad (2.13)$$

where

$$\gamma_{ij} = \frac{T_r}{\rho c_\varepsilon}\, \alpha_{ij}$$

$$x_{ijk\ell} = \beta_{ijk\ell} - \alpha_{ij}\gamma_{k\ell}$$

The Euler-Lagrange equations are

$$x_{ijk\ell}\sigma_{k\ell} + \frac{\gamma_{ij}}{T_r}\,(h - g' * q_{m,m}) - \Big(\frac{1}{\rho}\,(g * \sigma_{im,m} + f_i)\Big)_{,j} = 0 \qquad (2.14)$$

$$\Big(\gamma_{k\ell}\sigma_{k\ell} - \frac{1}{\rho c_\varepsilon}\,(h - g' * q_{m,m})\Big)_{,i} - r_{ij} q_j = 0 \qquad (2.15)$$

while at the boundaries

$$\frac{1}{\rho}\,(g * \sigma_{ij,j} + f_i) = u_i^* \qquad \text{on } A_u \qquad (2.16)$$

$$\frac{1}{\rho c_\varepsilon + T_r \alpha_{ij}\beta_{ij}}\,(h - g' * q_{i,i} - T_r \alpha_{ij}\sigma_{ij}) = \theta^* \qquad \text{on } A_T \qquad (2.17)$$

When u_i and θ satisfy the field equations (2.1) and (2.2), expressions (2.14) and (2.15) imply the strain-displacement relation. Finally. (2.16) and (2.17) are nothing but the displacement and the temperature boundary conditions (1.10) and (1.12).

Iesan as well as Sackman and Nickell have also derived canonical principles corresponding to Reissner's and Washizu's elasto-static principles.

Recently, Green and Lindsay[11] suggested that the internal energy and the stress are not only linear functions of θ and ε_{ij}, but depend in addition on $\partial_t\theta$. A variational principle encountering this case has been formulated by Rusu.[12]

3. Lagrangian thermoelasticity

3.1. Biot's principle. The above linear problem (1.17), (1.18) was also given a variational formulation by Biot.[4-6]

By analogy with the heat conduction problem, one introduces a thermoelastic potential

$$V = \int_\Omega (u_v - T_r s_v)\,d\Omega$$

$$= \frac{1}{2}\int_\Omega \left(\frac{\rho c_\varepsilon}{T_r}\,\theta^2 + C_{ijk\ell}\varepsilon_{ij}\varepsilon_{k\ell}\right)d\Omega \quad (*) \tag{3.1}$$

and a dissipation function defined by

$$\delta D = \int_\Omega r_{ij}\partial_t H_j \delta H_i\,d\Omega \tag{3.2}$$

Biot's principle states that the admissible displacements u_i and heat displacements H_i satisfying

(*) This result is demonstrated in the appendix. The advantage of V with respect to the classical Helmholtz free energy is that the former is additive and positive definite while the latter is not. It should be noted that the above definition of V differs from the corresponding thermal potential of Chap. III by a factor T^{-1}.

$$u_i = u_i^* \qquad \text{on } A_u \tag{3.3}$$

and

$$H_i = H_i^* \qquad \text{on } A_q \tag{3.4}$$

are solution of the thermoelastic problem if

$$\delta V + \frac{1}{T_r}\,\delta D + \int_\Omega (\rho \partial_t^2 u_i - F_i)\,\delta u_i\, d\Omega - \int_{A_\sigma} T_i^* \delta u_i\, dA + \int_{A_T} \frac{\theta^*}{T_r}\, n_i \delta H_i\, dA = 0 \tag{3.5}$$

To show it, let us derive the Euler-Lagrange equations. The variation of V is

$$\delta V = \int_\Omega \Big(\frac{\rho c_\varepsilon}{T_r}\,\theta\,\delta\theta - (C_{ijk\ell}\varepsilon_{k\ell})_{,j}\,\delta u_i\Big) d\Omega + \int_A C_{ijk\ell}\varepsilon_{k\ell} n_j \delta u_i\, dA \tag{3.6}$$

To express $\delta\theta$ in terms of δH_i, one applies the energy equation (1.18) which reads, after integration with respect to the time :

$$\rho c_\varepsilon \theta + T_r \beta_{ij}\varepsilon_{ij} = - H_{i,i} \tag{3.7}$$

Application of the δ operator yields

$$\frac{\rho c_\varepsilon}{T_r}\,\delta\theta = - \frac{1}{T_r}\,\delta H_{i,i} - \beta_{ij}\delta u_{i,j}$$

so that δV becomes, after integration by parts and use of Neumann-Duhamel's law,

$$\delta V = \int_\Omega \left(- \sigma_{ij,j}\delta u_i + \frac{1}{T_r}\theta_{,i}\,\delta H_i\right)d\Omega$$

$$+ \int_A \left((C_{ijk\ell}\varepsilon_{k\ell} - \beta_{ij}\theta)n_j\delta u_i - \frac{\theta}{T_r}n_i\delta H_i\right)dA$$

After substitution in (3.5), one arrives at

$$\int_\Omega \left((\rho\partial_t^2 u_i - \sigma_{ij,j} - F_i)\delta u_i + \frac{1}{T_r}(r_{ij}\partial_t H_j + \theta_{,i})\delta H_i\right)d\Omega$$

$$+ \int_{A_\sigma}(\sigma_{ij}n_j - T_i^*)\delta u_i dA - \int_{A_T}\frac{1}{T_r}(\theta - \theta^*)n_i\,\delta H_i dA = 0 \qquad (3.8)$$

This relation holds for arbitrary variations δu_i and δH_i if and only if the linearized equation of motion (1.12) and the Fourier law (1.3) are fulfilled. Moreover, as natural boundary conditions, one recovers the traction and the thermal conditions

$$\sigma_{ij}n_j = T_i^* \qquad \text{on } A_\sigma$$

$$\theta = \theta^* \qquad \text{on } A_T$$

For *static problems*, Biot's principle becomes

$$\delta V + \frac{1}{T_r}\delta D - \int_\Omega F_i\delta u_i d\Omega - \int_{A_\sigma} T_i^*\delta u_i dA + \int_{A_T}\frac{\theta^*}{T_r}n_i\delta H_i dA = 0 \qquad (3.9)$$

In the special case that q_i $(=\partial_t H_i)$ and F_i derive from two potential functions $\mathcal{X}$ and $\mathcal{F}$ such that

$$\frac{1}{T_r}\partial_t H_i = \frac{\partial \mathcal{X}}{\partial H_i} \quad ; \quad F_i = -\frac{\partial \mathcal{F}}{\partial u_i}$$

expression (3.9) can be viewed as a classical variational equation,

$$\delta I_1(u_i, H_i) = 0 \qquad (3.10)$$

where

$$I_1(u_i, H_i) = V + \int_\Omega (\mathcal{X} + \mathcal{J})\, d\Omega - \int_{A_\sigma} T_i^* u_i\, dA + \int_{A_T} \frac{\theta^*}{T_r} H_i n_i\, dA$$

In addition, since V is positive definite, (3.10) possesses a minimum property.

Lagrangian formulation. The derivation of Lagrange equations from the variational criterion (3.6) can be carried out in strict analogy with the heat conduction problem.

Express u_i and H_i in terms of generalized unknown coordinates $a_1(t) \ldots a_n(t)$:

$$u_i = u_i(a_1, a_2, \ldots a_n; x_j, t) \quad ; \quad H_i = H_i(a_1, a_2, \ldots a_n; x_j, t)$$

from which

$$\delta u_i = \sum_{\alpha=1}^{n} \frac{\partial u_i}{\partial a_\alpha} \delta a_\alpha \quad ; \quad \delta H_i = \sum_{\alpha=1}^{n} \frac{\partial H_i}{\partial a_\alpha} \delta a_\alpha$$

It follows directly that

$$\delta V = \sum_{\alpha=1}^{n} \frac{\partial V}{\partial a_\alpha} \delta a_\alpha \qquad (3.11)$$

while

$$\int_\Omega \rho\, \partial_t^2 u_i\, \delta u_i\, d\Omega = \int \rho \sum_{\alpha=1}^{n} \partial_t^2 u_i \frac{\partial u_i}{\partial a_\alpha} \delta a_\alpha\, d\Omega \qquad (3.12)$$

This latter term can be written in term of the kinetic energy

$$K = \frac{1}{2}\int_{\Omega} \rho\, \partial_t u_i\, \partial_t u_i \, d\Omega \tag{3.13}$$

To show it, start from the result

$$\partial_t u_i = \sum_{\alpha} \frac{\partial u_i}{\partial a_\alpha}\, \partial_t a_\alpha + \left.\frac{\partial u_i}{\partial t}\right|_{a_\alpha}$$

hence

$$\frac{\partial(\partial_t u_i)}{\partial(\partial_t a_\alpha)} = \frac{\partial u_i}{\partial a_\alpha}$$

and

$$\frac{\partial(\partial_t u_i)}{\partial\, a_\beta} = \sum_{\alpha} \frac{\partial^2 u_i}{\partial a_\alpha\, \partial a_\beta}\, \partial_t a_\alpha + \frac{\partial^2 u_i}{\partial t\, \partial a_\beta} = \partial_t \left(\frac{\partial u_i}{\partial a_\beta}\right)$$

Substitution of these results in the identity

$$\partial_t^2 u_i \frac{\partial u_i}{\partial a_\alpha} = \partial_t \left(\partial_t u_i \frac{\partial u_i}{\partial a_\alpha}\right) - \partial_t u_i \partial_t \left(\frac{\partial u_i}{\partial a_\alpha}\right)$$

gives

$$\partial_t^2 u_i \frac{\partial u_i}{\partial a_\alpha} = \partial_t \left(\partial_t u_i \frac{\partial(\partial_t u_i)}{\partial(\partial_t a_\alpha)}\right) - \partial_t u_i \frac{\partial(\partial_t u_i)}{\partial\, a_\alpha}$$

while (3.12) becomes

$$\int \rho\, \partial_t^2 u_i\, \delta u_i\, d\Omega = \sum_{\alpha} \left(\partial_t \frac{\partial K}{\partial(\partial_t a_\alpha)} - \frac{\partial K}{\partial a_\alpha}\right) \delta a_\alpha \tag{3.14}$$

The three last integrals of (3.6) can still be written as

$$-\int_{\Omega} F_i \delta u_i d\Omega + \int_{A_\sigma} T_i^{*} \delta u_i dA - \int_{A_T} \frac{\theta^{*}}{T_r} n_i \delta H_i dA = \sum_\alpha Q_\alpha \ \delta a_\alpha$$

where

$$Q_\alpha = \int_{A_\sigma} T_i^{*} \frac{\partial u_i}{\partial a_\alpha} d\Omega - \int_{A_T} \frac{\theta^{*}}{T_r} n_i \frac{\partial H_i}{\partial a_\alpha} dA$$

$$- \int_{\Omega} F_i \frac{\partial u_i}{\partial a_\alpha} d\Omega \tag{3.15}$$

Finally, in analogy with the heat conduction problem, put

$$\int r_{ij} \partial_t H_j \quad \delta H_i d\Omega = \sum_\alpha \frac{\partial D}{\partial(\partial_t a_\alpha)} \delta a_\alpha \tag{3.16}$$

Replacing (3.11), (3.14), (3.15) and (3.16) in (3.6) leads to the Lagrangian equation :

$$\sum_\alpha \left(\partial_t \frac{\partial K}{\partial(\partial_t a_\alpha)} - \frac{\partial K}{\partial a_\alpha} + \frac{\partial D}{\partial(\partial_t a_\alpha)} + \frac{\partial V}{\partial a_\alpha} \right) \delta a_\alpha = \sum_\alpha Q_\alpha \ \delta q_\alpha$$

Since the δa_α's are arbitrary,

$$\partial_t \frac{\partial K}{\partial(\partial_t a_\alpha)} - \frac{\partial K}{\partial a_\alpha} + \frac{\partial D}{\partial(\partial_t a_\alpha)} + \frac{\partial V}{\partial a_\alpha} = Q_\alpha \tag{3.17}$$

In absence of deformation ($u_i = \varepsilon_{ij} = 0$), equation (3.17) reduces to

$$\frac{\partial D}{\partial(\partial_t a_\alpha)} + \frac{\partial V}{\partial a_\alpha} = Q_\alpha$$

i.e. the Lagrangian heat conduction equation of chapter III. Notice that in order to obtain a relation like (3.17), it was necessary to express u_i and H_i in terms of the same unknown parameters a_α . If this is not done, no Lagrangian equation like (3.17) can be established.

Biot's principle has been generalized by Rafalski[14-16] for a large class of boundary conditions. Furthermore, Rafalski's analysis includes an integration over the time which allows to express the approximate function in terms of constant generalized coordinates instead of time dependent parameters as in Biot's original formulation. Rafalski used also the temperature as independent variational variable, which is much more convenient than the heat displacement field.

Despite its restricted character, Biot's criterion presents some advantages compared with Iesan-Nickell-Sackman's more rigorous formulation. From one side, the analytical expression of the principle is more simple and approximate solution can directly be derived from the Lagrangian form. Moreover, Biot was able to extend the above considerations to non-linear thermoelasticity,[7] fluid mechanics[17] and chemistry.[18,19]

3.2. A canonical principle for the static linear problem.[20] The linear equations of static thermoelasticity can be written as

$$\tau_{ij,j} + F_i - \beta_{ij} g_j = 0 \tag{3.18}$$

$$\rho\, c_\varepsilon \theta + H_{i,i} + T_r \beta_{ij} u_{i,j} = 0 \tag{3.19}$$

wherein g_j denotes the temperature gradient

$$g_j = \theta_{,j} \tag{3.20}$$

and τ_{ij} the part of the stress which is not due to the temperature and which is defined by

$$\varepsilon_{ij} = B_{ijk\ell} \tau_{k\ell} \tag{3.21}$$

Let us show that the field equations (3.18 - 3.21) and the appropriate boundary conditions can be derived from the variational equation

$$\delta I_2(\theta,\ H_i,\ u_i,\ \tau_{ij}) = 0 \tag{3.22}$$

where

$$I_2(\theta,\ H_i,\ u_i,\ \tau_{ij}) = \int_\Omega \left(\varepsilon_{ij}\tau_{ij} - F_i u_i - W_c^* + \frac{\rho c_\varepsilon}{T_r}\theta^2 - \frac{1}{T_r} H_i g_i - \frac{1}{T_r} D^*(g_i)\right) d\Omega$$

$$-\int_{A_\sigma} T_i^* u_i dA - \int_{A_u} T_i (u_i - u_i^*) dA + \frac{1}{T_r}\int_{A_q} \theta (H_i - H_i^*) n_i dA + \frac{1}{T_r}\int_{A_T} \theta^* H_i n_i dA \tag{3.23}$$

The quantity W_c^* appearing in (3.23) is a complementary deformation energy constructed from Biot's thermoelastic potential and given by

$$W_c^*(\tau_{ij}\ ,\ \theta) = \frac{1}{2} B_{ijk\ell}\tau_{ij}\tau_{k\ell} + \frac{1}{2}\frac{\rho c_\varepsilon}{T_r}\theta^2 \tag{3.24}$$

The specific dissipation function D^* is defined by

$$\partial_t D^* = \frac{1}{2} r_{ij} q_i q_j = \frac{1}{2} \lambda_{ij} g_i g_j$$

from which follows

$$\frac{\partial(\partial_t D^*)}{\partial g_i} = \lambda_{ij} g_j = -\ \partial_t H_i \tag{3.25}$$

and by integration with respect to the time,

$$\frac{\partial D^*}{\partial g_i} = -\ H_i \tag{3.26}$$

Application of the δ operator to (3.23) results in

$$\delta I_2(\theta, H_i, u_i, \tau_{ij}) = \int_\Omega \Bigg[(\tau_{ij} - \theta\beta_{ij})\delta u_{i,j} - F_i\delta u_i + (\varepsilon_{ij} - B_{ijk\ell}\tau_{k\ell})\delta\tau_{ij}$$

$$- \left(\frac{\partial W_c^*}{\partial\theta} - \frac{\rho c_\varepsilon}{T_r}\theta\right)\delta\theta - \frac{1}{T_r}\left(H_i - \frac{\partial D^*}{\partial g_i}\right)\delta\theta_{,i} - \frac{g_i}{T_r}\delta H_i - \frac{\theta}{T_r}\delta H_{i,i}\Bigg] d\Omega$$

$$- \int_{A_\sigma} T_i^*\delta u_i dA - \int_{A_u}\left((u_i - u_i^*)\delta T_i - T_i\delta u_i\right)dA + \frac{1}{T_r}\int_{A_q}\left((H_i - H_i^*)\delta\theta + \theta\delta H_i\right)n_i dA$$

$$+ \int_{A_T} \frac{1}{T_r}\theta^*\delta H_i n_i dA = 0$$

Integrating by parts and using (3.26), one obtains the Euler-Lagrange equations corresponding to arbitrary variations of u_i, τ_{ij}, θ and H_i, namely

$$\tau_{ij,j} + F_i - \beta_{ij}\theta_{,j} = 0 \tag{3.27}$$

$$\varepsilon_{ij} - B_{ijk\ell}\tau_{k\ell} = 0 \tag{3.28}$$

$$\frac{1}{T_r} H_{i,i} + \beta_{ij}u_{i,j} + \frac{\rho c_\varepsilon}{T_r}\theta = 0 \tag{3.29}$$

$$\theta_{,i} - g_i = 0 \tag{3.30}$$

while at the boundaries,

$$u_i - u_i^* = 0 \qquad \text{on } A_u$$

$$T_i - T_i^* = 0 \qquad \text{on } A_\sigma$$

$$H_i - H_i^* = 0 \qquad \text{on } A_q$$

$$\theta - \theta^{*} = 0 \qquad \text{on } A_T$$

This completes the proof.

3.3. A complementary principle for the static linear problem. Assume now that only τ_{ij} and θ are varied independently and impose that the momentum equation (3.18) as well as the relations (3.20) and (3.21) are satisfied. Suppose in addition that the traction and temperature boundary conditions are identically verified by the admissible functions:

$$\delta T_i = 0 \quad \text{on } A_\sigma \quad , \qquad \delta\theta = 0 \quad \text{on } A_T \tag{3.31}$$

The remaining field equations are then derived via a complementary variational principle. The details of its construction are the same as in elasticity and will not be repeated here. The principle takes the form

$$\delta I_3(\tau_{ij}\,,\,\theta) = 0$$

$$I_3(\tau_{ij}\,,\,\theta) = -\int_\Omega \left(W_c^{*} + \frac{1}{T_r} D^{*} + G\right) d\Omega + \int_{A_u} u_i^{*} T_i \, dA - \int_{A_q} H_i^{*} n_i \theta \, dA \tag{3.32}$$

where the potential function G is defined by

$$\frac{\partial G}{\partial F_i} = - u_i \tag{3.33}$$

The derivation of the Euler-Lagrange equations is classical. One has

$$\delta I_3(\tau_{ij}\,,\,\theta) = -\int_\Omega \left(\frac{\partial W_c^{*}}{\partial \tau_{ij}} \delta\tau_{ij} + \frac{\partial W_c^{*}}{\partial \theta} \delta\theta + \frac{1}{T_r} \frac{\partial D^{*}}{\partial g_i} \delta\theta_{,i} + \frac{\partial G}{\partial F_i} \delta F_i\right) d\Omega$$

$$+ \int_{A_u} u_i^{*} \delta T_i \, dA - \int_{A_q} \frac{1}{T_r} H_i^{*} n_i \, \delta\theta \, dA = 0 \tag{3.34}$$

But from (3.18),

$$\delta F_i = \beta_{ij}\delta\theta_{,j} - \delta\tau_{ij,j}$$

Substituting in (3.34) and integrating by parts, one gets

$$\delta I_3(\tau_{ij}, \theta) = -\int_\Omega \left[\left(\frac{\partial W_c^*}{\partial \tau_{ij}} - u_{i,j}\right)\delta\tau_{ij} + \left(\frac{\partial W_c^*}{\partial \theta} + \frac{H_{i,i}}{T_r} + \beta_{ij}u_{i,j}\right)\delta\theta\right] d\Omega$$

$$-\int_{A_u} (u_i - u_i^*)\delta T_i dA + \int_{A_q} \frac{1}{T_r}(H_i - H_i^*)n_i\delta\theta = 0 \tag{3.35}$$

from which follow the Euler-Lagrange equations :

$B_{ijk\ell}\tau_{k\ell} = \frac{1}{2}(u_{i,j} + u_{j,i})$ in Ω

$\rho\, c_\varepsilon \theta + H_{i,i} + T_r\beta_{ij}u_{i,j} = 0$ in Ω

$u_i = u_i^*$ on A_u

$H_i = H_i^*$ on A_q

The above principle represents an extension of the dual principle of elasticity ; since W_c^* is positive definite, it is a maximum principle. It is worthy to remember that the canonical criterion presents only stationary properties.

4. Lebon-Lambermont's principle for linear dynamic thermo-elasticity

Before deriving the principle, we introduce the following notations

$$F = \int_\Omega f_v d\Omega \tag{4.1}$$

with f_v given by (see appendix)

$$f_v = \frac{1}{2} C_{ijk\ell}\varepsilon_{ij}\varepsilon_{k\ell} - \beta_{ij}\,\theta\,\varepsilon_{ij} - \frac{1}{2}\rho\,\frac{c_\varepsilon}{T_r}\,\theta^2 \tag{4.2}$$

$$K = \frac{1}{2}\int_\Omega \rho\partial_t u_i\ \partial_t u_i d\Omega \qquad \text{(total kinetic energy)} \tag{4.3}$$

$$W = \int_\Omega F_i u_i d\Omega \qquad \text{(mechanical work)} \tag{4.4}$$

$$\mathcal{L} = F - K - W \qquad \text{(lagrangian)} \tag{4.5}$$

Assume that the admissible functions for u_i and H_i satisfy the boundary conditions on A_u and A_q respectively. The thermoelastic equations are then equivalent to the variational equation

$$\delta\left\{\int_{t_1}^{t_2}\mathcal{L}\,dt - \frac{1}{T_r}\int_{t_1}^{t_2}\int_\Omega H_{i,i}\,\theta\,d\Omega dt + \frac{1}{T_r}\int_{t_1}^{t_2}\int_{A_T}\theta^* n_i H_i dAdt - \int_{t_1}^{t_2}\int_{A_\sigma} T_i^* u_i dAdt\right\} + \frac{1}{T_r}\int_{t_1}^{t_2}\int_\Omega r_{ij}\partial_t H_j \delta H_i d\Omega dt - \int_\Omega \rho\,\partial_t u_i\ \delta u_i\Big|_{t_1}^{t_2} = 0 \tag{4.6}$$

The variations are to be taken with respect to u_i, θ and H_i. The corresponding Euler-Lagrange equations are

$$\rho\partial_t^2 u_i - F_i - C_{ijk\ell}\varepsilon_{k\ell,j} + \beta_{ij}\theta_{,j} = 0 \tag{4.7}$$

$$\rho c_\varepsilon\theta + T_r\beta_{ij}\varepsilon_{ij} + H_{i,i} = 0 \tag{4.8}$$

$$\theta_{,i} + r_{ij}\partial_t H_j = 0 \tag{4.9}$$

while on the boundaries, recalling that $\delta u_i = 0$ on A_u , $\delta H_i = 0$ on A_q ,

$$\sigma_{ij}n_j - T_i^* = 0 \qquad \text{on } A_\sigma$$

$$\theta - \theta^* = 0 \qquad \text{on } A_T$$

Since θ and H_i are considered as independent quantities, it is not allowed to inject from the start a relation between them. Therefore, it is not surprising to recover Fourier's law as Euler-Lagrange equation.

At this point, it is worth while to emphasize the differences with Biot's principle (3.5).

1. Lebon-Lambermont's functionals involve integration on the time-interval.
2. The quantities submitted to variation are u_i, H_i and θ while in Biot's, only u_i and H_i are varied.
3. It is no longer necessary to integrate the energy equation in order to express θ in terms of H_i .

Similarly with Biot's formalism, the variational equation (4.6) can be written in the form of a Lagrangian equation, namely

$$\partial_t \frac{\partial K}{\partial(\partial_t a_\alpha)} - \frac{\partial K}{\partial a_\alpha} + \frac{\partial D}{\partial(\partial_t a_\alpha)} + \frac{\partial F}{\partial a_\alpha} = Q_\alpha$$

Q_α and D are defined as earlier.

Lebon and Lambermont have also extended their principle to temperature dependent thermo-elastic properties ; for details, see reference.[8]

5. A generalized Hamilton's principle for finite deformations.

A few years ago, Parkus[10,22] proposed a generalization of Hamilton's principle to coupled thermoelasticity. Contrary to the previous criteria, it is applicable to large deformations. The version presented here differs slightly from Parkus' original formulation.

Introduce the following functionals :

$$K = \frac{1}{2}\int_{\Omega} \partial_t u_i \partial_t u_i \rho d\Omega \qquad \left(\partial_t u_i = \frac{\partial u_i}{\partial t}\Big|_X\right) \tag{5.1}$$

$$M = \int_{\Omega} (f + sT - \frac{1}{\rho_o} F_i u_i)\rho d\Omega - \int_{A_\sigma} T_i^* u_i dA \tag{5.2}$$

$$N = \int_{\Omega} (\psi + \frac{1}{2} T^2 \partial_t s)\rho d\Omega + \int_{A_q} q_i^* n_i dA \tag{5.3}$$

f is the free energy per unit mass, depending on the temperature T and the Green strain tensor E_{ij} , ψ is the so-called heat flux potential defined by

$$\rho_o \psi = \frac{1}{2} \lambda_{ij} \frac{\partial T}{\partial X_i} \frac{\partial T}{\partial X_j} \tag{5.4}$$

All the other quantities appearing in (5.1) to (5.3) have been specified earlier.

It is assumed that the body force F_i is a given function of X_i and t.

Consider all admissible variations of u_i and T, compatible with the displacement and temperature boundary conditions. It will be shown that the motion is governed by the following variational equation

$$\delta I(u_i , T) = \delta \int_{t_1}^{t_2} (K - M + N)dt$$

$$- \int_{\Omega} \rho \partial_t u_i \delta u_i d\Omega \Big|_{t_1}^{t_2} = 0 \tag{5.5}$$

wherein s and $\partial_t s$ are not submitted to variation. Clearly, (5.6) must be classified as a quasi-variational formulation.

To determine the Euler-Lagrange equations, apply the δ operator to the functionals K, M and N ; one obtains successively

$$\delta K = \int_{\Omega_o}\int_{t_1}^{t_2} \rho_o \partial_t u_i \, \partial_t \delta u_i d\Omega_o dt$$

$$= - \int_{\Omega_o}\int_{t_1}^{t_2} \rho_o \partial_t^2 u_i d\Omega_o dt + \int_{\Omega_o} \rho_o \partial_t u_i \delta u_i d\Omega_o \Big|_{t_1}^{t_2} \tag{5.6}$$

$$\delta M = \int_{\Omega_o}\int_{t_1}^{t_2} \left(\frac{\partial f}{\partial E_{ij}} \delta E_{ij} + \frac{\partial f}{\partial T} \delta T + s\delta T - \frac{1}{\rho_o} F_i \delta u_i \right) \rho_o d\Omega_o dt$$

$$- \int_{A_o}\int_{t_1}^{t_2} t_i^{*} \delta u_i dA_o dt \tag{5.7}$$

$$\delta N = \int_{\Omega_o}\int_{t_1}^{t_2} \left(\frac{\partial \psi}{\partial (\partial T/\partial X_i)} \delta \left(\frac{\partial T}{\partial X_i}\right) + \partial_t s T \delta T \right) \rho_o d\Omega_o dt$$

$$+ \int_{A_o}\int_{t_1}^{t_2} Q_i^o n_i^o \delta T dA dt \tag{5.8}$$

Use has been made of the mass conservation

$$\rho d\Omega = \rho_o d\Omega_o \tag{5.9}$$

and the general results

$$q_i n_i dA = Q_i n_i^o dA_o$$

$$T_i^* dA = t_i^* dA_o$$

t_i is the surface tension on the deformed body referred per unit undeformed surface. With the help of the stress-strain relation

$$S_{ij} = \rho_o \frac{\partial f}{\partial E_{ij}} \tag{5.10}$$

and the result

$$\delta E_{ij} = \left(\delta_{ki} + \frac{\partial u_k}{\partial X_i}\right) \delta \left(\frac{\partial u_k}{\partial X_j}\right) + \left(\delta_{kj} + \frac{\partial u_k}{\partial X_j}\right) \delta \frac{\partial u_k}{\partial X_i} \tag{5.11}$$

the first term in (5.8) writes as

$$\int_{\Omega_o} \rho_o \frac{\partial f}{\partial E_{ij}} \delta E_{ij} d\Omega_o = \int_{\Omega_o} S_{ij} F_{ki} \delta \frac{\partial u_k}{\partial X_j} d\Omega_o$$

$$= - \int_{\Omega_o} \frac{\partial}{\partial X_j} (S_{ij} F_{ki}) \delta u_k d\Omega_o + \int_{A_o} F_{ki} S_{ij} n_j^o \delta u_k dA_o \tag{5.12}$$

wherein one has introduced the notation

$$F_{ki} = \delta_{ki} + \frac{\partial u_k}{\partial X_i}$$

Using

$$s = - \frac{\partial f}{\partial T} \tag{5.13}$$

and integrating by parts the first term of (5.9), one obtains finally

$$\delta I(u_i , T) = \int_{\Omega_o}\int_{t_1}^{t_2} \Big(- \rho_o \, \partial_t^2 u_i + \frac{\partial}{\partial X_\ell} (F_{ik}S_{k\ell}) + F_i\Big)\delta u_i d\Omega_o dt$$

$$+ \int_{A_\sigma^o}\int_{t_1}^{t_2} (t_i^* - F_{ik}S_k \, n_\ell^o)\delta u_i dA_o dt + \int_{\Omega_o}\int_{t_1}^{t_2} (Q_{i,i} + \rho_o T \partial_t s)\delta T d\Omega_o dt$$

$$+ \int_{A_q^o}\int_{t_1}^{t_2} (Q_i - Q_i^*) n_i^o \delta T dA dt = 0$$

This holds for arbitrary variations of u_i and T at the condition that

$$\rho_o \partial_t^2 u_i = \frac{\partial}{\partial X_\ell} (F_{ik}S_{k\ell}) + F_i \qquad \text{(momentum balance)} \qquad (5.14)$$

$$\rho_o \partial_t s = - \frac{1}{T} \frac{\partial Q_i}{\partial X_i} \qquad \text{(entropy balance)} \qquad (5.15)$$

while on the boundaries,

$$t_i^* = F_{ik}S_{k\ell}n_\ell^o \qquad \text{on } A_\sigma^o \qquad (5.16)$$

$$Q_i = Q_i^* \qquad \text{on } A_q^o \qquad (5.17)$$

Remember that the displacement and the temperature fields are prescribed on the parts A_u^o and A_T^o of the surface.

In its original presentation, Parkus considered the principle as consisting of two separated variational equations

$$\delta\int(K - M)dt = 0 \qquad \text{and} \qquad \delta\int N \, dt = 0 \qquad (5.18)$$

with N given by

$$N = \int_{\Omega} \rho(\psi - s\,T\,\partial_t T)\,d\Omega + \int_{A} T\,\overset{*}{q}_i n_i\,dA \tag{5.19}$$

instead of (5.3).

In our opinion, this splitting is not justified because of the coupled character of the analysis. Moreover, by varying (5.18a) with respect to T, Parkus recovers one of the state equations, namely (5.13) as Euler-Lagrange, but not the other one (5.10) which is introduced a priori. Finally, in contrast with the above presentation, Parkus' analysis does not contain the last term of (5.6) and therefore, the admissible functions must coincide with the exact ones at t_1 and t_2 .

Appendix : some thermodynamic results

Let an anisotropic solid be characterized by constant thermo-mechanical properties and obey Neumann-Duhamel's linear constitutive equation. Start with the relation

$$ds_v = \left.\frac{\partial s_v}{\partial T}\right)_{\varepsilon_{ij}} dT + \left.\frac{\partial s_v}{\partial \varepsilon_{ij}}\right|_T d\varepsilon_{ij} \tag{A.1}$$

From the definition of the Helmholtz function per unit volume

$$f_v = u_v - Ts_v \tag{A.2}$$

where u_v is the internal energy per unit volume, one deduces :

$$df_v = \sigma_{ij} d\varepsilon_{ij} - s_v dT \tag{A.3}$$

Introducing the specific heat per unit mass c_ε and β_{ij} defined by

$$\rho\ c_\varepsilon = T \left.\frac{\partial s_v}{\partial T}\right|_{\varepsilon_{ij}} \quad \text{and} \quad \beta_{ij} = \left.\frac{\partial s_v}{\partial \varepsilon_{ij}}\right|_T \tag{A.4}$$

expression (A.1) may be written as

$$ds_v = \rho\ c_\varepsilon \frac{dT}{T} + \beta_{ij} d\varepsilon_{ij} \tag{A.5}$$

When T does not deviate very much from the reference temperature T_r, this equation can be linearized to

$$ds_v = \rho\ c_\varepsilon \frac{dT}{T_r} + \beta_{ij} d\varepsilon_{ij} \tag{A.6}$$

Integrating with respect to the time and setting $s_v = 0$ when $T = T_r$ and $\varepsilon_{ij} = 0$, results in

$$s_v = \rho \, c_\varepsilon \frac{T - T_r}{T_r} + \beta_{ij}\varepsilon_{ij} \tag{A.7}$$

A state equation involving u_v can be obtained from the Gibbs equation

$$du_v = Tds_v + \sigma_{ij}d\varepsilon_{ij} \tag{A.8}$$

from which follows, using (A.5)

$$du_v = \rho \, c_\varepsilon dT + (T\beta_{ij} + \sigma_{ij})d\varepsilon_{ij} \tag{A.9}$$

Substituting the Duhamel-Neumann equation

$$\sigma_{ij} = C_{ijk\ell}\varepsilon_{k\ell} - \beta_{ij}(T - T_r) \tag{A.10}$$

in equation (A.3) and integrating with respect to the strain, gives

$$f_v = \frac{1}{2} C_{ijk\ell}\varepsilon_{ij}\varepsilon_{k\ell} - \beta_{ij}(T - T_r)\varepsilon_{ij} + C(T) \tag{A.11}$$

where C(T) is a function of T. To evaluate C(T), observe that

$$s_v = - \left.\frac{\partial f_v}{\partial T}\right|_{\varepsilon_{ij}} = \beta_{ij}\varepsilon_{ij} - \frac{dC(T)}{dT} \tag{A.12}$$

Comparison of this relation with (A.7) yields after integrating with respect to T :

$$- C(T) = \frac{\rho c_\varepsilon}{2T_r} (T^2 - 2T \, T_r) + D \tag{A.13}$$

Introducing (A.13) in (A.11) and taking into account the condition that $f_v = 0$ when $\varepsilon_{ij} = 0$ and $T = T_r$, one obtains for the constant D :

$$D = \frac{\rho c_\varepsilon}{2T_r} T_r^2$$

so that finally the Helmholtz function per unit volume is obtained as

$$f_v = \frac{1}{2} C_{ijk\ell}\varepsilon_{ij}\varepsilon_{k\ell} - \beta_{ij}(T - T_r)\varepsilon_{ij} - \frac{\rho c_\varepsilon}{2T_r}(T - T_r)^2 \tag{A.14}$$

If f denotes the Helmholtz function per unit mass, eq. (A.14) writes as

$$\rho f = \frac{1}{2} C_{ijk\ell}\varepsilon_{ij}\varepsilon_{k\ell} - \beta_{ij}(T - T_r)\varepsilon_{ij} - \frac{\rho c_\varepsilon}{2T_r}(T - T_r)^2 \tag{A.15a}$$

For isotropic bodies,

$$\rho f = G\,\varepsilon_{ij}\varepsilon_{ij} + \frac{1}{2}\lambda_L\varepsilon_{kk}^2 - \beta(T - T_r)\varepsilon_{kk} - \frac{\rho c_\varepsilon}{2T_r}(T - T_r)^2 \tag{A.15b}$$

where λ_L and G are the Lamé coefficients.

The relation (A.14), or (A.15), is a fundamental one in the sense that it contains all thermodynamic information. In particular, the equations of state follow as derivatives of the above equation, i.e. :

$$\sigma_{ij} = \left.\frac{\partial(\rho f)}{\partial\varepsilon_{ij}}\right|_T = C_{ijk\ell}\varepsilon_{k\ell} - \beta_{ij}(T - T_r) \tag{A.16}$$

and

$$\rho s = -\left.\frac{\partial(\rho f)}{\partial T}\right|_{\varepsilon_{ij}} = \beta_{ij}\varepsilon_{ij} + \frac{\rho c_\varepsilon}{T_r}(T - T_r) \tag{A.17}$$

The expression of the thermoelastic potential

$$V = \int_\Omega (u_v - T_r s_v) d\Omega \tag{A.17}$$

is directly derived. Indeed, one has

$$u_v - T_r s_v = f_v + (T - T_r) s_v \tag{A.18}$$

or using (A.13) and (A.15),

$$u_v - T_r s_v = \frac{1}{2} C_{ijk\ell} \varepsilon_{ij} \varepsilon_{k\ell} + \rho \frac{c_\varepsilon (T - T_r)^2}{2T_r} \tag{A.19}$$

so that

$$V = \frac{1}{2} \int_\Omega \left(C_{ijk\ell} \varepsilon_{ij} \varepsilon_{k\ell} + \rho \frac{c_\varepsilon (T - T_r)^2}{T_r} \right) d\Omega \tag{A.20}$$

References

1. Iesan, D., Principes variationnels dans la théorie de la thermoélasticité, *Ann. St. Univ. Al.I.Cuza Iasi*, 12, 439, 1966.
2. Iesan, D., On some reciprocity theorems and variational theorems in linear dynamic theories of continuum mechanics, *Mem. Acc. Sciences Torino*, serie 4, n°17, 1974.
3. Nickell, R. and Sackman, J., Variational principles for linear coupled thermoelasticity, *Quart. Appl. Math.*, 26, 11, 1968.
4. Biot, M., Thermoelasticity and irreversible thermodynamics, *J. Appl. Phys.*, 27, 240, 1956.
5. Biot, M., Linear thermodynamics and the mechanics of solids, Proc. Third U.S. Nat. Congress Appl. Mec., 1, 1958.

6. Biot, M., *Variational principles in heat transfer*, Oxford Math. Mono., Oxford, 1970.
7. Biot, M., Non-linear thermoelasticity, irreversible thermodynamics and elastic instability, *Indiana Univ. Math. J.*, 23, 309, 1973.
8. Lebon, G. and Lambermont, J., Some variational principles pertaining to non-stationary heat conduction and coupled thermoelasticity, *Acta Mech.*, 15, 121, 1972.
9. Lebon, G., Une nouvelle formulation variationnelle du problème de la thermoélasticité, *Proc. of A.I.M.E.T.A.*, 1, 249, 1972.
10. Parkus, H., Variational principles in thermo- and magneto-elasticity, *C.I.S.M. Lectures*, N°58, Springer-Verlag, Berlin, 1970.
11. Green, A.E. and Lindsay, K., Thermoelasticity, *J. of Elastiticy*, 2, 1, 1972.
12. Rusu, E., On some theorems in a generalized theory of linear thermo-elasticity, *Bul. Inst. Pol. Iasi*, 20, 75, 1974.
13. Rusu, E., A reciprocity theorem and a variational theorem for coupled mechanical and thermoelectric fields, 21, 83, 1975.
14. Rafalski, P., Lagrangian formulation of dynamic thermoelastic problem, *Bull. Ac. Pol. Sci.*, 16, 25, 1968.
15. Rafalski, P., The lagrangian formulation of the dynamic thermoelastic problem for mixed boundary conditions, in *Proceedings of vibrations problems*, Warsaw, 1, 9, 1968.
16. Rafalski, P., A variational principle for the coupled thermoelastic problem, *Int. J. Engng. Sci.*, 6, 465, 1968.
17. Biot, M., Variational thermodynamics of viscous compressible heat-conducting fluids, *Quart. Appl. Math.*, 4, 323, 1977.
18. Biot, M., New chemical thermodynamics of open systems, thermobaric potential, a new concept. *Bull. Class. Sci. Acad. Belgique*, 62, 239, 1976.
19. Biot, M., Variational lagrangian thermodynamics of evolution of collective chemical systems, *Chem. Phys.*, 29, 97, 1978.
20. Hermann, G., On variational principles in thermoelasticity and heat conduction, *Quart. Appl. Math.*, 21, 151, 1963.

21. Hermann, G., On a complementary energy principle in linear thermo-elasticity, *J. Aero-space Sci.*, 25, 660, 1958.

22. Parkus, H., Uber eine Erweiterung der Hamilton'schen Prinzipes auf thermoelastische Vorgänge, *Federhofer-Girkmann Festschrift*, Wien, 1950.

GENERAL APPENDIX : CONVERGENCE AND ERROR ESTIMATES OF THE VARIATIONAL AND GALERKIN METHODS

1. Definitions

In this appendix, the problems of convergence and error estimates of the variational and Galerkin methods are discussed. For practical calculations, numerical convergence of the various approximations is a sufficient test but in some circumstances, it is interesting to derive convergence criteria and error estimates on mathematical bases. The purpose is not to give here a fully account of the problem but rather to present a scope of some useful criteria. Most of the results will be given without demonstration ; more details can be found in specialized treatises.[1,2]

Consider a bounded domain Ω with a surface A in a N dimensional Euclidean space $x_1, x_2, \ldots x_N$. The following definitions are introduced.

The scalar inner product (u,v), the norm $\|u\|$ and the distance (or metric) d(u,v) of two functions u and v relative to the domain Ω are respectively defined by

$$(u,v) = \int_\Omega uvdx \tag{1.1}$$

$$\|u\| = (u,u)^{1/2} \tag{1.2}$$

$$d(u,v) = \|u-v\| = (u-v\ ,\ u-v)^{1/2} \tag{1.3}$$

Two functions are orthogonal if

$$(u,v) = 0$$

Let L be a linear operator. Its inner product is defined by

$$(u,Lv) = \int_{\Omega} uLvdx \tag{1.4}$$

L is symmetric if

$$(u,Lv) = (v,Lu) \tag{1.5}$$

L is positive definite if

$$(u,Lu) \geq 0 \tag{1.6}$$

it is positive bounded below if

$$(u,Lu) > \gamma(u,u) \qquad (\gamma > 0) \tag{1.7}$$

where γ is a positive constant.

Consider a functional F(u) and a sequence of function $\{u_n\}$; if

$$F(u_n) \geq \min F(u) \qquad (=d) \tag{1.8}$$

the set u_n is a minimizing sequence for the functional F and d is the greatest lower bound.

Introduce now the notion of *convergence*. As a matter of fact, there exists several definitions of convergence.

Uniform or pointwise convergence : given $\varepsilon > 0$, there exists a number n such that

$$|u_n(x) - u(x)| < \varepsilon \tag{1.9}$$

Convergence in the mean : the condition (1.9) is replaced by

$$\|u_n - u\| < \varepsilon \tag{1.10}$$

Convergence in energy : the condition (1.9) is substituted by

$$|||u_n - u||| < \varepsilon \tag{1.11}$$

where, by definition,

$$|||u||| = (u,Lu)^{1/2} \tag{1.12}$$

Convergence of a functional $I(u_n) \rightarrow I(u)$ does not necessarily imply convergence of the function itself.[4]

A set of functions ϕ_i is *complete* if any function u can be expanded in terms of the set. In particular, a set of functions is complete, with respect to convergence in the mean, if for n sufficiently large, one has

$$\|u - \sum_{i=1}^{n} a_i\phi_i\| < \varepsilon \tag{1.13}$$

A set of functions is complete in energy if

$$|||u - \sum_{i=1}^{n} b_i\phi_i||| < \varepsilon \tag{1.14}$$

Theorem 1. If L is positive bounded below, convergence in energy implies convergence in the mean.

Theorem 2. A set of functions is complete, if the only function orthogonal to each element of the set is the null function. This theorem is the basis of the method of weighted residuals.

As a rule, the trial functions should be selected as members of a complete serie. Examples of complete functions for the Sturm-Liouville problem

$$L(u) \equiv - \left(p(x)\, u_{,x}\right)_{,x} + q(x)u = f(x), \quad u(0)= u(1) = 0$$

are

$$\phi_i = x^i/(1-x) \qquad \phi_i = \sin i\pi x$$

For the two dimensional problem

$$L(u) \equiv (A_{ij}u_{,j})_{,i} + Cu = f(x) \qquad (i,j = 1,2)$$

$$u = 0 \text{ on } A$$

a system of complete functions is

$$\phi_{ij} = \sin (i\pi x) \sin (j\pi y)$$

As shown by Courant and Hilbert,[1] eigenvalue problems furnish a convenient source of complete sets.

2. Convergence and error estimates in boundary value problems

As pointed out by Finlayson,[3] the usefulness of theorems on convergence and error estimates depends on the characteristics of the problem, like its dimensions, the nature of the boundary conditions and the existence of a variational principle. For instance, a pointwise convergence theorem can be established for ordinary and two-dimensional differential equations while three-dimensional problems are related to convergence in the mean. Likewise, error bounds can be found for problems with classical variational principles ; exceptionally, some are available for non-linear or non-self adjoint problems. In this section, we shall briefly examine the problems of convergence and error estimates for the Rayleigh-Ritz, the Galerkin and the least square methods.

2.1. <u>The Rayleigh-Ritz method.</u> Consider the problems governed by

$$L\,u = f(x) \qquad B(u) = 0 \tag{2.1}$$

where L is linear and symmetric. If moreover L is positive bounded below,

$$(Lu,u) > \gamma \|u\|^2 \tag{2.2}$$

the unicity of the exact solution u_o is ensured.[2]

The corresponding variational equation is

$$\delta I(u) = \delta\{(Lu,u) - 2(u,f)\} = 0 \tag{2.3}$$

According to Rayleigh-Ritz's method, select as approximate function

$$u_n = \sum_{i=1}^{n} a_i \phi_i(x) \tag{2.4}$$

The constants a_i, depending on n, are determined by requiring that $I(u_n)$ is stationary :

$$\frac{\partial I(u_n)}{\partial a_i} = 0$$

Convergence. By using a sequence of trial functions which are complete in energy and linearly independent, it can be shown[2] that

1. the approximate solution (2.4) forms a minimizing sequence for I(u). Denoting by d the least value of I(u)

$$d = \min I(u) \ ,$$

one has

$$I(u_n) > d. \tag{2.5}$$

2. any approximate solution u_n converges in energy to the exact solution u_o :

$$\|u_n - u_o\| < \varepsilon \tag{2.6}$$

If L is a positive bounded below operator, one has simultaneously

$$|u_n - u_o\| < \varepsilon$$

The error estimate. If u_o denotes the exact solution, one has

$$Lu_o = f(x) \tag{2.7}$$

and formally

$$u_o = L^{-1}f \tag{2.8}$$

The error can be written as

$$u_o - u_n = L^{-1}L(u_o - u_n) = L^{-1}(f - Lu_n)$$

If L^{-1} is bounded, one has[2]

$$\|L^{-1}(f - Lu_n)\| < \|L^{-1}\| \quad \|f - Lu_n\| < \frac{1}{\gamma}\|f - Lu_n\|$$

from which results

$$\|u_o - u_n\| < \frac{1}{\gamma}\|f - Lu_n\| \tag{2.9}$$

This relation furnishes only a crude estimation, because in general Lu_n does not tend to f.

A better method of evaluating the error is the following. Consider the functional

$$I(u) = (u,Lu) - 2(u,f)$$

According to (2.7) one has

$$I(u) = (u,Lu) - 2(u,Lu_o)$$

$$= \big((u-u_o),L(u-u_o)\big) - (u_o,Lu_o) \tag{2.10}$$

and, with the notation (1.12),

$$I(u) = |||u-u_o|||^2 - |||u_o|||^2 \tag{2.11}$$

Similarly, one has

$$I(u_n) = |||u_n-u_o|||^2 - |||u_o|||^2 \tag{2.12}$$

It is clear from (2.11) that

$$\min I(u) = -|||u_o|||^2 \quad (=d) \tag{2.13}$$

In virtue of (2.12), the error of u_n is

$$|||u_n-u_o||| = \big(I(u_n) - d\big)^{1/2} \tag{2.14}$$

The problem reduces of finding a number δ lesser than d, in order that

$$|||u_n-u_o||| \leq \big(I(u_n) - \delta\big)^{1/2} \tag{2.15}$$

Such a construction is particularly easy when there exists a dual principle

$$\delta J(v) = 0 \tag{2.16}$$

such that

$$\max J(v) = \min I(u) \quad (=d) \tag{2.17}$$

If one constructs a maximizing sequence $\{v_m\}$ for $J(v)$

$$J(v_m) \leq \max J(v) \quad (=d) \tag{2.18}$$

one can take

$$\delta \equiv J(v_m) < d$$

Finally, the estimate of the error is given by

$$\| u_n - u_o \| \leq \left(I(u_n) - J(v_m) \right)^{1/2} \tag{2.19}$$

For m sufficiently large, the quantity (2.19) tends to the exact value

$$\left(I(u_n) - d \right)^{1/2}$$

The more general problem consisting of finding a number $\delta < d$ has been solved by using the method of Trefftz or the method of orthogonal projections. These methods resort to functional analysis and are largely expounded in ref.[2]

2.2. Galerkin method. Suppose that it is required to solve the non self-ajoint equation

$$Lu = f(x) \tag{2.20}$$

Galerkin's methods amounts to write an approximate solution in the form :

$$u_n = \sum_{i=1}^{n} a_i \phi_i (x) \tag{2.21}$$

and to solve the set of algebraic equations in the unknowns a_i

$$\sum_{i=1}^{n} (L\phi_i \, , \, \phi_j) a_i = (f, \, \phi_j) \qquad j = 1, 2, \ldots, n \tag{2.22}$$

Convergence has been proved by Mikhlin[2] for various types of differential equations. As an example, consider the second order linear ordinary differential equations

$$- u_{,xx} + p(x) u_{,x} + q(x) u = f(x) \tag{2.23}$$

with

$$u(0) = u(1) = 0$$

It was shown[2] that by using trial functions forming a complete set, the approximate solution u_n converges uniformly to the exact solution u_o :

$$|u_n(x) - u_o(x)| < \varepsilon \tag{2.24}$$

This result has been extended to a more general class of linear ordinary differential equations of the form

$$(-)^m u_{,2m} - Lu = f(x) \tag{2.25}$$

$$u = u_{,x} = u_{,xx} = \ldots u_{,m-1} = 0 \quad \text{at} \quad x = 0, \;\; x = 1.$$

Assuming that the solution of the problem is unique, it was proved that for $k < m$ $\quad |(u_n)_{,k} - (u_o)_{,k}| < \varepsilon$ (uniform convergence)

for $k = m$ $\quad \|(u_n)_{,m} - (u_o)_{,m}\| < \varepsilon$ (convergence in the mean)

Convergence of some partial differential equations has also been studied.[2] As illustration, consider the steady state transport equation

$$-\left(A_{ij}(x)u_{,j}\right)_{,i} + B_i(x)u_{,i} + C(x)u = f(x) \quad (i,j = 1,2,3) \tag{2.26}$$

with $u = 0$ at the boundary A. This equation is a second order elliptic equation, it is not self-adjoint except if $B_i(x) = 0$. Suppose that the solution is unique and put

$$Ku = -(A_{ij}u_{,j})_{,i}$$

Operator K has been proved to be positive bounded below for the set of functions vanishing at the boundary :

$$(Ku,u) > \gamma \|u\|^2$$

Under these conditions, it was shown[2] that by using trial functions which are complete in energy of K , linearly independent and satisfying $u = 0$ on A , the Galerkin method leads to approximate solution u_n converging in energy to the true solution u_o.

This result is applicable to heat and mass transfer problems with known coefficients A_{ij} , B_i and C in a finite domain. When (2.26) is non linear, convergence of the Galerkin method can also be assured.[3]

Error estimates for the Galerkin method can be derived in terms of the mean-square residual, as shown on the next problem

$$Lu = g(u,x) \tag{2.27}$$

where L is a nth order differential operator and g a non-linear function of u. Formally the exact solution can be expressed as

$$u_o = L^{-1}g(u_o,x) \tag{2.28}$$

The approximate solution gives raise to a residual R_n, defined by

$$Lu_n - g(u_n,x) = R_n \tag{2.29}$$

from which follows

$$u_n = L^{-1}g(u_n,x) + L^{-1}R_n \tag{2.30}$$

By substraction of (2.28) and (2.30), one obtains

$$u_n - u_o = L^{-1}\left(g(u_n,x) - g(u_o,x)\right) + L^{-1}R_n \tag{2.31}$$

Introducing a so-called Lipschitz constant k defined by

$$\|g(u_n,x) - g(u_o,x)\| \leq k \|u_n - u_o\| \tag{2.32}$$

(2.31) yields

$$\|u_n - u_o\| \leq k\|L^{-1}\| \|u_n - u_o\| + \|L^{-1}\| \|R_n\|$$

or

$$\|u_n - u_o\| \leq \frac{\|L^{-1}\| \|R_n\|}{1 - k\|L^{-1}\|} \tag{2.30}$$

This result gives the error in term of the mean square residual $\|R_n\|$.

2.3. The least square method. Consider the linear equation

$$Lu = f(x), \quad u = 0 \text{ on } A \tag{2.31}$$

and the associated least square functional

$$I(u) = \int_\Omega (Lu - f)^2 d\Omega \tag{2.32}$$

By selecting as trial function

$$u_n = \sum_{i=1}^{n} a_i \phi_i(x) \tag{2.33}$$

the coefficients a_i are obtained from

$$\sum_{i=1}^{n} a_i \,(L\phi_i \,, L\phi_j) = (f \,, L\phi_j) \quad j=1,2,\ldots, n \tag{2.34}$$

Convergence : The method gives a sequence of approximate solution which converges in the mean to the exact solution if[2]

1. equation (2.31) is soluble
2. there exists a constant k such that

$$\|u\| \leq k \,\|Lu\| \tag{2.35}$$

3. the sequence of trial functions ϕ_i is L complete. By L complete is meant that for a given $\varepsilon > 0$, it is possible to find an integer n and constants a_i such that

$$\|Lu - Lu_n\| < \varepsilon \tag{2.36}$$

It should be observed that the least-square method converges more slowly than Rayleigh-Ritz's technique. To show it, apply the two methods with the same set of trial functions ϕ_i. The variational principle associated to (2.31) is

$$\delta J = 0$$

Let u_n denote the solution with the least square and v_n the solution with the Rayleigh-Ritz method. Since the Rayleigh-Ritz method minimizes the functional J, one has

$$I(u_n) \geq J(v_n)$$

which implies

$$|||u_n - u_o||| \geq |||v_n - u_o|||$$

This indicates that Rayleigh-Ritz's method converges faster because it has the smallest energy norm.

Error estimate. The error of an approximate solution obtained from the least square method is estimated by the inequality

$$\|u_n - u_o\|^2 \leq k\left(\|f\|^2 - (Lu_n , f)\right) \tag{2.37}$$

For the two-dimensional heat conduction problem

$$T_{xx} + T_{yy} = f(x,y) \tag{2.38}$$

error bounds have been calculated and given by[2]

$$|T_n - T_o|^2 < \int_\Omega G^2(x,y,\xi,\eta)d\xi d\eta \int (T_{xx} + T_{yy} - f)^2 dxdy \tag{2.39}$$

G is the Green function ; the first integral is bounded and the second is minimized by the least-square method. When more terms are incorporated in the trial function, the error bounds are improved.

To sum up, it has been seen that to prove convergence, it is necessary that the set of expansion functions be complete. For partial differential equations, convergence in energy is observed for positive definite operators, convergence in the mean is observed for positive bounded below operators. The error estimates make generally use of the special properties of the equations, like the existence of a variational equation or a Green function.

3. Convergence and error estimates in initial value problems.

A great variety of examples[3] have indicated that the Galerkin method furnishes solutions which converge to the exact solution.

As illustration, consider the linear problem

$$Lu \equiv \left(\frac{\partial^2}{\partial x^2} - \frac{\partial}{\partial t} - g(x,t)\right)u = f(x,t) \qquad 0 \leq x \leq \pi \tag{3.1}$$

$$u(x,0) = u(0,t) = u(\pi,t) = 0 \tag{3.2}$$

In addition,

$$f(0,0) = f(\pi,0) = 0 \tag{3.3}$$

The approximate solution is represented by

$$u_n(x,t) = \sum_{i=1}^{n} a_i(t) \sin ix \tag{3.4}$$

wherein the a_i are solution of

$$\int_o^{\pi} (L\, u_n - f) \sin jx \, dx = 0 \qquad j = 1,2,\ldots,n \tag{3.5}$$

$a_i(0) = 0$

Green[5] proves that

1. the sequence u_n converges uniformly to the solution u_o ; $\partial_t(u_n)$ and $(u_n)_{,x}$ converge uniformly to $\partial_t u_o$ and $u^o_{o,x}$ respectively ; $(u_{n,xx})$ converges in the mean to $u_{o,xx}$.
2. Error bounds are given in terms of the mean square residual

$$\int_o^\pi (u_n - u_o)^2 dx \leq t^2 B_n^2(t) \tag{3.6}$$

where

$$B_n(t) = \max_{\tau \leq t} \left\{ \int_o^\pi \left(Lu_n(x,\tau) - f(x,\tau) \right)^2 dx \right\}^{1/2} \tag{3.7}$$

The convergence of Galerkin's method has also been established for non linear equations, like the non linear heat conduction equation. For more details, the reader is referred to the books listed as references.

References

1. Courant, R. and Hilbert, D., *Methods of mathematical physics*, Wiley, New-York, 1953.
2. Mikhlin, S.G., *The variational methods in mathematical physics*, Pergamon Press, New-York, 1964.
3. Finlayson, B., *The method of weighted residuals*, Academic Press, New-York, 1972.
4. Kantorovitch, L. and Krilov, V., *Approximate methods in higher analysis*, Wiley, New-York, 1958.
5. Green, J.W., An expansion method for parabolic partial differential equations, *J. Res. Natl. Bur. Std.*, 51, 27, 1953.

CONTENTS

INTRODUCTION 221

I. BASIC CONCEPTS OF THE CALCULUS OF VARIATION 224

1. Basic definition and theorems 224
   1.1. The variation operator δ 224
   1.2. The Euler-Lagrange equations 226
   1.3. Constrained integrals 228
   1.4. Natural boundary conditions 229
   1.5. Dual variational principles 231
2. Variational methods of approximation 235
   2.1. The Rayleigh-Ritz method 235
   2.2. The Kantorovitch method 237
   2.3. The method of weighted residuals 238
3. The inverse problem 243
   3.1. Adjoint and self-adjoint operators 243
   3.2. The self-adjoint problem 244
   3.3. The non-self-adjoint problem 246
   3.4. General condition of existence of a variational principle 247
4. Non classical variational principles 251
   4.1. Quasi variational principles 251
   4.2. Restricted variational principles 252
Appendix : The finite element technique 255
References 262

II. VARIATIONAL PRINCIPLES IN CLASSICAL MECHANICS AND IN ELASTICITY 264

1. The law of varying action and Hamilton's principle in classical mechanics 264
   1.1. Formulation of the principles 264
   1.2. An example 269
2. General results in elasticity and thermoelasticity 272
   2.1. Notation and definitions 272
   2.2. Laws of balance 276
   2.3. Thermodynamic description 278
3. Basic variational principles in elastostatics 283
   3.1. The minimum energy principle for small deformations . . 283
   3.2. The Reissner and the complementary energy principles for small deformations 289
   3.3. An extension of Reissner's principle : H_u - Washizu's principle 296
   3.4. Illustrative examples 297
      i) Bending of a beam under static loading 297
      ii) A beam with a clamped and a supported end, loaded at its middle 300
      iii) Complementary energy principle 302
4. Variational principles for large deformations 305
   4.1. Principle of stationary potential energy 305
   4.2. A canonical principle 307
   4.3. A complementary energy principle 309
5. The elastodynamic problem 310
   5.1. The law of varying action and Hamilton's principle . . . 310
   5.2. Gurtin's principle of linear elasticity 314
   5.3. Numerical applications of the law of varying action to the vibrations of a beam 319

References 328

III. VARIATIONAL THEORY OF HEAT CONDUCTION 330

1. Introduction 330
2. Steady heat conduction 331
   2.1. The linear problem 331
   2.2. The non-linear problem 333
   2.3. Canonical and dual principles 335
3. Unsteady linear heat conduction 338
   3.1. The parabolic equation 338
   3.2. The hyperbolic equation 342
4. Unsteady non-linear heat conduction 343
   4.1. Biot's principle 343
   4.2. Vujanovic's principle 350
   4.3. Glansdorff-Prigogine's local potential 353
   4.4. Lebon-Lambermont's principle 359
References 363

IV. COUPLED THERMOELASTICITY 366

1. Basic equations 366
2. The Iesan-Nickell-Sackman principle 370
3. Lagrangian Thermoelasticity 373
   3.1. Biot's principle 373
   3.2. A canonical principle for the static linear problem . . 379
   3.3. A complementary principle for the static linear problem 382
4. Lebon-Lambermont's principle for linear dynamic thermoelasticity 383
5. A generalized Hamilton's principle for finite deformations . . 385
Appendix : Some thermodynamic results 391
References 394

GENERAL APPENDIX : CONVERGENCE AND ERROR ESTIMATES OF THE VARIATIONAL AND GALERKIN METHODS 397

1. Definitions 397
2. Convergence and error estimates in boundary value problems . . 400
   2.1. The Rayleigh-Ritz method 400
   2.2. The Galerkin method 404
   2.3. The least square method 408
3. Convergence and error estimates in initial value problems . . 410
References 411

Printed in the United States
By Bookmasters